# PATENTS IN THE KNOWLEDGE-BASED ECONOMY

Wesley M. Cohen and Stephen A. Merrill, Editors

Committee on Intellectual Property Rights in the Knowledge-Based Economy
Board on Science, Technology, and Economic Policy
Policy and Global Affairs

NATIONAL RESEARCH COUNCIL
*OF THE NATIONAL ACADEMIES*

THE NATIONAL ACADEMIES PRESS
Washington, D.C.
**www.nap.edu**

THE NATIONAL ACADEMIES PRESS • 500 Fifth Street, N.W. • Washington, DC 20001

NOTICE: The project that is the subject of this report was approved by the Governing Board of the National Research Council, whose members are drawn from the councils of the National Academy of Sciences, the National Academy of Engineering, and the Institute of Medicine. The members of the committee responsible for the report were chosen for their special competences and with regard for appropriate balance.

This study was supported by Contract No. NASW-99037, Task Order 103, between the National Academy of Sciences and the National Aeronautics and Space Administration, the Department of Commerce, Andrew W. Mellon Foundation, Center for Public Domain, Pharmacia Corporation, Merck & Company, Procter & Gamble, and IBM. Any opinions, findings, conclusions, or recommendations expressed in this publication are those of the author(s) and do not necessarily reflect the views of the organizations or agencies that provided support for the project.

International Standard Book Number 0-309-08636-1 (Book)
International Standard Book Number 0-309-50941-6 (PDF)

Limited copies are available from:

Board on Science, Technology, and Economic Policy
National Research Council
500 Fifth Street, N.W.
Washington, D.C. 20001
Phone: 202-334-2200
Fax: 202-334-1505

Additional copies of this report are available from the National Academies Press, 2101 Constitution Avenue, N.W., Lockbox 285, Washington, D.C. 20055; (800) 624-6242 or (202) 334-3313 (in the Washington metropolitan area); Internet, http://www.nap.edu

Copyright 2003 by the National Academy of Sciences. All rights reserved.
Printed in the United States of America

The cover design incorporates original illustration from the following U.S. patents issued over a nearly 160-year period:

U.S. Patent 6,506,554; Core structure of gp41 from the HIV envelope glycoprotein; Chan, David C. (Brookline, MA); Fass, Deborah (Cambridge, MA); Lu, Min (New York, NY); Berger, James M., (Cambridge, MA); Kim, Peter S. (Lexington, MA); Granted January 14, 2003.
U.S. Patent 6,423,583; Methodology for electrically induced selective breakdown of nanotubes; Avouris, Phaedon (Yorktown Heights, NY); Collins, Philip G. (Ossining, NY); Martel, Richard (Peekskill, NY); Granted July 23, 2003.
U.S. Patent 6,313,562; Microelectromechanical ratcheting apparatus; Barnes, Stephen M. (Albuquerque, NM); Miller, Samuel L. (Albuquerque, NM); Jensen, Brian D. (Albuquerque, NM); Rodgers, M. Steven (Albuquerque, NM); Burg, Michael S., (Albuquerque, NM); Granted November 6, 2001.
U.S. Patent 821,393; Flying machine; Wright , Orville (Dayton, OH) and Wright, Wilbur (Dayton, OH); Granted May 22, 1906.
U.S. Patent 223,898; Electric lamp; Edison, Thomas A. (Menlo Park, NJ); Granted January 27, 1880.
U.S. Patent 4750; Improvement in sewing machines; Howe, Jr., Elias, (Cambridge, MA); Granted September 10, 1846.

# THE NATIONAL ACADEMIES
*Advisers to the Nation on Science, Engineering, and Medicine*

The **National Academy of Sciences** is a private, nonprofit, self-perpetuating society of distinguished scholars engaged in scientific and engineering research, dedicated to the furtherance of science and technology and to their use for the general welfare. Upon the authority of the charter granted to it by the Congress in 1863, the Academy has a mandate that requires it to advise the federal government on scientific and technical matters. Dr. Bruce M. Alberts is president of the National Academy of Sciences.

The **National Academy of Engineering** was established in 1964, under the charter of the National Academy of Sciences, as a parallel organization of outstanding engineers. It is autonomous in its administration and in the selection of its members, sharing with the National Academy of Sciences the responsibility for advising the federal government. The National Academy of Engineering also sponsors engineering programs aimed at meeting national needs, encourages education and research, and recognizes the superior achievements of engineers. Dr. Wm. A. Wulf is president of the National Academy of Engineering.

The **Institute of Medicine** was established in 1970 by the National Academy of Sciences to secure the services of eminent members of appropriate professions in the examination of policy matters pertaining to the health of the public. The Institute acts under the responsibility given to the National Academy of Sciences by its congressional charter to be an adviser to the federal government and, upon its own initiative, to identify issues of medical care, research, and education. Dr. Harvey V. Fineberg is president of the Institute of Medicine.

The **National Research Council** was organized by the National Academy of Sciences in 1916 to associate the broad community of science and technology with the Academy's purposes of furthering knowledge and advising the federal government. Functioning in accordance with general policies determined by the Academy, the Council has become the principal operating agency of both the National Academy of Sciences and the National Academy of Engineering in providing services to the government, the public, and the scientific and engineering communities. The Council is administered jointly by both Academies and the Institute of Medicine. Dr. Bruce M. Alberts and Dr. Wm. A. Wulf are chair and vice chair, respectively, of the National Research Council.

**www.national-academies.org**

## Committee on Intellectual Property Rights in the Knowledge-Based Economy

## Board on Science, Technology, and Economic Policy

## National Research Council

Richard Levin (Co-chair)
President
Yale University

Mark Myers (Co-chair)
Visiting Executive Professor of
  Management
The Wharton School
University of Pennsylvania

John Barton
George E. Osborne Professor of Law
Stanford University

Robert Blackburn
Vice President and Chief Patent Counsel
Chiron Corporation
and Distinguished Scholar
Berkeley Center for Law and Technology

Wesley Cohen
Professor of Economics and Management
Fuqua School of Business
Duke University

Frank Collins
Senior Vice President for Research
ZymoGenetics

Rochelle Dreyfuss
Pauline Newman Professor of Law
New York University

Bronwyn Hall
Professor of Economics
University of California, Berkeley

Eugene Lynch
Judge (retired)
Federal District Court of Northern
  California

Daniel McCurdy
President & CEO
ThinkFire, Ltd.

Gerald J. Mossinghoff
Senior Counsel
Oblon, Spivak, McClelland, Maier &
  Neustadt, P.C.

Gail Naughton
Dean, School of Business
San Diego State University

Richard Nelson
George Blumenthal Professor of
  International and Public Affairs
Columbia University

James Pooley
Partner
Milbank, Tweed, Hadley & McCloy LLP

William Raduchel
Great Falls, Virginia

Pamela Samuelson
Professor of Law and Information
  Management
University of California, Berkeley

## LIAISONS WITH OTHER NRC PROGRAMS

*Science, Technology, and Law Program*
David Korn
Senior Vice President for Biomedical and Health Science
Association of American Medical Colleges

*National Cancer Policy Board*
Pilar Ossorio
Assistant Professor of Law and Medical Affairs
University of Wisconsin Law School

## STAFF

Stephen A. Merrill
Project Director

Craig Schultz
Research Associate

Camille Collett
Project Associate (until September 2002)

George Elliott
Commerce Science and Technology Fellow (until September 2001)

# Board on Science, Technology, and Economic Policy
# National Research Council

*Chairman*

Dale Jorgenson
Samuel W. Morris University Professor
Harvard University

*Vice Chairman*

Bill Spencer
Chairman Emeritus
International Sematech

M. Kathy Behrens
Managing Partner
RS Investments

Bronwyn Hall
Professor of Economics
University of California, Berkeley

James Heckman
Henry Schultz Distinguished Service
 Professor of Economics
University of Chicago

Ralph Landau
Senior Fellow
Stanford Institute for Economic Policy
 Research
Stanford University

Richard Levin
President
Yale University

David Morgenthaler
Founding Partner
Morgenthaler Ventures

Mark Myers
Visiting Executive Professor of
 Management
The Wharton School
University of Pennsylvania

Roger Noll
Morris M. Doyle Centennial Professor of
 Economics and Director, Public Policy
 Program
Stanford University

Edward E. Penhoet
Director, Science and Higher Education
 Programs
Gordon and Betty Moore Foundation

William Raduchel
Great Falls, Virginia

Alan Wm. Wolff
Managing Partner
Dewey Ballantine, DC

*Staff*

Stephen A. Merrill
Executive Director

Charles Wessner
Deputy Director

Russell Moy
Senior Program Officer

Sujai Shivakumar
Program Officer

Craig Schultz
Research Associate

Tabitha Benney
Program Associate

David Dierksheide
Program Associate

Chris Hayter
Program Associate

# Preface

Over the past 25 years a series of court decisions, legislative and administrative actions, and international agreements has extended and strengthened intellectual property rights (IPRs) in the United States and elsewhere. In turn these policy changes contributed to more zealous acquisition and vigorous exercise and defense of IPRs. Curiosity about the effects of these developments on innovation and economic performance led the National Academies' Board on Science, Technology, and Economic Policy (STEP) in 2000 to embark on an extended inquiry that encompassed workshops, conferences, commissioned research, and committee deliberations focused on the operation of the patent system, especially in two areas, information technology (IT) and biotechnology.

The need for specialized legal and technical expertise to carry out a study leading to policy recommendations in these areas led the STEP Board to propose to the Academies the creation of the Committee on Intellectual Property Rights in the Knowledge-Based Economy, composed of economists specializing in intellectual property and technology development, legal scholars, practitioners from corporations and private law practice, a former federal judge and a former Commissioner of the U.S. Patent and Trademark Office (USPTO), biomedical scientists, and managers of research and business development in the IT sector. We were asked to co-chair this committee, whose report, *A Patent System for the 21st Century*, accompanies this volume.

At the same time the need for additional analysis and data to inform recommendations in these areas led the STEP Board to commission eight research projects. With one additional chapter, the results of this work comprise this volume, edited by Wesley Cohen, professor of economics and management at Duke and a member of the study committee, and Stephen Merrill, executive director of the STEP program and director of the project.

The process of selecting the topics and authors of this collection was unusual for STEP and for the Academies. We decided to solicit proposals via a formal request for proposals that in March 2000 was widely circulated to the academic, consulting, and legal communities. The solicitation specified four policy-related areas of interest—the patent examination process and its bearing on patent quality, the incidence of patent litigation and its costs, and patent acquisition and use in two technologies—software and biotechnology. It further stipulated that the work involve original empirical research or data analysis, that it fall within a narrow range of costs, and that it be completed within approximately 12 months.

We received more than 80 proposals, necessitating a more elaborate review and selection process than we had originally planned. After an initial screening by staff we recruited a group of economists and legal scholars to help review more than 60 proposals. The reviewers included Bronwyn Hall, Berkeley economist and member of the STEP Board, Wesley Cohen, then at Carnegie Mellon University; John Barton, Stanford professor of law; and Robert Merges, professor of intellectual property law at Boalt Hall, Berkeley. Each proposal was read by at least one economist and one legal scholar. Three reviewers recused themselves from considering proposals on which they were listed as principal or co-investigator. The evaluation criteria included 1) policy relevance and conformity to the issues discussed in the request for proposals, 2) quality of issue framing and methodology, and 3) feasibility of the research. As co-leaders of the project for the STEP Board, we assumed responsibility for the final selection from among 25 highly ranked proposals. Although the final selection included proposals by Hall and Cohen, other proposals by these two investigators were not selected. Negotiations with individuals and their institutions consumed several weeks so that the work commenced in the late summer of 2000.

Preliminary results were presented at a Washington conference in October 2001, where attorneys, judges and former PTO officials, and corporate managers were able to comment on the methodology and the findings. Audio tracks, slide presentations, and transcripts of this and two other STEP conferences are available on a CD accompanying this volume. Following the meeting, papers were reviewed by the editors and in most cases by the external reviewers listed below and were in all cases revised before publication in this volume. In the meantime they were available at various stages on the project website and to the Committee on Intellectual Property Rights in the Knowledge-Based Economy. The Committee has found them useful and in a few cases directly relevant to its findings and recommendations but in no way has been constrained from considering other research and commentary.

The generosity of two foundations made the research element of the project possible. The Andrew W. Mellon Foundation supported the February 2000 STEP conference, *Intellectual Property Rights: How Far Should They Be Extended?*, at which many of these ideas germinated as well as the subsequent preparation of the papers and their publication. In addition to supporting the research, The Center

PREFACE                                                                                                    ix

for the Public Domain enabled us to develop a dedicated website that has been indispensable to our efforts to keep a wide community of interested people informed of our progress and enabled them to express their views. Naturally, most of the contributions to this volume represent parts of larger and longer research projects supported by other sources, including the National Science Foundation, Brookings Institution, and the Wharton School's Reginald H. Jones Center for Management Policy, Strategy, and Organization. These organizations' sponsorship of work in this important area deserves to be highlighted and commended.

Two contributions to this volume received no funding from the STEP Board. Jonathan King's chapter, although selected as a result of the solicitation and review described above, is part of important ongoing analytical work on intellectual property policy at the Economic Research Service of the U.S. Department of Agriculture. We are grateful to the Service for making it available to us. The chapter exploring the theoretical benefits of a patent opposition process by Richard Levin and Jonathan Levin evolved independently of the STEP project but is included here because of its close relevance to the empirical comparison of European oppositions and U.S. patent re-examinations by Graham, Harhoff, Hall, and Mowery. Both papers were nevertheless subject to the Academy review process.

Individual chapters in this volume have been reviewed in draft form by people chosen for their technical expertise, in accordance with procedures approved by the National Research Council's (NRC) Report Review Committee. The purpose of this independent review is to provide candid and critical comments that will assist the institution in making its published report as sound as possible and to ensure that the report meets institutional standards for quality. The review comments and draft manuscript remain confidential to protect the integrity of the process. We wish to thank the following individuals for their review of selected papers: Mildred Cho, Stanford University; Robert Cook-Deegan, Duke University; Jeffrey Kushan, Sidley, Austin, Brown, and Wood LLC; Joshua Lerner, Harvard Business School; Arti Rai, University of Pennsylvania Law School; F.M Scherer, Harvard University (emeritus); and Brian Wright, University of California, Berkeley.

Although the reviewers listed above have provided constructive comments and suggestions, they were not asked to endorse the content of the papers. Responsibility for the final content of the papers rests with the authors, and statements made in them do not necessarily represent positions of the National Research Council or the Committee.

Finally, we want to thank all of the authors and the editors, who worked successfully to produce these results under rather severe constraints of time and limited budgets. Their collective work not only aided the Committee but, more importantly, advances our common knowledge of the patent system and demonstrates the value of continuing efforts to understand its operation and effects. We are grateful to Craig Schultz, who administered the contracts for these papers and oversaw the review process and the production of this volume.

Richard Levin  
President  
Yale University

Mark B. Myers  
Visiting Executive Professor of Management  
University of Pennsylvania

# Contents

**Introduction**
  Wesley M. Cohen and Stephen A. Merrill .......................... 1

**Patent Quality**

  *Are All Patent Examiners Equal? Examiners, Patent
  Characteristics, and Litigation Outcomes* ....................... 19
  Iain M. Cockburn, Samuel Kortum, and Scott Stern

  *Patent Examination Procedures and Patent Quality* ................ 54
  John L. King

  *Patent Quality Control: A Comparison of U.S. Patent
  Re-examinations and European Patent Oppositions* ............... 74
  Stuart J.H. Graham, Bronwyn H. Hall, Dietmar Harhoff, and
  David C. Mowery

  *Benefits and Costs of an Opposition Process* ..................... 120
  Jonathan Levin and Richard Levin

**Patent Litigation**

  *Enforcement of Patent Rights in the United States* ................ 145
  Jean O. Lanjouw and Mark Schankerman

*Patent Litigation in the U.S. Semiconductor Industry* .............. 180
    Rosemarie Ham Ziedonis

## Patents in Software and Biotechnology

*Intellectual Property Protection in the U.S. Software Industry* ....... 219
    Stuart J.H. Graham and David C. Mowery

*Internet Business Method Patents* .............................. 259
    John R. Allison and Emerson H. Tiller

*Effects of Research Tool Patents and Licensing on Biomedical
Innovation* ................................................. 285
    John P. Walsh, Ashish Arora, and Wesley M. Cohen

# Introduction

Wesley M. Cohen
Duke University

Stephen A. Merrill
National Research Council

Since 1980, successive changes in patent policy, one of the oldest elements of U.S. technology policy, have expanded intellectual property rights and strengthened the position of patent owners. The establishment of the Court of Appeals for the Federal Circuit (1982), which consolidated all appeals from patent case decisions of federal district courts in a single specialized court, led to a sharp increase in plaintiff success rates in patent infringement law suits. A number of large, widely publicized judgments in infringement cases also suggested a marked rise in the value of patents, at least at the upper end of the distribution.[1] Encouraged by a series of court decisions, patenting has been extended to new scientific and technological domains such as life forms, genes, software, and methods of doing business. A federal statute enacted in 1980 encouraged universities and other nonprofit institutions conducting research with public funds to obtain and license patents. Partly as a result of these changes in the policy environment, business strategy in some sectors has placed a greater premium on acquiring and using patents (cf. Kortum and Lerner, 1999). This is especially the case in biomedicine and information technology.

At the same time these policy changes have raised concerns about their impact on innovation and the factors driving innovation. Along with the growth in patenting itself, there has been an increase in patent litigation, which some consider to be an unproductive increment to the cost of innovation. Others see in the

---

[1] For example, in 1991 Kodak was compelled to pay Polaroid $873,158,971 for infringement of the latter's instant photography patents as well as to cease production of its own instant camera.

proliferation of software and business method patents a weakening of the standards of novelty and non-obviousness, thereby undermining the purpose of patents to provide an incentive to those who innovate in a genuine way. The proliferation of patents in biotechnology, especially those involving DNA sequences, has raised a different set of concerns—whether intellectual property rights are becoming so fragmented that assembling the rights necessary to commercialize a new therapy or drug is prohibitively costly and whether some promising lines of research are abandoned prematurely.

This volume assembles papers commissioned by the National Research Council's Board on Science, Technology, and Economic Policy (STEP) to inform judgments about some of these institutional and policy developments made over the past two decades. The chapters fall into three areas. The first four chapters consider the determinants and effects of changes in patent "quality." Quality refers to whether patents issued by the U.S. Patent and Trademark Office (USPTO) meet the statutory standards of patentability, including novelty, non-obviousness, and utility. The fifth and sixth chapters consider the growth in patent litigation, which may itself be a function of changes in the quality of contested patents. The final three chapters explore controversies associated with the extension of patents into new domains of technology, including biomedicine, software, and business methods.

The style of these contributions varies. Several are based on descriptive and in some cases qualitative data. Others develop and test hypotheses in the manner of empirical economics. And one chapter is a theoretical exploration of the costs and benefits of a proposed institutional change intended to improve patent quality. These contributions are discussed below.

We are interested in questions of patent quality, litigation, and extension into new areas of technology because we are concerned with how the patent system affects the rate and direction of technological change. The trouble with trying to understand the import of the changes in patent policy over the past two decades is that we have a limited understanding of the effects of the patent system to begin with. There has been little systematic empirical analysis of the impact of patents on innovation. Even the narrower question of whether patenting stimulates research and development investment has only recently begun to be studied.

There are theoretical as well as empirical reasons to question whether patent rights advance innovation in a substantial way in most industries. The rationale for patent protection is to augment the incentives to invent by conferring the right to exclude others from making, using, or selling the invention in exchange for the disclosure of its details. Although the prospect of monopoly rents should induce inventive effort, the costs of disclosure can in some circumstances more than offset the prospective gains to patenting (Horstmann et al., 1985). "Strengthening" patent protection enhances the value of not only a given firm's patents but also those of its rivals who may be able to constrain the original firm's ability to commercialize its innovations (Jaffe, 2000; Gallini, 2002).

Merges and Nelson (1990) and Scotchmer (1991) argue that where technological advances build upon one another cumulatively, as is increasingly the case, broad patent protection on upstream discoveries may slow the rate of technical change by impeding subsequent innovations. Heller and Eisenberg (1998) suspect that in the domain of genetics patenting has been extended to such fine-grained inventions that the intellectual property covering any new drug or therapy may now be so complex and dispersed that heterogeneous patent owners may not agree on the licensing terms necessary to bring a product to market. Cohen and colleagues (2000) point out that in industries such as microelectronics there can be hundreds of patentable elements in one product, with the consequence that typically no single firm ever holds all of the rights necessary for its commercialization. In complex product industries generally and in the semiconductor industry in particular (Hall and Ziedonis, 2001) such mutual dependence commonly spawns extensive cross-licensing. In these cases the kind of breakdown hypothesized by Heller and Eisenberg does not often occur, but the need for patents as bargaining chips to arrive at satisfactory cross-licensing agreements may stimulate expensive patent portfolio races among industry incumbents. Especially if powerful incumbents insist on trading like-for-like in licensing arrangements (Shapiro, 2001), firms with modest or negligible patent holdings may be barred from entry. To the extent that entrants are a vehicle for innovation and their entry is obstructed, technical advance may suffer.

Empirical work by a number of economists over nearly fifty years suggests that patents play a prominent role in stimulating invention in only a few manufacturing industries (Scherer et al., 1959; Taylor and Silberston, 1973; Mansfield, 1986). Surveys of R&D managers by Levin and colleagues (1987) and, more recently, Cohen and colleagues (2000) found that in most industries patents are judged to be less important means of protecting innovations than, for example, being first to market or retaining know-how as trade secrets.[2]

Although we should therefore not assume that patents invariably induce innovation, neither should we assume the contrary. Firms may rely more heavily on other means of protecting innovations, but patents may still yield a return. Arora and colleagues (2003) recently showed that patents do appear to stimulate R&D across the manufacturing sector, although the magnitude of the stimulus varies greatly from industry to industry. Levin and colleagues (1987), Mansfield (1986), and Cohen and colleagues (2000) all find that pharmaceutical and medical equipment R&D benefits the most from patenting.

The literature on the impact of patents on innovation must be considered emergent. One reason is that the effect of patent policy has many dimensions, some fundamental to understanding the determinants of innovation generally,

---

[2]A related line of analysis suggests that compulsory licensing, which abrogates patent protection, need not be detrimental to innovation (Scherer, 1977; Bresnahan, 1985).

and these continue to challenge scholars both theoretically and empirically. For example, although the literature has identified offsetting impacts of intra-industry R&D knowledge flows on R&D incentives (Cohen and Levin, 1989), it is not clear how the protection patents afford and the information they disclose contribute to those flows and associated incentives. Another reason is insufficient data. Although patent and patent citation data are readily available and extensively used in the study of innovation (Jaffe and Trajtenberg, 2002), information on the uses and impacts of patenting is quite limited. For example, without data on the incidence and terms of patent licensing and associated fees and royalties, it is difficult to assess the efficiency and social welfare effects of markets for technology whose growth can depend on the allocation and strength of patent rights (Arora et al., 2001).

Other important data limitations relate to conflicts over patent rights. Although the studies in this volume and others are informative about trends in, parties to, costs of, and determinants of outcomes in formal patent lawsuits, litigation is only one aspect of firms' maneuvering to exploit patent positions against rivals and to enforce their intellectual property rights. Letters of notification claiming infringement and demanding licensing agreements are far more common than lawsuits and may have a significant effect on firm behavior, including R&D activity and entry. Moreover, the direct cost of prosecuting or defending cases does not represent the full range of litigation costs affecting innovation. Although we know anecdotally that participating in legal strategizing and discovery can consume considerable time of corporate managers and technical staff, we have no data on the associated opportunity costs, let alone on the innovative paths not taken as a result of actual or threatened litigation. In short, scholars gravitate to available data, leaving broad economic impacts only partially examined.

We now turn to reviewing the contents of this volume. Although sharing many of the limitations discussed above, these contributions advance our understanding of the determinants and social welfare implications of patent quality, litigation, and the extension of patenting into new technological domains—some of them in novel ways.

## PATENT QUALITY

Over the past decade the quality of issued patents has come under attack. The claim that quality has declined in a broad or systematic way has not been empirically tested, although Quillen and Webster's (2001) claim that patent approval rates are much higher (on the order of 80 to 90 percent or more) than officially reported is consistent with the hypothesis. The conjecture that patent quality is declining has been characterized in two ways. First, Barton (2000) and others have suggested that the standards for patentability—especially the non-obviousness standard—have been relaxed largely as a result of court interpretations. Second, other critics have suggested that the USPTO frequently issues patents for inventions that do not conform to the standards for patentability, especially in technology areas that are newly patentable, notably genomics, software, and business methods. Although separable in principle, the notion that standards for patentability are slipping and the notion that the USPTO is failing to apply the legal standards appropriately are difficult to distinguish in practice.

The first two chapters in this volume focus on whether the existing standards have been appropriately applied by the USPTO, not whether the standards themselves have changed. Although Griliches (1990) attempted to estimate a patent office "production function" and concluded that the number of patent examiners is the major determinant of the *number* of patents issued, these studies are the first to examine the impact of characteristics of the patent examination process and the examiners themselves on different indices of patent *quality*. This is an avenue of research that needs to be pursued vigorously if we are to have even a modest research basis for making informed judgments about how USPTO organization, management, and resources affect the quality and level of its output.

In both of the first two chapters the measure of quality is the likelihood that a patent's validity is challenged in litigation and is upheld or overturned by the courts. Merges (1999) suggests that the quality of patent examination, associated with what he calls "front-end costs," can have an important effect on the likelihood of whether a patent will be litigated, entailing "back-end costs." In modeling the choice between settlement versus litigation in the case of infringement, Meurer (1989) elaborates the intuition behind this assertion, namely that greater uncertainty over patent validity will lead to a higher incidence of both infringement and litigation. The chapter in this volume by Jonathan Levin and Richard Levin highlights the costs in addition to litigation that are associated with poor quality patents. They argue, "the holders of dubious patents may be unjustly enriched and the entry of competitive products and services that would enhance consumer welfare may be deterred." Further, they point out that "...uncertainty about what is patentable in an emerging technology may discourage investment in innovation and product development until the courts clarify the law, or, in the alternative, inventors may choose to incur the cost of product development only to abandon the market years later when their technology is deemed to infringe."

John King in his chapter provides the only evidence to date suggesting a link between the care with which patents are examined and subsequent litigation. In a simple regression controlling for both cost of litigation and USPTO examination (technology-based) groups, he shows a strong negative effect of average examination hours per examination group on the rate at which issued patents were involved in legal complaints. Although not providing evidence of a direct effect of quality on lawsuits, the result suggests that greater effort dedicated to examination does have a shielding effect. With a breakdown by examination group of the annual rate of patents involved in complaints per patent allowed, and estimates of median out-of-pocket legal costs, he even suggests, under plausible assumptions, that the social costs of increasing the resources dedicated to examination may be more than offset by the savings in "back-end" legal costs.

The study by Iain Cockburn, Samuel Kortum, and Scott Stern shifts the focus from the effects of characteristics of the examination process to the effects of characteristics of the examiners themselves. First, they demonstrate considerable heterogeneity across examiners, even within USPTO examination groups, with respect to tenure in the Office, the number of patents they have examined over time, and the degree to which their patents are subsequently cited in other patents. The authors then explore the link between those characteristics and subsequent validity decisions by the Court of Appeals for the Federal Circuit (CAFC) for the small set of patents that are ultimately litigated at that level. Perhaps their most noteworthy results are negative ones. They find that neither the length of experience of examiners, measured either in years or cumulative number of patents issued, or examiner workload, represented by the count of issued patents in the three months prior to issue date of a given patent, appears to have any impact on whether an issued patent is judged valid by the CAFC.

If the result means that fewer examination hours per patent, corresponding to a higher workload, has no impact on measures of patent quality, then we have a finding contrary to King's, described above. As the authors themselves acknowledge, however, these results warrant skepticism. They are based, for example, on what may be an unrepresentative sample, namely patent suits selected by the CAFC to give the court opportunities to decide significant points of law. To the degree that more representative validity decisions are made at the trial court level, it may be useful to repeat the analysis with a sample of those cases, especially in light of the importance of the relationship being studied for the management of the USPTO.

Cockburn and colleagues arrive at one clear result, that patents issued by examiners who tend to issue patents generating more citations in subsequent patents are more likely to be judged invalid by the CAFC. Although the robustness of their result remains to be seen, the authors interpret it as showing that some examiners systematically approve claims that are broader in scope and that such claims tend to be more vulnerable to invalidity judgments. The authors further interpret their result as suggesting that the courts provide a needed check on the predisposition of some examiners to issue patents with broader claims. This finding also underscores the observation of Gallini (2002), among others, that patent "strength" is an ambiguous and possibly misleading concept. What are commonly considered to be two dimensions of "strength"—claim breadth and enforceability—may well be at odds.

Reflecting growing concerns over patent quality, Merges (1999) has suggested that the United States consider adopting an administrative procedure similar to the post-grant patent opposition process employed in Europe and, more recently, Japan.[3] This volume provides two complementary perspectives on post-grant review of patent validity—first, an empirical comparison by Stuart Graham, Bronwyn Hall, Dietmar Harhoff, and David Mowery of certain features of the European opposition system and the current U.S. patent re-examination procedure and, second, Levin's and Levin's theoretical analysis of the social welfare effects of post-grant opposition. In what is almost certainly the first cross-national comparison of patent institutions and procedures, Graham and colleagues show that the European and U.S. systems for reviewing issued patents are very different from one another. The chief differences have to do with the roles of both challengers and patent holders in the proceedings, the grounds for challenge, and whether the outcome limits subsequent litigation. Among other things, the U.S. re-examination process makes it advantageous for the patent holder to request review to accommodate newly discovered prior art and disadvantageous for challengers to initiate re-examination (with the result that the owner-initiated cases constitute 40 percent of the total). This may account for the authors' finding that only 0.3 percent of U.S. patents were re-examined in the 1981-1998 period whereas 8.3 percent of European patents were subject to opposition. Graham and colleagues also show that in Europe oppositions focus on more commercially important patents and do not appear to be used by large established firms as a competitive weapon against smaller firms.

The most compelling reason to have a post-grant opposition procedure, according to both sets of authors, may be a reduction in patent litigation and its associated costs. An opposition proceeding that lowers uncertainty about a given patent's validity should make it less likely that the patent will be later litigated, assuming that both infringement and a mutual reluctance to settle patent disputes are associated with uncertainty about patents' validity. Further assuming that oppositions are on average resolved more quickly and inexpensively than lawsuits, Levin and Levin show that an opposition procedure can offer significant social welfare advantages over the current reliance on litigation to resolve issues of patent validity, especially in newly patented technologies where patent quality is most uncertain.

Graham and colleagues are unable to confirm the Levins' prediction that the use of opposition should substitute for subsequent litigation over validity if the process is speedier and cheaper. That is partly because European oppositions are relatively unconstrained by deadlines and tend to drag on for extended periods of time. Furthermore, the authors were not able to collect data on the litigation histories of patents that have gone through European opposition proceedings in time to

---

[3]Adoption of an opposition system is endorsed in principle by the leadership of the USPTO, in the agency's 21st Century Strategic Plan, and with more detail by the National Academies' Committee on Intellectual Property Rights in the Knowledge-Based Economy. See National Research Council. (2003). *A Patent System for the 21st Century*, Washington, D.C.: The National Academies Press.

incorporate in their analysis. Consequently, we do not yet know whether the European experience supports the plausible arguments for an opposition system.

There is another source of uncertainty regarding the social welfare impact of an opposition procedure. If prospective challengers can invalidate patents without incurring the high costs of litigation, there will be more challenges and, conceivably, a greater aggregate cost to society. That may well be the case for pharmaceutical and biotechnology patent disputes. Harhoff and Reitzig (2002) calculate that in Europe over 8 percent of such patents are opposed, while in the United States the litigation rate is just over 1 percent (Lanjouw and Schankerman, 2001). If the price elasticity of demand for post-grant validity checks via either opposition or litigation is roughly unity, then the argument for instituting an opposition procedure must be based on some benefit other than savings on litigation. Levin and Levin acknowledge the point and suggest that there are other substantial gains from a vigorous system of post-grant review of validity. In particular, it would help ensure that society realizes the benefits of conferring monopoly profit only upon those patented inventions representing a genuine technical advance and deserving of encouragement while minimizing the consumer welfare losses that invalid patents may impose.

Because the value distribution of patents is highly skewed (Scherer et al., 1959; Scherer and Harhoff, 2000) *and* the majority of patents are not even commercialized, an opposition procedure may well be more efficient than devoting additional resources to examining all patent applications more rigorously. Does the European opposition process tend to select for close scrutiny prospectively valuable patents? Consistent with Harhoff and colleagues' findings (1999, 2002) on the determinants of opposition for European Patent Office (EPO) biotechnology and pharmaceutical patents, Graham and colleagues find that the likelihood of opposition does indeed increase with forward citations, which Trajtenberg (1990) shows to be an indicator of social value.

## PATENT LITIGATION

The almost tenfold growth in patent litigation over the past two decades (Merz and Pace, 1994; Moore, 2001) and the escalating cost of prosecuting and defending an infringement suit (AIPLA, 1997, 2001) have raised concerns about the effects of the cost of patent litigation on R&D incentives and innovation (Barton, 2000). Although some observers have labeled these costs a tax on innovation, it may be that litigation is an essential complement to patenting itself and therefore part of the investment in innovation. Even if that is the case, it is appropriate to ask if the costs of litigation can be contained or reduced, possibly to the benefit of investment in innovation. The two chapters on patent litigation in this volume, by Jean Lanjouw and Mark Schankerman and by Rosemarie Ziedonis, take some initial, complementary steps to addressing these questions.

Building on their earlier work (2001), Lanjouw and Schankerman consider patent litigation in the United States as a whole from 1978 through 1995 and probe the features of patents and their owners that affect decisions to file suits and their outcomes. First they show that litigation rates—defined as decisions to file a suit normalized by numbers of patents—vary substantially across technology fields and that, once disaggregated in that way, litigation rates have not changed over the two decades of rapid growth in patent law suits. In other words, litigation has simply kept pace with patenting. Next they relate a range of characteristics of patents and their owners to decisions to file suit. Having a larger portfolio of patents or a history of repeated interaction with other established firms in the industry seems to reduce the probability of becoming involved in a suit. Consistent with that finding, asymmetry in size affects the probability of filing a suit; it is less likely that a firm that is large relative to likely disputants will file a suit. The authors also find that patents with a larger number of forward citations, indicative of greater value, are more likely to be the objects of suits while patents with more backward citations, reflecting more derivative or incremental innovation, are less likely to be litigated. Patents subject to greater rates of self-citation, interpreted as signaling more cumulative technologies, are more likely to be the subjects of suits, suggesting that the probability of conflict over intellectual property increases when pioneers and followers need to come to terms. This finding is consistent with the arguments of Scotchmer (1991) and Merges and Nelson (1990).

These various characteristics of patents and owners appear to affect only decisions to file suits, not their outcomes, including the likelihood of settlements prior to verdicts. Lanjouw and Schankerman interpret this finding as suggesting that the patent and firm characteristics that reduce suits—asymmetric firm size, large portfolios, and repeat interactions—may lower the social cost imposed by patent litigation. The analysis raises further issues, however. Larger firm size or possession of a large patent portfolio may reduce the direct costs of patent litigation but these characteristics may be associated with other costs. For example, large firms' threat letters (i.e., letters of notification of infringement) may chill smaller firms' incentives to undertake innovation in selected markets and may even discourage entry. Although the analysis of firm behavior regarding formal litigation and costs arising from it is extremely useful, we are still far from a complete understanding of the costs and benefits of conflict over intellectual property rights—as far as we are from fully understanding the costs and benefits of licensing.

Building on her earlier work with Hall (2001), Ziedonis's chapter in this volume presents a detailed picture of litigation trends in a single industry—semiconductors—between 1973 and 2000. The semiconductor industry has experienced some of the most visible patent settlements over the past two decades as well as a rapid growth in patents per dollar invested in R&D, rising from 0.3 in

1982 to 0.8 in 1997. Some scholars have suggested that this is the result of patent portfolio races in which incumbents aggressively amass patents for use in cross-licensing and to fend off patent infringements suits (Cohen et al., 2000; Hall and Ziedonis, 2001).

A key question is what the pattern of behavior in semiconductors tells us about whether patenting is serving its constitutionally mandated purpose of stimulating innovation. One possibility is that the sharp increase in patent filings reflects little more than a non-cooperative bargaining game wherein rival incumbents are amassing larger and larger portfolios for defensive reasons and the prospect of obtaining patents provides little incentive for R&D beyond that provided by other means of protecting inventions, such as secrecy or the exploitation of lead time advantages. Hall and Ziedonis (2001) call into question whether patenting stimulates R&D in semiconductors to any significant degree. On the other hand, the recent analysis by Arora and colleagues (2003) of the impact of patenting on R&D in major U.S. manufacturing sectors suggests that patenting does stimulate R&D in semiconductors, although not as much as in most other industries.

In describing litigation rates in the industry, Ziedonis emphasizes a different metric than that of Lanjouw and Schankerman. She acknowledges that the fraction of patents involved in legal disputes in the industry has not risen over time. To assess the social costs of patent litigation, however, Ziedonis calculates a litigation rate normalized by R&D expenditures. She shows that, relative to R&D spending, the patent litigation rate in semiconductors has risen 93 percent during the 1986-2000 period. Her interpretation is that: "...semiconductor firms have been directing a larger share of their innovation-related resources towards defending, enforcing, and challenging patents in courts since the mid-1980s...." To say that patent litigation has made innovation more costly is not to suggest that it has actually been a net drag on innovation in the industry, and Ziedonis does not go that far. To address that important question, one would have to estimate the net contribution to innovation that patents *and* their enforcement have made.

A prominent concern about the contemporary use of patents has to do with barriers to entry in industries such as semiconductors in which large patent portfolios are acquired and used as the basis for cross-licensing and that licensing takes the form of trading "like for like" (Shapiro, 2001). Barriers to becoming an integrated semiconductor manufacturer are no doubt very high, more as a consequence of the huge capital requirements of production than of incumbents' patent portfolios. But Ziedonis suggests that since the 1980s patent protection has underpinned the entry and growth of fabless chip design firms, an important component of the industry. Interestingly, fabless design firms tend to be more R&D-intensive and more prone to litigation than integrated manufacturers. The question of the overall impact of patents and their enforcement on the semiconductor industry's growth, structure, and technological advance remains open.

## CHALLENGES POSED BY NEWLY PATENTED TECHNOLOGIES

Apart from institutional changes, an important development in the patent system over the last generation has been the expansion of patentable subject matter. Patenting was extended to life forms with the landmark Supreme Court case of *Diamond v. Chakrabarty* in 1980 and subsequently to genes and gene fragments. The CAFC also endorsed the limited patentability of software, as an adjunct to a physical process, in its 1981 decision in *Diamond v. Diehr.* Seventeen years later, in *State Street Bank and Trust v. Signature Financial Group,* the same court rejected arguments against patents on "methods of doing business" and appeared to dispense with virtually all limitations on software-related subject matter.

The extensions of patents to biotechnology, software, and business methods have aroused controversy on a variety of grounds—the ethical implications of patenting life forms, especially human genetic material, the alleged burden on research and product development of patents on upstream research tools and foundational discoveries, and the proximity of some software and business method developments and DNA discoveries to ideas and information theoretically outside the scope of the patent system.

Three chapters in this volume consider aspects of American experience with the patenting of software, business methods, and biotechnology. They do not go very far in addressing the broad concerns raised by the extension of patenting into these areas, but they do illuminate some challenges they have posed for the patent system and its impact on innovation.

### Software and Business Methods

Stuart Graham and David Mowery describe the evolution of software from a "relatively open intellectual property regime to one in which formal protection, especially patents, figures prominently." They attribute this shift to the diffusion since the 1980s of microcomputing, giving rise to the growth of packaged software and, more recently, the internet. The authors document the rapid growth in patenting activity of packaged software companies but acknowledge that, historically, manufacturers of computers and other electronic systems have been the most aggressive software patenting companies. Their analysis highlights the corresponding decline in the "copyright propensity" of the largest packaged software firms, and they speculate that the shift is a function of the strengthening of patent relative to copyright protection in recent years.

Graham's and Mowery's analysis raises several questions for future work. For example, how will the use of software patents continue to evolve? What will be the effect on industry entry of aggressive software patenting on the part of larger incumbents? As software tends to develop cumulatively, what will be the effect of contemporary upstream patents on subsequent innovation in the industry? Finally, a good deal of software development and patenting stands outside of the well defined software technology categories examined by Graham and Mowery and, for the time being, eludes description and analysis.

Both the chapter by Graham and Mowery and the chapter by Allison and Tiller attempt to address the controversial question of patent quality in software generally and business methods in particular. Graham and Mowery use a relative measure of the "importance" of software patents—the forward citations to their sample of the patents obtained by large packaged-software firms relative to citations to all software patents for the technology areas that they examine. They find that citations to the large firms' patents exceeded those to all software firms' patents and that this ratio moved modestly upward through 1996. Although not reflective of the importance or quality of software patents in general, their data suggest that the relative importance of patents issued to large producers of personal computer software has not declined during a period in which their patenting rate has accelerated. A similar measure of relative importance suggests the software patents of large electronics firms also have not declined during this period.

Many business methods patents have come under attack for either being obvious or not novel in light of the contemporary practice and teaching of business methods. With regard to novelty, critics have claimed that the USPTO is either inattentive or lacks access to relevant non-patent prior art in business literature. In the first attempt to evaluate this claim empirically, Allison and Tiller compare the number of non-patent prior art references (i.e., backward citations) in a sample of internet business method patents to those found in a sample of all other patents. They find that there were substantially more total references, patent references, and non-patent references in the business methods patents than in the general sample of patents. Nevertheless, Allison and Tillers' data cannot answer several intriguing questions. For example, is the body of non-patented prior art in the area of business methods so large or diverse that examiners are still missing a good share of it? Does the examination process overlook some business methods that are in common use but not documented in written sources? Notwithstanding these uncertainties, the USPTO appears to be paying more attention to non-patented prior art in the examination of business method patent applications than is widely assumed to be the case.

### Biomedicine

Economic research has made a convincing case that in at least one area—pharmaceuticals—patents have played a critical role in stimulating technical advance (Scherer et al., 1959; Mansfield, 1986; Levin et al., 1987; Cohen et al., 2000). In recent years, however, a few scholars have speculated that in some circumstances the opposite may be the case—patents may now be impeding drug discovery and development. Two related concerns about the patenting and licensing of biomedical innovations have been articulated. First, Heller and Eisenberg

(1998) posited what they termed a "tragedy of the anticommons," resulting when heterogeneous players assert numerous property rights claims to separate building blocks for some product or line of research. In these circumstances negotiations to assemble the rights may fail, blocking otherwise promising lines of research or product development. Concern focused initially on access to "research tools" (i.e., inputs into the discovery of new drugs, diagnostics, and therapies), which some firms and many public research institutions were beginning to patent extensively. A related argument, previously developed in general terms by Merges and Nelson (1990) and Scotchmer (1991), is that patents on upstream discoveries if sufficiently broad in scope can impede follow-on discoveries and development if access to the foundational intellectual property is restricted.

In this volume, John Walsh, Ashish Arora, and Wesley Cohen present the first empirical evidence regarding the impact of research tool patenting and licensing on biomedical innovation. The authors draw upon 70 interviews with firms, intellectual property practitioners, and university and government personnel to address two questions: 1) whether an emergent anticommons is in fact impeding the development and commercialization of new drugs, diagnostics, and other therapies; and 2) whether restricted access to patents on upstream, foundational discoveries is blocking important follow-on research and innovation. The preconditions for both results appear to exist. There are now more patents associated with any new therapeutic product. Furthermore, since the passage in 1980 of the Bayh-Dole Amendment, encouraging nonprofit research institutions and small businesses to acquire title to inventions developed with public support, many research universities, the locus of fundamental upstream discoveries, have been patenting and licensing more aggressively.

Walsh and colleagues do not find, however, that these developments are yet impeding the development of drugs or other therapies in a significant way. First, the number of patents required for most projects remains manageable. Most importantly, firms and other institutions have developed a number of "working solutions" that limit the effects of the intellectual property complexities that exist. These range from the normal responses of licensing and occasional litigation to other less visible solutions, including fairly pervasive infringement of patents in the course of laboratory research at a pre-product stage. Such infringement seems to be common in both public research institutions and firms and is informally rationalized as causing no commercial harm and, in any event, shielded from infringement liability by the court-interpreted "research exemption."[4] Finally, the

---

[4]This prevalent and questionable assumption has been clearly contradicted by an October 2002 decision by the Court of Appeals for the Federal Circuit, in *Madey v. Duke University,* 64 USPQ2d 1737 (CAFC 2002). Although agreeing that research "solely for amusement, to satisfy idle curiosity, or for strictly philosophical inquiry," is protected, the court ruled that the protection does not extend to organized scientific research activity pursued as part of the legitimate business of an institution, whether nonprofit or for-profit. As Walsh and colleagues observe, this decision undermines one of the working solutions that has contributed to the progress of biomedical research.

National Institutes of Health (NIH), other influential research funders, and some scientific publications have encouraged ease of access to important research materials and tools.

An exception to this general finding involves cases where the intellectual property required for follow-on research is the same intellectual property associated with diagnostic tests for genetic predisposition to particular diseases. Walsh and colleagues also suggest that restricted access to upstream discoveries could substantially impede subsequent research and development for particular disease categories and therapies in the future and therefore recommend careful monitoring. Were significant impediments to emerge, a solution would, however, be difficult to devise given the importance of patents to biomedical innovation generally.

## CONCLUSION

Over the past two decades, policy and court decisions have moved patent policy toward a regime of stronger enforcement and extended patents into new domains of technology. At the same time, firms and public research institutions in technology sectors important to the economy now and in the future —biotechnology and pharmaceuticals and computer hardware components and software— have embraced patenting aggressively. These changes have proceeded, however, with a limited understanding of their consequences. The National Academies' STEP Board's initiative to support original research on the patent system represents a modest step to illuminate these consequences. The chapters that follow focus on the issue of patent quality, the transactions costs imposed by patent enforcement through litigation, and some of the challenges posed by the extension of patentability to new domains, a process that almost certainly will be repeated indefinitely. Few of these contributions point to particular policy prescriptions and those that are prescriptive are not definitive, but they have informed the findings and recommendations of the STEP Board's Committee on Intellectual Property Rights in the Knowledge-Based Economy and should be useful in other policy deliberations.

## REFERENCES

AIPLA. (1997, 2001). "Report of the Economic Survey." Arlington, VA: American Intellectual Property Law Association.

Arora, A., M. Ceccagnoli, and W. M. Cohen. (2003). "R&D and the Patent Premium." NBER Working Paper No. 9431.

Arora, A., A. Fosfuri, and A. Gambardella. (2001). *Markets for Technology: Economics of Innovation and Corporate Strategy.* Cambridge, MA: MIT Press.

Barton, J. (2000). "Intellectual Property Rights. Reforming the Patent System." *Science* 287: 1933-34.

Bresnahan, T. (1985). "Post-entry Competition in the Plain Paper Copier Market." *American Economic Review.* 75(2): 15-19.

Cohen, W. M., and R. C. Levin. (1989). "Empirical Studies of Innovation and Market Structure." In R. Schmalensee and R. Willig, eds., *Handbook of Industrial Organization*. Amsterdam: North-Holland Publishers.

Cohen, W. M., R. R. Nelson, and J. P. Walsh. (2000). "Protecting Their Intellectual Assets: Appropriability Conditions and Why U.S. Manufacturing Firms Patent (or not)." NBER Working Paper 7522.

Gallini, N. (2002). "The Economics of Patents: Lessons from Recent U.S. Patent Reform." *Journal of Economic Perspectives* 16(2): 131-154.

Griliches, Z. (1990). "Patent Statistics as Economic Indicators: A Survey." *Journal of Economic Literature*. 28(4): 1661-1707.

Hall, B. H., and R. H. Ziedonis. (2001). "The Patent Paradox Revisited: An Empirical Study of Patenting in the U.S. Semiconductor Industry, 1979-1995." *RAND Journal of Economics*. 32(1): 101-128.

Harhoff, D., and M. Reitzig. (2002). "Determinants of Opposition Against EPO Patent Grants: The Case of Biotechnology and Pharmaceuticals." CEPR Discussion Paper DP3645.

Harhoff, D., F.M. Scherer, and K. Vopel. (1999). "Citations, Family Size, Opposition and the Value of Patent Rights." Discussion paper at http://emlab.berkeley.edu/users/bhhall/harhoffetal99.pdf

Heller, M. A., and R. S. Eisenberg. (1998). "Can Patents Deter Innovation? The Anticommons in Biomedical Research." *Science* 280(May 1): 698-701.

Horstmann, I., J. M. MacDonald, and A. Slivinski. (1985). "Patents as Information Transfer Mechanisms: To Patent or (maybe) Not to Patent." *Journal of Political Economy* 93: 837-858.

Jaffe, A. (2000). "The U.S. Patent System in Transition: Policy Innovation and the Innovation Process." *Research Policy* 29: 531-558.

Jaffe, A., and M. Trajtenberg. (2002). *Patents, Citations, and Innovations: A Window on the Knowledge Economy*. Cambridge, MA: MIT Press.

Kortum, S., and J. Lerner. (1999). "What Is Behind the Recent Surge in Patenting?" *Research Policy* 28: 1-22.

Lanjouw, J., and M. Schankerman. (2001). "Characteristics of Patent Litigation: A Window on Competition." *RAND Journal of Economics* 32(1): 129-151.

Levin, R. C., A. K. Klevorick, R. R. Nelson, and S. G. Winter. (1987). "Appropriating the Returns from Industrial R&D." *Brookings Papers on Economic Activity* 3: 783-820.

Mansfield, E. (1986). "Patents and Innovation: An Empirical Study." *Management Science* 32: 173-181.

Merges, R. (1999). "As Many as Six Impossible Patents Before Breakfast: Property Rights for Business Concepts and Patent System Reform." *Berkeley Technology Law Journal* 14: 578-615.

Merges, R., and R. R. Nelson. (1990). "On the Complex Economics of Patent Scope." *Columbia Law Review* 90: 839-916.

Merz, J., and N. Pace. (1994). "Trends in Patent Litigation: The Apparent Influence of Strengthened Patents Attributable to the Court of Appeals for the Federal Circuit." *Journal of the Patent and Trademark Office Society* 76: 579-590.

Meurer, M. (1989). "The Settlement of Patent Litigation." *RAND Journal of Economics* 20: 77-91.

Moore, K. (2001). "Forum Shopping in Patent Cases: Does Geographic Choice Affect Innovation?" *North Carolina Law Review* 79: 889.

Quillen, C. D., and O. H. Webster. (2001). "Continuing Patent Applications and Performance of the U.S. Patent Office." *Federal Circuit Bar Journal* 11(1): 1-21.

Scherer, F. M. (1977). *The Economic Effects of Compulsory Patent Licensing*. New York: New York University Graduate School of Business Administration.

Scherer, F. M., and D. Harhoff. (2000). "Technology Policy for a World of Skew-Distributed Outcomes." *Research Policy* 29: 559-566.

Scherer, F. M., S. Herzstein, Jr., A. Dreyfoos, W. Whitney, O. Bachmann, C. Pesek, C. Scott, T. Kelly, and J. Galvin. (1959). *Patents and the Corporation: A Report on Industrial Technology Under Changing Public Policy.* 2nd Edn. Boston, MA: Harvard University, Graduate School of Business Administration.

Scotchmer, S. (1991). "Standing on the Shoulders of Giants: Cumulative Research and the Patent Law." *Journal of Economic Perspectives* 5: 29-41.

Shapiro, C. (2001). "Navigating the Patent Thicket: Cross Licenses, Patent Pools, and Standard Setting." In A. Jaffe, J. Lerner, and S. Stern, eds., *Innovation Policy and the Economy*, Vol. 1. Cambridge, MA: National Bureau of Economic Research and MIT Press.

Taylor, C. T., and Z. A. Silberston. (1973). *The Economic Impact of the Patent System: A Study of the British Experience.* Cambridge: Cambridge University Press.

Trajtenberg, M. (1990). "A Penny for Your Quotes: Patent Citations and the Value of Innovations." *Rand Journal of Economics* 21(1): 172-187.

# Patent Quality

# Are All Patent Examiners Equal? Examiners, Patent Characteristics, and Litigation Outcomes[1]

Iain M. Cockburn
Boston University and NBER

Samuel Kortum
University of Minnesota and NBER

Scott Stern
Northwestern University, Brookings Institution, and NBER

### ABSTRACT

*We conducted an empirical investigation, both qualitative and quantitative, on the role of patent examiner characteristics in the allocation of intellectual property rights. Building on insights gained from interviewing administrators and patent examiners at the U.S. Patent and Trademark Office (USPTO), we collected and analyzed a novel data set of patent examiners and patent litigation outcomes. This data set is based on 182 patents for which the Court of Appeals for the Federal Circuit (CAFC) ruled on validity between 1997 and 2000. For each patent, we identified a USPTO primary examiner and collected historical statistics derived from the examiner's entire patent examination history. These data were used to conduct an exploratory investigation of the connection between the patent examination process and the strength of ensuing patent rights. Our main findings are as follows: (i) Patent examiners and the patent examination process are not homogeneous. There is substantial variation in observable characteristics of patent examiners, such as their tenure at the USPTO, the number of patents they have exam-*

---

[1] We thank USPTO personnel for offering their time and insight, members of the STEP Committee on the Intellectual Property Rights in a Knowledge-Based Economy, Wesley Cohen, George Elliott, and an anonymous reviewer for their comments and suggestions. Tariq Ashrati provided excellent research assistance. All errors, however, remain our own.

ined, and the degree to which the patents that they examine are later cited by other patents. (ii) There is no evidence in our data set that examiner experience or workload at the time a patent is issued affects the probability that the CAFC will find a patent invalid. (iii) Examiners whose patents tend to be more frequently cited tend to have a higher probability of a CAFC invalidity ruling. Although we interpret these results cautiously, our findings suggest that all patent examiners are not equal and that one of the roles of the CAFC is to limit the impact of discretion and specialization on the part of patent examiners.

## INTRODUCTION

Recent years have seen a worldwide surge in interest in intellectual property rights, particularly patents, in academia, in policy circles, and in the business community. This heightened level of interest has produced a substantial body of research in economics ranging from analyses of decisions to use patents rather than alternative means of protecting intellectual property (Cohen et al., 2000) to studies of the ways in which patents are used and enforced once granted (see, for example, Hall and Ziedonis, 2001; Lanjouw and Schankerman, 2001; Lanjouw and Lerner, 2001). However, little systematic attention has been paid to the *process* of how patent rights are created.

Indeed, only recently have researchers begun to develop a systematic understanding of the differences in intellectual property regimes across countries and over time (Lerner, 2002). Moreover, except for some preliminary aggregate statistics (Griliches, 1984; 1990), there are no published studies of the empirical determinants of patent examiner productivity, or of linkages between characteristics of patent examiners and the subsequent performance of the patent rights that they issue.[2] This chapter offers a preliminary evaluation of the role that some aspects of the examination process may have in determining the allocation of patent rights, in particular the consequences of specialization of examiners in specific technologies and their exercise of discretion in examining patent applications.

Filling in this gap in our knowledge may yield a number of benefits. First, and perhaps most importantly, it is difficult to assess the likely impact of changes in the funding or operation of the U.S. Patent and Trademark Office (USPTO) without some understanding of the "USPTO production function." For example,

---

[2]King in this volume offers an examination complementary to the one conducted here, in which he undertakes a detailed analysis of the impact of resource allocation per se on "art unit" performance, whereas our quantitative research focuses on how examiner characteristics and workload might impact litigation outcomes. The overall literature on the use of patent statistics and the impact of patents on innovation is far too large to be summarized here, but see Levin et al. (1987), Griliches (1990), Cohen et al. (2000), and Hall et al. (2001) for an introduction.

at various points in the past there have been shifts in the resources available to the USPTO as well as in the incentives and objectives provided to examiners, recently focused on reducing the time taken between initial filing of a patent application and final issuance. At the same time, court rulings and revisions in USPTO practice have broadened intellectual property protection into new areas, such as genomics and business methods, where the novelty and obviousness of inventions and the scope of awarded claims may be difficult to assess. These developments raise several important policy concerns. How do the structure and process of patent examination impact the allocation of intellectual property rights? How might changes in the structure and process of examination, from the provision of new incentives to the establishment of new examination procedures, impact patent application and litigation outcomes?

Our analysis has both qualitative and quantitative components. In the first part of the chapter, we review our qualitative investigation, in which we developed an informal understanding of the process of patent examination and investigated potential areas for differences among patent examiners to impact policy-relevant measures of the performance of the patent system. The key insight from our qualitative analysis is that "there may be as many patent offices as patent examiners." On the basis of this insight, we hypothesize that there may be substantial—and quantifiable—heterogeneity among examiners and that this heterogeneity may affect the outcome of the examination process. In the remainder of the chapter we develop some exploratory tests of this hypothesis.

To perform our quantitative analysis we constructed a novel data set linking USPTO "front page" information for issued patents with data based on the U.S. Court of Appeals for the Federal Circuit (CAFC) record between 1997 and 2000. We considered a sample of 182 patents: those on which the CAFC issued a ruling on validity during this period. For each patent, we identified the primary and secondary examiners associated with the patent and collected the complete set of patents issued by that examiner during his or her tenure at the USPTO. We then constructed measures based on this examiner-specific patent collection, including the examiner's experience with examination, workload, and measures based on the citation patterns associated with issued patents.

Our sample of "CAFC-tested" patents comes with several limitations. First, it is fairly small, giving us relatively little statistical power for testing some of our hypotheses, such as the effect of examiner experience on subsequent judgments of validity. Second, the sample excluded all patent litigation that had been settled before appeal or had not been appealed from the District Court level. It is quite possible that a lot of the more apparent validity decisions were taken care of below the CAFC level. With these caveats in mind, however, our data set does offer a valuable first look at the characteristics of examiners associated with those patents receiving a high level of judicial scrutiny. We hope that follow-up research will be undertaken to examine whether our findings are confirmed using broader samples of court-tested patents.

We present our key findings in several steps. First, we show that patent examiners differ on a number of observable characteristics, including their overall experience at the USPTO (both in terms of years as well as total number of issued patents), their degree of technological specialization, their propensity to cite their own patents, and their propensity to issue patents that are highly cited. Indeed, a significant portion of the overall variance among patents in measures such as the number and pattern of citations received, the number and pattern of citations made, and the approval time can be explained by the identity of the examiner—in the language of econometrics, "examiner fixed effects." These examiner effects are significant even after controlling for the patent's technology field and its cohort (i.e., the year the patent was issued).

We then turn to an examination of whether observable characteristics of our sample of CAFC-tested patents, such as their citation rate or approval time, can be tied to observable characteristics of examiners, such as their experience or the rate at which "their" patents receive citations. Here we find intriguing evidence for the impact of examiners. For example, there is a significant positive relationship between the citations received by a subsequently litigated patent and the "propensity" of its examiner to issue patents that attract a large number of citations. We then tie these relationships to patent validity rulings. Our econometric results provide evidence of a linkage between the patent examination procedures and litigation outcomes. Although the outcome of a test of validity by the CAFC is unrelated to the number of citations received by that particular patent, validity *is* related to the portion of the citation rate explained by the examiner's idiosyncratic propensity to issue patents that receive a high level of citations. This examiner-specific citation rate may reflect a number of aspects of the patent examination process, and it may therefore be difficult to attach an unambiguous interpretation to this measure. On the one hand, examiner-specific differences in the propensity of "their" patents to receive future citations may capture differences in the "generosity" of examiners in allowing claims. On the other hand, this variable may capture the impact of examiner specialization, as a consequence of an examiner concentrating on an especially "hot" technology area where patents attract large numbers of future citations. Nonetheless, our empirical findings suggest that USPTO patent examination procedures do allow for significant differences across examiners in the nature and scope of patent rights that are granted. This finding points to an important role for litigation and judicial review in checking the impact of discretion and specialization in the patent examination process.

The remainder of the chapter is organized as follows. In the next section, we review our qualitative data gathering, and motivate the evidence for our key testable hypotheses, which we state in the third section. The fourth section describes the novel data set we have constructed, and the fifth section reviews the results. A final section offers a discussion of our findings and identifies areas for future empirical research in this area.

# THE PATENT EXAMINATION PROCESS

## Methodology

This section reviews the initial stage of our research, a qualitative investigative phase in which we sought to understand the process of patent examination and the potential role of patent examiner characteristics in that process. This type of investigation is precisely what has been lacking from much academic and policy discussion of the impact of patent office practices, procedures, and personnel on the performance of the intellectual property rights system. Although practitioners and USPTO personnel are intimately acquainted with these procedures, there has been little attempt to identify which aspects of the examination process can be linked through rigorous empirical analysis to the key policy challenges facing USPTO.

Overall, our qualitative research phase included interviews with approximately 20 current or former patent examiners and an equal number of patent attorneys with considerable experience in patent prosecution. This phase involved three distinct stages. First, we informally interviewed former patent examiners and patent attorneys outside the USPTO to develop a basic grounding in the process and procedures of the USPTO and to evaluate some of our initial hypotheses on the impact of patent examiner characteristics and USPTO practice on the allocation of intellectual property rights. We developed a proposal based on this working knowledge to undertake systematic interviews within the USPTO, and with the assistance of the National Academies' STEP Board, we met with senior USPTO managers to discuss administering a survey linking detailed information about examiner history with information that could be gleaned from patent statistics about differences among patent examiners. We were unable to obtain approval to distribute a systematic survey of our own design to a broad cross section of current and former examiners, but USPTO management generously allowed us to conduct informal interviews and question-and-answer sessions during several visits with a small number of examiners, mostly those in a supervisory role. These conversations were very helpful in developing more subtle, precise, and econometrically testable hypotheses. In the third stage of qualitative research, we confirmed the viability of our hypotheses with individuals external to the USPTO.

## The Examination Process

Here we describe the patent examination process in general terms, focusing on the aspects for which we identified potential sources of heterogeneity in examination practice. The USPTO is one of the earliest and among the most visible agencies of the federal government, receiving more certified mail per day than any other single organization in the world. Located in a single campus of connected buildings, the USPTO is staffed by over 3,000 patent examiners and has more than 6,000 total full-time equivalent employees. In recent years the examiner corps has been responsible for over 160,000 patent approvals per year. The federal government raises nearly $1 billion in revenue per year from the fees and other revenue streams associated with the USPTO.

The work flow and procedures associated with patent approval are quite systematic and well-determined.[3] After arriving at a central receiving office, and passing basic checks to qualify for a filing date, patent applications are sorted by a specialized classification branch[4] that allocates them to one of 235 "Art Units"— a group of examiners who examine closely related technology and constitute an administrative unit. Within the Art Unit, a "Supervisory Patent Examiner" (a senior examiner with administrative responsibilities) looks at the technology claimed in the application and assigns it to a specific examiner. Once the patent is allocated to a given examiner, that examiner will, in most cases, have continuing responsibility for examination of the case until it is disposed of—through rejection, allowance, or discontinuation. The examination process therefore typically involves an interaction between a single examiner and the attorneys of the inventor or assignee. Although the stages associated with this process are relatively structured (and exhaustively documented in the *Manual of Patent Examining and Procedure*), they leave substantial discretion to the examiner in how to deal with a particular application.

The examination of an application begins with a review of legal formalities and requirements and an analysis of the claims to determine what the claimed invention actually is. The examiner also reads the description of the invention (part of the "specification") to ensure that disclosure requirements are met. The next step is a search of prior art to determine whether the claimed invention is anticipated by prior patents or nonpatent references and whether the claimed invention is obvious in view of the prior art. There is considerable scope for heterogeneity in this search procedure. The prior art search typically begins with a review of existing U.S. patents in relevant technology classes and subclasses, either through computerized tools or by hand examination of hard copy stacks of issued patents, and may then proceed to a word search of foreign patent documents, scientific and technical journals, or other databases and indexes. USPTO's Scientific and Technical Information Center maintains extensive collections of reference materials. Word searches typically require significant skill and time to conduct effectively.

---

[3] In this short discussion, we do not cover the legal requirements for patentability, because these are covered in great detail elsewhere. Indeed, the departure point between our analysis and more of the prior literature in this area is that we are principally concerned with the actual process of examination rather than the standards as defined by the patent law.

[4] This sorting function identifies and appropriately treats applications with national security implications.

The applicant may also include significant amounts of material documenting prior art with the application. The extent to which examiners review this nonpatent material may be a function of the nature of the technology, the maturity of the field, and the ease with which it can be searched. For example, in science-intensive fields like biotechnology where much of the relevant prior art is in the form of research articles published in the scientific literature and indexed by services such as Medline, examiners may rely extensively on nonpatent materials. In very young technologies, or in areas where the USPTO has just begun to grant patents, there may be very limited patent prior art. In more mature technologies examiners may have only a moderate interest in nonpatent materials and a limited ability to easily or effectively search them. Although the scope of patent examination prior art searches has been criticized, our interviews of USPTO personnel suggest that senior USPTO management are keenly aware of external critiques of the examination process and that a variety of initiatives have been set in motion to address some of these limitations.

Once relevant prior art has been identified, the examiner obtains and reads relevant documents. Again, different examiners and different Art Units may use substantially different examination technologies. For example, although many of the mechanical Art Units have historically relied on the "shoes" (the storage bins for hard copy patent documents), and may search for prior art primarily by viewing drawings, a typical search in the life sciences can involve detailed algorithmic searches by computer to evaluate long genetic sequences and review of tens or hundreds of research articles and other references. Some examiners may develop and keep close to hand their own specialized collections of prior art to facilitate searching. Indeed, patent examiners identify and frequently refer to "favorite" examples of prior art that usefully describe ("teach") the technology area and the bounds of prior art in a way that facilitates the examination of a wide range of subsequent applications.[5]

After the specification is reviewed to ensure that it provides an adequate "enabling disclosure" and an appropriate wording of claims, the initial examination is complete. The examiner then arrives at a determination of whether or not the claimed invention is patentable and composes a "first action" letter to the applicant (or, normally, the applicant's attorney) that accepts ("allows"), or rejects, the claims. Some applications may be allowed in their entirety upon first examination. More commonly, some or all of the claims are rejected as being anticipated by the prior art, obvious, not adequately enabled, or lacking in utility, and the examiner will write a detailed analysis of the basis for rejection. The

---

[5]Many of these "favorites" are university or public sector patents, which may be written less strategically than those for private firms. In part, this may help to explain the finding that university patents are more highly cited than control patents by private firms (Jaffe, Henderson, and Trajtenberg, 1998).

applicant then has a fixed length of time to respond by amending the claims and/ or supplying additional evidence or argument. After receiving and evaluating this response, the examiner can then "allow" the application if it is satisfactory (the most common stage in the process at which an application proceeds on to final issuance of a patent), negotiate minor changes with the attorney, or write a "second action" letter, which maintains some or all of the initial rejections. In this letter the examiner is encouraged to point out what might be done to overcome these rejections. Although at this stage the applicant's ability to further amend the application is formally somewhat restricted, in effect, additional rounds of negotiation between the examiner and applicant may ensue. The applicant also has the opportunity to appeal decisions for re-examination or evaluation within an internal USPTO administrative proceeding. However, such actions are quite rare; most applications are allowed (or not) on the second or third action letter.

USPTO operates various internal systems to ensure "quality control" through auditing, reviewing, and checking examiners' work. This includes the collection and analysis of detailed statistics about various measures of examiner work product flow. For example, Supervisory Patent Examiners, as well as their supervisors, routinely evaluate data relating to the distribution of times to action and the number of actions required before "disposal" of an application through allowance, abandonment, or appeal. These measurements are one of the many tools that USPTO uses to refine the internal management of the examination process.

It is also useful to note that examiners are allocated fixed amounts of time for completing the initial examination of the application and for disposal of the application. However, examiners are free to average these time allotments over their caseload. Moreover, there are differences in these time allocations across technology groups, and there also have been changes over time. Although we do not explore this variation in the current study, exploiting these changes in USPTO practice across technology groups and over time could give some leverage for understanding the relationship between time constraints and patent quality.

## Examiner Training and Specialization

Variation among examiners in their conduct of the examination process may arise from several sources. We focus here on two possibilities suggested by our interviews. First, at a given point in time, or for a particular patent cohort, examiners necessarily vary substantially in their experience. Experience may affect the quality of patent examination, and this has been a source of concern in recent years as the rate of hiring into the USPTO has increased, particularly into art areas with little in-house expertise. On the other hand, our qualitative research greatly emphasized the role of the systematic apprenticeship process within the USPTO, which is likely to reduce errors made by junior examiners. For the first several years of their career, examiners are denoted as Secondary Examiners and their work is routinely reviewed by a more senior Primary Examiner. Over time, the Secondary Examiner takes greater control over his/her caseload and the Primary Examiner focuses on teaching more subtle lessons about the practice of dealing with applicants and their attorneys and instilling the delicate "not too much, not too little" balance that the USPTO is trying to achieve in the patent examination process.

Second, as alluded to above, Art Units may vary substantially in their organization and functioning. In the most traditional group structure, the allocation of work promotes a maximal amount of specialization by individual examiners. For example, in many of the mechanical Art Units, an individual examiner may be responsible for nearly all of the applications within specific patent classes or subclasses. In other Art Units, however, the approach is more team-oriented. In these groups, there is less technological specialization (multiple subclasses are shared by multiple examiners) and there is likely a higher degree of discussion and knowledge sharing among examiners. In the more specialized organization, there are far fewer checks and balances on the practices of a given examiner. When the examiner has all of the relevant technological information; the cost for an auditor to effectively review his/her work becomes very high. By contrast, in less specialized environments, there are likely to be greater opportunities for monitoring, although, obviously, decreased specialization may reduce examiners' level of expertise in any specific area.

In part because of specialization, primary examiners maintain substantial discretion in their approach to individual applications. Our qualitative interviews suggest that this latitude may result in variation among examiners in how they *balance* multiple USPTO objectives. Consider the impact of the Clinton administration program (headed up by Vice President Gore) to establish the USPTO as a "Performance-Based Organization" (National Partnership for Reinventing Government, 2000).[6] Among other goals, this initiative encouraged examiners to treat applicants as *customers* and to *cooperate* with applicants' attorneys to define and allow (legitimate) claims. Although not changing the *formal* standards for claims assessment, this program encouraged examiners to use their discretion to increase the applicants' ability to receive at least some protection for inventions. In our qualitative interviews, there were significant differences among examiners in how they claimed to respond to this new "customer" orientation. Although some acknowledged that it changed their approach to interactions with applicants' attorneys, others claimed that it had "made no difference" for the day-to-day "balancing act" associated with allowing claims. This heterogeneous response to a single well-defined change in USPTO policy supports our hypothesis that examiners may vary in their approach to the examination process.

---

[6]The interviews for this project took place between June 2000 and June 2001.

## Qualitative Findings

Our qualitative investigation of the patent examination process both generated a number of insights central to our hypothesis development and raised some flags about potential hazards for empirical research in this area.

The first key finding from the qualitative evaluation of patent examination can be summarized in the phrase of one of our informants: "There may be as many patent offices as there are patent examiners." In other words, although the examination process is relatively structured, and USPTO devotes considerable resources to quality control, substantial discretion is provided to examiners in how they deal with applications, and the extent to which they exercise this discretion can potentially vary substantially across examiners. Several features contribute to this potential for heterogeneity, including the formal emphasis on specialization, variation among Art Units and individual examiners in their approach to searching prior art, the fact that much learning is through an apprenticeship system with only a small number of mentors, and the existence of differences across groups and examiners in the time allocated to specific tasks and examination procedures.

This heterogeneity might manifest itself in several ways. First, there may be substantial variation across examiners in the breadth of patent grants—some examiners may have a propensity to systematically allow a more restrictive or more expansive set of claims. One potential consequence of this use of discretion may be that patents issued by examiners who tend to allow broader claims will impinge on a greater number of follow-on inventions and therefore receive more citations over time. Although prior research has emphasized the degree to which the number of citations received by a patent is an indicator of its underlying inventive significance, it is important to recognize that a given patent's propensity to receive future citations may also be related to the "generosity" of the examiner in allowing a broad patent, relative to an average examiner's practice.

Second, examiners differ as a result of specialization. Perhaps the key consequence of the organizational structure of the USPTO is the existence of only a handful of examiners within a narrowly defined technological field at a point in time. Specialization confers several benefits, most notably the development of "deep" human capital in established technology areas. At the same time, specialization can bring its own challenges. By construction, specialization raises the costs of monitoring, because it is difficult to disentangle whether the "practice" of a given examiner reflects the nature of the art under his or her purview or reflects idiosyncratic aspects of that examiner that are independent of the art. For example, examiners may vary in their observed propensity for self-citation. (Self-citation is the practice by which examiners tend to include citations to "their" patents, i.e., patents for which they were the examiner.) A high degree of self-citation may reflect an examiner's reluctance to search beyond a narrow set of prior art with which he is already familiar. But it may equally be driven by the technology area in which the examiner works. Our interviews suggest that a high degree of self-citation is particularly likely for examiners working in technology areas that are highly compartmentalized, with little communication across examiners, and that are highly reliant on hard copy technologies for the prior art search process.

Another impact of specialization may be to reduce the sensitivity of the USPTO to *new* technology areas. Before the establishment and development of norms for new Art Units, patent applications in a new technology area may be "shoehorned" into existing Art Units. As a result, in the earliest stages of a new technology (a time when the standards of patentability are being established), the examination process depends heavily on the idiosyncratic knowledge base of a small group of examiners with limited expertise in the new technology area. Although the establishment of new Art Units and the development of new standards can address such problems over time, relying on highly specialized examiners in the earliest stages of a new technology area may slow the rate at which USPTO can establish and implement such norms and procedures.

Third, examiners may vary substantially in their effective average "approval time," the length of time between initial application and the date at which the patent issues. Although a large fraction of the lag between application and approval will, of course, be driven by external forces—the speed at which applicants respond to office actions, for example—differences across Art Units and across examiners in their workload and the type of applications they receive will likely lead to differences in average approval time. It is an interesting question whether this involves a trade-off with other dimensions of quality, specifically the ability to withstand judicial scrutiny.

At the same time that this qualitative analysis formed the basis for our hypotheses concerning how examiners might influence the allocation of patent rights, it also suggested several limitations to any empirical work and some challenges that must be overcome before drawing policy conclusions from it. First, and perhaps most importantly, the analysis highlighted the importance of taking account of variation across technologies and patent cohorts in any empirical analysis. Our investigation suggests that there are large differences across Art Units in examination practice, and these technology effects must be controlled for. In addition, examination practice, resources, and management processes have changed over time, so it is also necessary to control in a detailed way for the cohort in which a particular patent was granted.

Second, we were prompted to be cognizant of how noisy the underlying data generation process is likely to be. Much of the variation in any observable patent characteristic is likely to reflect the nature of the invention, the behavior of the applicant, and other unobserved factors. Our guarded interpretation of the econometric results presented below reflects our recognition that we are investigating rather subtle relationships, in which the impact of examiner effects may be difficult to evaluate in light of the overall noisiness of the data-generating process.

Finally, our qualitative analysis clearly indicated that our econometric analysis should recognize and incorporate the fact that the USPTO has multiple objectives and that there is no single "silver bullet" measure of performance, particularly among easily available statistics. Although, all else being equal, shorter approval times are socially beneficial (particularly in the era when disclosure did not occur until the patent was issued), speed is not a virtue in and of itself; achieving shorter approval times may require trade-offs with other objectives, such as enforceability. With these caveats in mind, we now turn to a fuller development of testable (though exploratory) hypotheses associated with examiner characteristics.

## HYPOTHESIS DEVELOPMENT

Our empirical analysis is organized around two sets of hypotheses, those reflecting the relationship between patent characteristics and examiner characteristics and those reflecting the relationship of patent litigation outcomes to patent and/or examiner characteristics.

### The Impact of Examiners on Patent Characteristics

One of the key insights from our qualitative analysis is the potential for heterogeneity across examiners in their discretion and specialization to affect observable outcomes of the examination process. First, as a result of their exercise of discretion, examiners may differ in the average scope of the claims in patents issued under their review. Inventors who receive patent rights with substantial scope will, on average, have been allowed more valuable rights. Identifying the impact of examiner "generosity" is subtle. Patents with broader claims are more likely to constrain the claims granted to future inventors. As a result, beyond their innate inventive importance, patents with broader allowed claims will tend to be more highly cited. Conversely, if all examiners use discretion similarly, and all receive applications with a similar distribution of inventive importance, then the average level of citation should not vary across examiners.[7]

However, examiner specialization may result in differences across examiners in terms of the distribution of inventive performance under their review. For example, some examiners may work in particularly "hot" technology areas where there is a rapid rate of progress; as a result, "their" patents receive large numbers of citations simply because of the larger size of the future "risk set"—i.e., the

---

[7]Although a "generous" grant is a boon to the inventor associated with the application, such treatment may reduce incentives for future inventors, as the hurdle associated with achieving a significant inventive step increases with the breadth and scope offered to inventors from the past. From the perspective of these follow-on inventors, one mechanism to earn a higher return on their own inventions is to seek to invalidate the broad scope associated with a given patent, resulting in specific instances of litigation among the population of an examiner's patents.

number of patents that could potentially cite that examiner's patents, regardless of their breadth. Thus specialization of examiners may result in variation across examiners in the "average" number of citations received by patents issued by each examiner. Moreover, although the effects of specialization can be conditioned by statistical controls for technology area, it is possible that this specialization effect operates in a more nuanced way or at a level of detail that is not easy for us to control for.

Of course, a number of additional factors determine the number of citations received by a patent or even the average level of citations received by patents associated with a patent examiner, including the particular type of technology and the amount of time that has passed since the application. However, after controlling for technology and cohort effects, variation in the exercise of discretion and specialization may still lead to different citation levels, yielding our first hypothesis:[8]

H1: Even after controlling for broad technology class, patent examiners will vary in terms of the average level of citations received by the patents they examine.

In addition to this variation among examiners in their discretion and specialization, there is likely to be variation among examiners in their ability to use search technologies that identify the broadest range of possible prior art. Furthermore, differences in the organization of different Art Units will likely result in different levels of communication and monitoring among examiners and among examiners and their supervisors. As discussed above, one of the consequences of this heterogeneity among examiners is that some examiners may tend toward a more autarkic approach to examination, principally relying on their past experience examining in a particular technological field, whereas others will draw on a wider range of resources. This discussion motivates our second hypothesis:

H2: Even after controlling for technology area, examiners will vary in their level of self-citation. Self-citation should be decreasing with the adoption of more advanced prior art search procedures and increasing with the technological specialization of the examiner.

Finally, examiners will vary in the workload they are given and in the allocations of time for particular tasks associated with the examination process. As several examiners related to us, however, this variation may be in place to allow

---

[8]It is possible that variation in citations received by an examiner reflects selectivity in the assignment of applications to examiners (e.g., Supervising Patent Examiners (SPEs) tend to allocate particularly important inventions to particularly able examiners). We discuss this hypothesis further when considering the impact of average citations received on litigation outcomes.

examiners to more effectively achieve other objectives of the examination process, such as precision or effective communication with the patent bar community in their technological specialty. Thus we offer a third hypothesis about the role of the approval time:

H3: Examiners will vary in their average approval time, above and beyond what can be attributed to the technology of the patents examined. Slower approval will be positively correlated with other dimensions of performance.

### The Impact of Examiners on Patent Litigation Outcomes

Ultimately, we are interested in tying examiner characteristics to more objective measures of the performance of the examination process. We organize this portion of the analysis around patent litigation outcomes. Specifically, we are interested in the possibility that the type of heterogeneity implicit in hypotheses H1, H2, and H3 (as well as other examiner characteristics) will manifest itself in imperfections in the scope of patent rights that are allowed by examiners. As a preliminary foray into this area, we focus on findings of invalidity by the CAFC.[9] Although the CAFC is not the "ideal" setting in which to study validity (because "obvious" invalidity cases are resolved through settlement or at the District Court level), CAFC decisions do provide a useful exploratory window into how examiner characteristics vary (and matter for litigation outcomes) for patents receiving a very high level of judicial scrutiny. Furthermore, by focusing on *invalidity*, we develop hypotheses relating to the role that heterogeneity among examiners might play in leading to the "excess" allocation of patent rights (as adjudicated by the CAFC); however, in future work, we hope to explore the converse possibility that this same heterogeneity may also occasionally manifest itself as underprovision.

Perhaps the most obvious potential source of variation among examiners is their overall level of examination experience. In recent years, various commentators have hypothesized that the rapid growth in patent applications and the concomitant rise in the number of examiners have reduced the experience of the average examiner, particularly in technology areas such as business methods, which have only recently begun to receive patent rights. Implicit in this argument is the proposition that less experienced examiners are more likely to inappropriately allow patent rights that should not be granted. Although it is likely true that

---

[9]We discuss how this particular sampling choice may impact our results in the fourth section, where we present the data and our sampling scheme in more detail. In future work, we hope to redirect analysis toward earlier stages of the litigation process, including the probability of an initial suit, settlement outcomes, and District Court decisions.

experience is helpful in the examination process, the procedures of the USPTO explicitly recognize the value of experience through practices such as the division of responsibilities between primary and secondary examiners and the strong culture of internal promotion. There may therefore be competing effects that mitigate the impact of experience on litigation outcomes. However, to be precise about the specific theory that has been put forth, we offer a testable hypothesis about the impact of examiner experience:

H4: The probability of a litigated patent being ruled valid will be increasing with the experience of the examiner.

In addition, hypotheses H1, H2, and H3 offer at least three potential sources of heterogeneity that may be associated with excess allocation of patent rights and therefore with invalidity findings. First, hypothesis H1 states that some examiners may vary in the degree to which they exercise discretion and the extent to which they are specialized within technology areas and that this variation should be associated with variation in the level of citations received by their patents. Claims allowed by examiners whose exercise of discretion results in allowance of relatively broad claims or who are specialized in "hot" technology areas undergoing rapid changes (either in terms of the technology or in the underlying norms of patentability) may be more likely to be found invalid by the CAFC. As a result, the probability of validity should be declining with examiners' average level of citations received. Similarly, to the extent that it may be easier to overturn the validity of patents based on less thorough searchers of the prior art, the probability of a ruling of validity may be declining with the self-citation of the examiner. Finally, if there is a trade-off between the speed of approval and the quality of the examination, then the probability of validity will likely be increasing with the approval time of the examiner. This discussion motivates the following hypothesis:

H5: The probability of a litigated patent being ruled valid will be declining with the examiner's average citations received per patent, declining with the self-citation rate of the examiner, but increasing with the examiner's average approval time.

It is important to recognize that the relationship between validity and average citations per examiner is subtle, and difficult to interpret, because it measures the combined impact of discretion *and* specialization (i.e., the two distinct forces leading the average citations per patent to vary among examiners). In our empirical work, we therefore explicitly compare how the relationship between validity and average citations per examiner changes when we include detailed technology class controls. To the extent that including controls for each technology class does not reduce the impact of average citations per examiner, this suggests that the impact of specialization and/or discretion occurs at a relatively fine-grained level within the USPTO. In addition, we emphasize a more general point: Regardless of whether discretion or specialization is the driver of variation among examiners, the judicial system may provide a "check" on patent examination patterns that result from the organization of the USPTO.[10]

Finally, the richness of patent data allows us to explore the impact of examiner heterogeneity more precisely. When determining validity, the CAFC will, of course, only consider the merits of the patent under review rather than the historical record of a particular examiner. To determine the impact of examiner characteristics on the probability of validity, only that portion of the citations received by the particular patent that are due to the examiner's overall patterns are relevant. If hypothesis H1 is true, i.e., the number of citations received by the litigated patent is a function of the examiner's average number of citations per patent, then this relationship allows us to estimate this portion econometrically.[11] This reasoning motivates the following hypothesis:

H6: The probability of validity should be declining with the predicted number of citations received by a patent, where the prediction is based on the examiner-specific citation rate.

Together, these hypotheses provide several potential observable consequences of examiner heterogeneity. Consider, for example, the perennial policy issue of patent "disposal" times. By linking approval to other outcomes (such as validity rulings), these hypotheses offer potential insight into the potential for trade-offs associated with speeding up the examination process. To empirically test these propositions, we must tie these hypotheses to a specific set of data, which we now describe.

---

[10]To the extent that the assignment of patent applications to examiners is subject to *selectivity* (i.e., particularly important technologies, associated with higher citation rates, are assigned to more able examiners), our ability to find evidence for the impact of discretion and specialization becomes more difficult.

[11]Specifically, hypothesis H6 can be tested by using an "instrumental variables" estimator where the validity ruling is regressed on the predicted level of citations associated with the litigated patent and the excluded exogenous variable in the validity equation is the examiner's average citation per patent. Intuitively, this procedure is equivalent to a two-stage estimation procedure. In the first stage, the total number of citations received by each patent is regressed on the examiner's average citations per patent and other controls. Predicted values of the total citations variable, i.e., the portion of citations attributable to the examiner's specialization and discretion, are then used in the second-stage validity regression.

## THE DATA

Data for this study were derived from the USPTO's public access patent databases and from the Lexis-Nexis database of decisions of the CAFC.

We began by searching for CAFC decisions in cases where the validity of a patent was contested. In the years 1997-2000, there were 216 such cases, of which 34 were excluded from further consideration because they involved plant patents or re-examined patents or other complicating factors were present. For each of the remaining 182 "CAFC-tested" patents, we determined whether the CAFC found the patent to be valid or invalid and on what grounds: novelty, subject matter, obviousness, procedural errors, etc. Note that in many instances the CAFC found the patent invalid for more than one reason. In just over 50 percent of these 182 cases, the patent was found to be invalid. Of these, the CAFC found problems with novelty (Section 101) in 37 percent of cases, with obviousness (Section 102) in 47 percent of cases, and with the specification of the patent (Section 112) in 15 percent of cases. Although examining CAFC-tested cases does bias the sample away from "obvious" validity issues [i.e., cases with little uncertainty about outcome are likely settled before trial at the District Court level or are less likely to be appealed to the CAFC (Waldfogel, 1995)], this sampling does allow us to assess the characteristics of examiners associated with cases receiving a high level of judicial scrutiny.[12]

Having obtained this list of CAFC-tested patents, we then used it to construct a sample of "CAFC-tested" examiners.[13] To do so we identified the 196 individuals listed as either the primary or secondary examiner for each of the 182 CAFC-tested patents.[14]

For each CAFC-tested examiner, we searched for all patents granted in the period 1976-2000 on which the individual was listed as a primary or secondary examiner. This search was conducted using a fairly generous "wild card" procedure to allow for typographical errors in the source data and variations in the spelling or formatting of names. Results were then carefully screened by hand to ensure that individuals were correctly identified. For example, our procedure would recognize "Merrill, Stephen A.," "Merril, Stephen," "Merrill, S.A.," and "Merrill, Steve" as being the same person, but would exclude "Meril, S." or

---

[12]As well, by excluding "obvious" cases (because they are settled before the CAFC stage), we are simply reducing the amount of underlying variation in the data set. By looking at CAFC cases we are therefore building in a bias *against* finding any effect.

[13]In using the phrase "CAFC-tested" we certainly do not mean to imply that a ruling of invalidity by the CAFC necessarily implies any shortcoming on the part of the examiner.

[14]In future work it would be possible to conduct parts of our analysis of patent examiners on a much wider sample of individuals who performed this function at the USPTO. A useful feature of our small sample, however, is that each examiner in the sample has examined at least one patent that was "tested" for validity by the CAFC.

"Merrill, Stavros A." If anything, this process erred on the side of caution, so that we may be slightly undercounting examiners' output. The initial search returned just over 316,000 candidate patents, from which we excluded about 6 percent misidentified patents to arrive at a base data set of 298,441 patents attributable to the 196 CAFC-tested examiners.[15]

Using the data set of 298,441 patents we constructed complete histories of each CAFC-tested examiner's patent output during the sample period, as well as various measures of their productivity, experience with examination, workload, and examining practice. Each of these patents was matched to the NBER Patent Citation Data File (Hall, Jaffe, and Trajtenberg, 2001) to obtain data on each patent's technology classes, citations made, and citations received, as well as variables computed from these data, which measure the breadth of citations.[16]

In the empirical analysis that follows, we focus on the primary examiner for each of the 182 CAFC-tested patents. Because the same primary examiner may show up several times in the sample, we actually have 136 CAFC-tested primary examiners. In computing statistics across examiners, we weight examiner characteristics by the number of times that each examiner shows up in our data. Table 1 gives variable definitions, and Table 2A presents descriptive statistics for our linked data sets on CAFC-tested patents, CAFC-tested primary examiners, and patent histories of these examiners.

Again, we stress that the set of 182 CAFC-tested patents is a highly selective sample; these patents are not at all representative of the population of all granted patents. Table 2B compares mean values of some key variables for the 182 CAFC-tested patents with those typical of a utility patent applied for in 1980s.[17] On average, the CAFC-tested patents contain more claims, make more citations, receive more citations, and take longer to issue. This is not surprising, because litigants who pursue CAFC review likely perceive a high value for intellectual property over a given technology. As well, given that the litigants have not settled, these patents are likely associated with a higher level of ambiguity than an average patent (perhaps an additional reason for the longer time to approval).

---

[15]Because we have not been able to obtain a definitive matching of examiner ID numbers with issued patents, and have had to work from published data sources, this search misses a small number of patents. We are confident, however, that missing observations are missing at random and therefore do not bias our results.

[16]Jaffe, Henderson and Trajtenberg (1998) computed two measures of the breadth of citations across technology classes: "originality," which captures the extent to which citations made by a patent are spread across technology classes, and "generality," which captures the extent to which citations received by a patent are spread across technology classes. See Table 1 for definitions.

[17]The statistics for a typical patent are based on the tables and figures in Hall, Jaffe, and Trajtenberg (2001).

**TABLE 1** Variables and Definitions

| Variable | Definition |
|---|---|
| **Validity** | |
| Valid | Valid = 1 if patent validity upheld by CAFC; 0 else |
| **CAFC Patent Characteristics** | |
| Citations Received | No. of Citations to CAFC Patent from grant date through 6/2001 |
| Claims | No. of Distinct Claims for CAFC Patent |
| Approval Time | Patent Issue Date—Patent Application Date (Days) |
| Generality | Jaffe-Henderson-Trajtenberg "Generality" index: $$1 - \sum_j \left( \frac{Cites\ Received_j}{Total\ Cites\ Received} \right)^2 \quad j = \text{technology classes}$$ |
| Originality | Jaffe-Henderson-Trajtenberg "Originality" index: $$1 - \sum_j \left( \frac{Cites\ Made_j}{Total\ Cites\ Made} \right)^2 \quad j = \text{technology classes}$$ |
| **Primary Examiner Characteristics** | |
| Experience (no. of patents) | Cumulative Patent Production by Examiner, both primary and secondary (see Figure 1) |
| Examiner Citations | Cumulative Citations to Examiner Patents (through July, 2001) |
| Examiner Cites Per Patent | EXAM CITATIONS divided by EXPERIENCE (NO. OF PATENTS) (see Figure 4) |
| Secondary Experience | Cumulative Patent Production as Secondary Examiner |
| Self-Cite | Share of All Citations to Own Prior Patents (see Figure 5) |
| Examiner Tech. Experience | Number of broad technology classes of patents on which the examiner has experience |
| Examiner Specialization | Herfindahl-type measure of distribution of examiner's patents across broad technology classes |
| Experience (years) | Cumulative Years Observed as Issuing Examiner (both primary and secondary) |
| 3-Month Volume | Count of Issued Patents in Three Months Immediately Before Issue Date |
| **Control Variables** | |
| Tech Class Fixed Effects | 6 Distinct Technology Categories Based on Patent Classes (see Figure 6) |
| Technology Subclass Fixed Effects | 35 Distinct Technology Subclasses Based on Patent Subclasses (see Hall-Jaffe-Trajtenberg) |
| Cohort Fixed Effects | 20 Individual Year Dummies Based on CAFC Patent Issue Date |
| Assignee Fixed Effects | 4 Dummies for Type of Assignee (see Figure 8) |

**TABLE 2A** Means and Standard Deviations

|  | Mean | Standard Deviation |
|---|---|---|
| **Validity** | | |
| Valid | 0.48 | 0.50 |
| **CAFC Patent Characteristics** | | |
| Citations Received | 16.74 | 21.47 |
| Claims | 20.52 | 26.05 |
| Approval Time | 804.60 | 799.80 |
| Generality | 0.41 | 0.27 |
| Originality | 0.39 | 0.28 |
| **Examiner Characteristics** | | |
| Experience (no. of patents) | 2180.38 | 1395.65 |
| Examiner Citations | 14201.68 | 12673.34 |
| Examiner Cites Per Patent | 6.32 | 3.49 |
| Secondary Experience | 207.05 | 137.67 |
| Self-Cite | 0.10 | 0.06 |
| Examiner Specialization | 0.75 | 0.20 |
| Experience (years) | 18.67 | 5.67 |
| 3-Month Volume | 41.52 | 35.66 |

Although the CAFC-tested patents are quite selective, there is little reason to believe that the CAFC-tested primary examiners are very different from the population of all examiners.[18] On one hand, because of the way we have constructed our sample, the probability of an examiner being in our data set is likely proportional to the examiner's experience (measured in terms of total patents examined) at the USPTO. Thus, relative to the set of examiners working at the USPTO on

**TABLE 2B** Patent Characteristics—CAFC Sample Compared to "Universe"

|  | CAFC Sample (182) | Typical Patent (1980s application yr) |
|---|---|---|
| Claims | 20.5 | 9-14 |
| Citations Received | 14.0 | 6-8 |
| Citations Made | 16.7 | 6-8 |
| Originality | 0.36 | 0.3-0.4 |
| Generality | 0.41 | 0.3-0.4 |
| Approval Time (years) | 2.21 | 1.76-2.05 |

---

[18] We intend to test this proposition more carefully by randomly sampling examiners. Initial comparisons with the small set of examiners caught in the wildcard search, but rejected as poor matches, find no substantive differences between them and the sample of CAFC-tested primary examiners in terms of experience and other characteristics.

any given day, we are undersampling inexperienced examiners. As well, our sample may underrepresent variation in the degree of generosity; patents associated with the least generous examiners are less likely to be subject to an appellate validity claim. As such, our empirical design is providing a lower bound of the impact of examiner experience or generosity on patent litigation outcomes. Because examiner patent histories in our data set begin in 1976, the measures of experience are slightly downward biased. About 30 percent of the examiners in our sample first appear in the data set in 1976, some fraction of whom must be assumed to have begun their careers somewhat earlier. Similarly, citations to patents granted before 1976 cannot be evaluated as self-citations (or not) because we do not have information on who the examiner was, and information based on citations received by patents granted in recent years is limited by the truncation of the data set in 2001.

## RESULTS

We present our results in several steps. First, we review evidence of the existence of heterogeneity among examiners and show that an important component of the overall variation in commonly used patent statistics can be explained by examiner "fixed effects." Having established the existence of observable examiner heterogeneity, we then examine the sensitivity of various characteristics of CAFC-tested patents to observable examiner characteristics. We then turn to a discussion of the determinants of patent validity. Consistent with our discussion in the third section, we evaluate a reduced-form model of the sensitivity of validity to examiner characteristics as well as a more nuanced instrumental variables estimation that only allows examiner characteristics to impact validity through their predicted impact on characteristics unique to the CAFC-tested patent.

### The Nature of Examiner Heterogeneity

Our analysis begins with a set of figures that display the heterogeneity among examiners along four distinct dimensions: experience, the level of citations received per patent, the degree of self-citation, and the degree of technological specialization in the patents examined. Figure 1 plots Experience (number of patents) across examiners. We see that although the average examiner in our sample has a lifetime experience of over 2,000 patents, a large number are associated with over 4,000 patents, with a few outliers of over 7,000 patents. This distribution is consistent with the substantial variation we see in the examiners' length of tenure at the USPTO. For example, nearly a third of the CAFC-tested examiners have over 24 years' experience at the USPTO.[19]

---

[19]This may be somewhat biased upward because patent examiners may not "exit" in the way we currently compute this particular statistic.

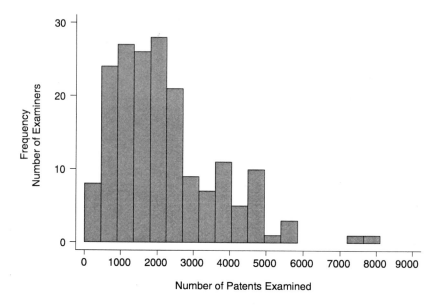

**FIGURE 1** Experience of examiners.

We next turn to an evaluation of the extent to which examiners specialize in particular technology classes over the course of their career. One simple way to measure this specialization is to compute the number of distinct technology classes appearing among the patents examined by a particular examiner. Using six broad technology classes, this measure (Examiner Tech. Experience) is displayed in Figure 2. We see that it is most common to have examined patents in nearly all of the six classes. Yet even if an examiner has dealt with all types of patents, he or she may still be highly specialized within a single technology category with only an occasional patent elsewhere. A more sophisticated approach to deal with this issue is to compute a Herfindahl-type index of the dispersion of an examiner's patents over technology classes.[20] This measure (Examiner Specialization) is plotted in Figure 3. Although some noise is inherent in this measure because of the nature of the technology classification system, its mean level across examiners (0.75) indicates a high average degree of specialization. As Figure 3 indicates, however, there is also considerable variation: Although the modal examiner is highly specialized, with a specialization index near 1, there are still a

---

[20]A Herfindahl index is a commonly used measure of concentration, based on the sum of the squares of the share of a variable across categories.

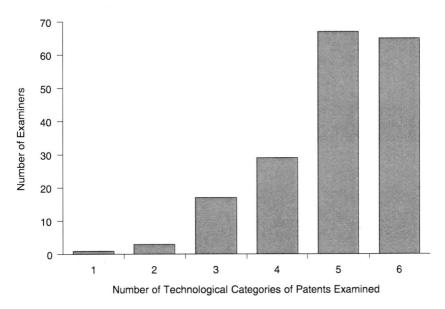

**FIGURE 2** Technological experience of examiners.

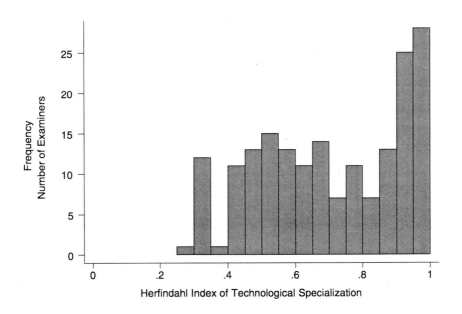

**FIGURE 3** Technological specialization of examiners.

significant number of examiners with a much greater degree of dispersion of patents across technology classes.

Perhaps more interestingly, there is also substantial variation among examiners in the characteristics of "their" patents. Figure 4 shows the distribution of the average number of citations received overall for all patents issued by each examiner (Examiner Cites Per Patent). The distribution is highly skewed. The coefficient of variation associated with examiner cites received per patent issued is over 0.5; and over 10 percent of examiners have citation rates more than double the average citation rate. Similarly, as shown in Figure 5, although the average self-citation rate (Self-Cite) is relatively low, particularly given the technological specialization of examiners, some examiners have self-citation rates more than three times the sample mean. Another method for understanding the importance of heterogeneity across examiners is to use ANOVA analysis to formally test for the presence of examiner effects in several key statistics associated with the examination process. An advantage of this statistical approach is that we can condition on other variables that might explain the observed differences across examiners, such as the technological areas of the patents they examine. Recall that in H1, H2, and H3 we hypothesized that the differences across examiners were not simply a reflection of the technological area of the patents they examined.

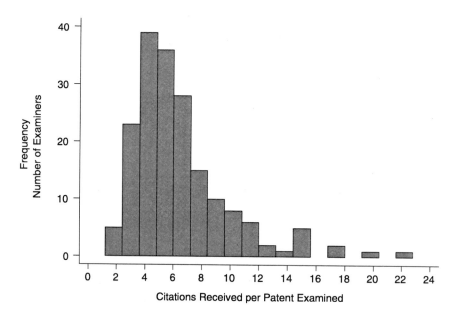

**FIGURE 4** Citations received by examiners.

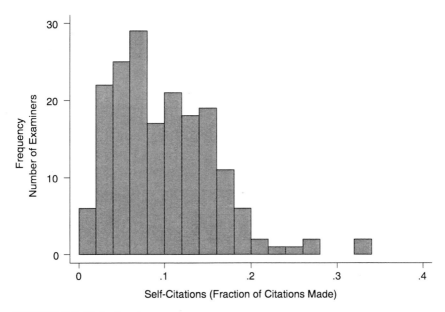

**FIGURE 5** Self-citations by examiners.

In Table 3A, we present a simple ANOVA analysis based on our complete sample of 298,441 patents attributed to the 196 CAFC-tested examiners. The results indicate that examiners matter: A significant share of the variance in this sample in the four variables capturing the volume and pattern of citations by and to a particular patent (Citations Made, Citations Received, Originality, and Generality) is accounted for by fixed examiner effects, with a particularly strong ef-

**TABLE 3A** Analysis of Variance of Patent Characteristics ($N = 289{,}441$; 196 Examiner Effects; 36 Technology Subclass Effects; 24 Cohort Effects)

| Variable | Fraction of Variance Explained by Examiner Effects | $F$-Statistic for No Examiner Effect | $F$-Statistic for No Examiner Effect, Controlling for Detailed Technology Class and Cohort |
|---|---|---|---|
| Citations Made | 0.077 | 121.71 | 52.64 |
| Citations Received | 0.117 | 193.40 | 51.07 |
| Approval Time | 0.083 | 131.77 | 78.92 |
| Claims | 0.030 | 44.83 | 16.06 |
| Generality | 0.079 | 105.56 | 38.97 |
| Originality | 0.069 | 104.23 | 61.30 |

fect in the ANOVA of Citations Received. A similar result is obtained for the length of time between application and grant: About 8 percent of the variance in this measure can be attributed to differences among examiners. A much smaller share of variance is explained for the number of claims on each patent. These results are robust to controlling for differences across technology classes. As Table 3B shows, there are visible differences across technology classes in the fraction of variance explained by examiner effects. There appears to be much more homogeneity across examiners in examination of mechanical patents, with significantly less homogeneity in Citations Made for chemical patents, and in the approval time for electrical/electronic patents. Overall, these results confirm the intuition we developed in our qualitative investigation: There is substantial heterogeneity across examiners, even after controlling for the important technology and cohort effects.

The above analysis suggests that examiners vary, particularly in terms of the rate at which their patents tend to receive citations. But how does this variation, which we have suggested as a proxy for examiner discretion and/or specialization, affect our set of CAFC-tested patents? Table 4 presents regressions relating Citations Received by the CAFC-tested patents to a set of examiner characteristics and, in particular, Examiner Cites Per Patent. One result is particularly striking: There is a very strong relationship between Examiner Cites Per Patent and Citations Received by CAFC-tested patents. The effect is slightly reduced, but still quite significant, after conditioning on the patent's detailed technology subclass, cohort, and assignee type. In each of the specifications in Table 4, increasing Examiner Cites Per Patent by one patent (less than one-third of a standard deviation) increases the predicted number of citations of the CAFC-tested patent by more than one (recall that CAFC-tested patents have much higher overall citation rates). Other observable examiner characteristics have a less clear relationship with Citations Received. The overall level of self-citation, experience (both in terms of years as well as the total level of issued patents), and a measure

**TABLE 3B** Analysis of Variance of Patent Characteristics by Technology Class

| Variable | Fraction of Variance Explained by Examiner Effects | | | | | |
|---|---|---|---|---|---|---|
| | Chemical | ICT | Drug/Med | Electronic | Mechanical | Other |
| Citations Made | 0.123 | 0.054 | 0.104 | 0.078 | 0.054 | 0.059 |
| Citations Received | 0.058 | 0.099 | 0.110 | 0.066 | 0.076 | 0.072 |
| Approval Time | 0.098 | 0.083 | 0.074 | 0.116 | 0.053 | 0.053 |
| Claims | 0.033 | 0.027 | 0.028 | 0.022 | 0.031 | 0.037 |
| Generality | 0.084 | 0.112 | 0.078 | 0.086 | 0.055 | 0.081 |
| Originality | 0.087 | 0.964 | 0.044 | 0.063 | 0.069 | 0.082 |

**TABLE 4**  Citations-Received Equation

|  | Citations Received | | | |
| --- | --- | --- | --- | --- |
| Dependent Variable | (4-1) | (4-2) | (4-3) | (4-4) |
| **Examiner Characteristics** | | | | |
| Examiner Cites Per Patent | 2.68 | 1.83 | 1.82 | 1.69 |
|  | (0.41) | (0.53) | (0.54) | (0.58) |
| Self-Cite |  |  | 25.80 | 40.56 |
|  |  |  | (26.18) | (28.03) |
| Experience (years) |  |  | 0.13 | 0.32 |
|  |  |  | (0.27) | (0.30) |
| 3-Month Volume |  |  | 0.01 | 0.01 |
|  |  |  | (0.04) | (0.04) |
| **Patent Characteristics** | | | | |
| Generality |  |  |  | 11.28 |
|  |  |  |  | (6.70) |
| Originality |  |  |  | 1.90 |
|  |  |  |  | (6.33) |
| **Control Variables** | | | | |
| Cohort Fixed Effects |  | Sig. | Sig. | Sig. |
| Technology Subclass Fixed Effects |  | Sig. | Sig. | Sig. |
| Assignee Fixed Effects |  | Insig. | Insig. | Insig. |
| **Regression Statistics** | | | | |
| Adj. R-squared | 0.19 | 0.44 | 0.45 | 0.45 |
| No. of Observations | 182.00 | 182.00 | 170.00 | 170.00 |

Note: Standard errors are in parentheses.

of near-term work flow (3-Month Volume) are all insignificant in their impact on Citations Received.

Many factors may affect how many citations a patent receives. Citations received are frequently thought to reflect the technological significance of the claimed invention. Pioneering inventions with broad claims and no closely related prior art will tend to be cited frequently as follow-on inventors improve on the original invention. Citations may also reflect the quality or scope of the disclosure accompanying the claims. We cannot directly measure either of these factors here. Nonetheless, these results do indicate that a significant fraction of the variation in citations received by any particular patent is driven by a single aspect of examiner heterogeneity, the average propensity of "their" patents to attract citations. This is true even after controlling for other important attributes of the patent such as the technology class, the year when it was approved, and the type of assignee.

## The Impact of Examiner and Patent Characteristics on Litigation Outcomes

We now turn to the final part of our analysis—linking examiner characteristics to litigation outcomes. Although the overall probability of validity being upheld is approximately 50 percent, there is substantial variation in this percentage across technological area, year of patent approval, and even the type of assignee (see Figures 6-8). For example, although pharmaceutical and medical patents are more likely than not to be upheld, a substantial majority of computers and communications equipment patents are overturned. As well, the age of a patent seems to be an important predictor of validity—pre-1990 approvals are much more likely to be upheld by the CAFC than post-1990 approvals. As we emphasized in the second section of this chapter, these findings suggest the importance of controlling for detailed technology classes and cohorts in our analysis as we seek to evaluate the sensitivity of validity findings to examiner characteristics.

We begin our analysis in Tables 5 and 6, which compare the means of examiner characteristics and patent characteristics, conditional on whether the CAFC ruled the patent valid. Several issues stand out. First, the conditional means associated with most of the patent characteristics are roughly the same. It is useful to note that there is less than a 10 percent difference in the level of Citations Received between the two groups. The only striking difference is in Approval Time, where the time taken to approve *invalid* patents is significantly higher than the time taken to approve those that were found to be valid. Although it is hard to establish a "negative" result in the context of our highly selected CAFC-tested sample of patents,[21] this finding does offer evidence against a simplistic relationship between approval times and validity rulings. Turning to the mean examiner characteristics by validity (Table 6), the striking differences are in terms of Examiner Cites Per Patent and 3-Month Volume. There is no significant difference in the means according to experience level; if anything, invalid patents are associated with examiners with higher mean levels of experience, both in terms of

**TABLE 5** Patent Characteristics: Means Conditional on CAFC Validity Ruling

|                    | Invalid | Valid  |
|--------------------|---------|--------|
| Claims             | 20.73   | 20.28  |
| Citations Received | 17.38   | 16.04  |
| Originality        | 0.36    | 0.41   |
| Generality         | 0.41    | 0.41   |
| Approval Time      | 845.51  | 760.90 |

---

[21] As argued above, the selectivity effect biases against finding such a result.

**TABLE 6** Examiner Characteristics: Means Conditional on CAFC Validity Ruling

|  | Invalid | Valid |
| --- | --- | --- |
| Experience (no. of patents) | 2276.40 | 2077.81 |
| Experience (years) | 18.82 | 18.51 |
| Examiner Cites Per Patent | 6.89 | 5.72 |
| Self-Cite | 0.10 | 0.10 |
| 3-Month Volume | 45.56 | 37.19 |

volume and tenure. This stands in useful contrast to the most naïve interpretation of hypothesis H4, which predicts that Experience should be positively correlated with Validity. In contrast, consistent with the suggestion in hypothesis H5, invalid patents do seem to be associated with examiners who have a higher average citation rate.

Of course, these conditional means ignore the important differences across technologies (Figure 6), cohorts (Figure 7), or assignees (Figure 8) and the poten-

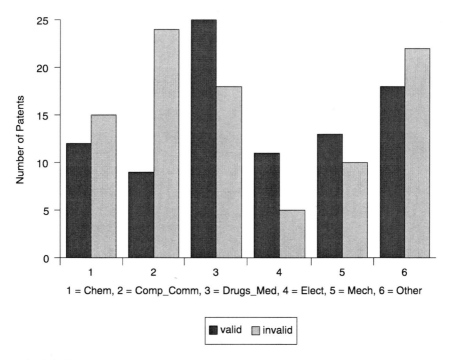

**FIGURE 6** CAFC patents by technology.

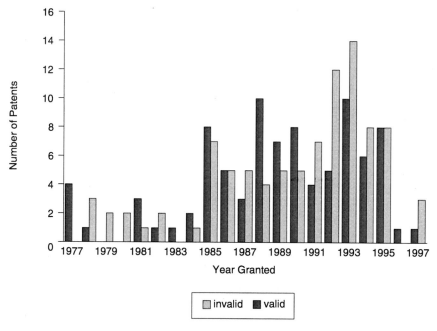

**FIGURE 7** CAFC patents by year issued.

tial for correlation among the examiner characteristics themselves. We therefore turn to a more systematic set of regression analyses in Table 7. The dependent variable in the regressions takes the value 1 if the CAFC-tested patent is ruled valid, 0 otherwise.[22] The first two columns of Table 7 provide a test for hypothesis H4, the sensitivity of the probability of a validity finding to the experience of the examiner. Whether or not detailed controls are included, there is no significant relationship between any measure of experience and the probability of a ruling of validity. Indeed, we have experimented with a wide variety of specifications relating to these experience measures and there is no systematic relationship between validity and these measures in these data. Once again, these "negative" results must be interpreted with caution given the special nature of the sample; they might suggest that a mechanical relationship between examiner experience and the outcomes of validity rulings, if it exists, may be more subtle than some would argue.

---

[22]Both Tables 7 and 8 employ a linear probability model, either OLS or IV. The coefficients are therefore easily interpretable and comparable with each other, and we avoid the technical subtleties associated with an implementing instrumental variables probit in the context of a small sample. We experimented with a probit model for the reduced-form OLS results, and the results remain quantitatively and statistically significant.

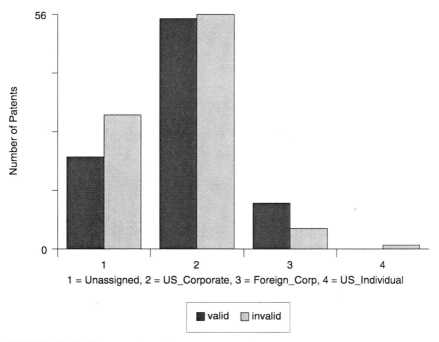

**FIGURE 8** CAFC patents by assignee type.

In the last two columns of Table 7, we turn to hypothesis H5, the sensitivity of a validity ruling to other examiner characteristics. The only significant relationship is with Examiner Cites Per Patent, which has a significant and large negative coefficient. Moreover, this coefficient increases in absolute value when detailed technology and cohort controls are included. According to (7-4), by increasing the Examiner Cites Per Patent by one standard deviation (3.49), the probability of validity is predicted to decline by over 14 percentage points, from a mean of 48 percent. In other words, the probability of validity is strongly associated with the average rate at which that examiner's patents have received citations. As we have emphasized in our hypothesis development, the interpretation of this result is subtle. On one hand, the results provide evidence that even high-level litigation outcomes are related to examiner characteristics that result from two key features of the organization of the USPTO—specialization of examiners in narrow technology areas and exercise of discretion by examiners in allowing claims. Moreover, when we control for specialization, the results become stronger, providing a hint that variation among examiners in their "generosity"—a phenomenon we observed in our qualitative interviews—may be important for understanding the allocation of intellectual property rights that result from patent examination procedures.

**TABLE 7** Reduced-Form OLS Validity Equation

| | Valid | | | |
|---|---|---|---|---|
| Dependent Variable | (7-1) | (7-2) | (7-3) | (7-4) |
| **Examiner Characteristics** | | | | |
| Experience (no. of patents) | −2.97 E-05 | −4.97 E-05 | 2.43 E-05 | −7.60 E-05 |
| | (3.26 E-05) | (3.52 E-05) | (4.57 E-05) | (5.53 E-05) |
| Experience (years) | −0.002 | −0.005 | −0.0003 | −0.002 |
| | (0.008) | (0.008) | (0.008) | (0.009) |
| Self-Cite | | | −0.41 | −0.190 |
| | | | (0.67) | (0.797) |
| 3-Month Volume | | | −0.002 | 0.001 |
| | | | (0.0014) | (0.001) |
| Examiner Cites Per Patent | | | −0.024 | −0.041 |
| | | | (0.011) | (0.015) |
| **Control Variables** | | | | |
| Cohort Fixed Effects | | Insig. | | Insig. |
| Technology Subclass Fixed Effects | | Sig. | | Sig. |
| Assignee Fixed Effects | | Insig. | | Insig. |
| **Regression Statistics** | | | | |
| Adj. R-squared | 0.000 | 0.113 | 0.017 | 0.143 |
| No. of observations | 182.00 | 182.00 | 182.00 | 182.00 |

Note: Standard errors are in parentheses.

This finding motivates our final set of regressions, using the instrumental variables procedure, in Table 8. As discussed in the third section of this chapter, we investigate the mechanism by which Examiner Cites Per Patent might affect patent validity rulings by restricting its impact to the citation rate of the litigated patent. In other words, we impose the exclusion restriction that, but for its impact on Citations Received, Examiner Cites Per Patent is exogenous to the validity decision. The results of this IV analysis are striking. On the one hand, the OLS relationship between validity and Citations Received is insignificant (8-1). However, the coefficient on Citations Received in the instrumental variables equations is significant, large, and negative. Although validity is unrelated to the total number of citations received by a patent, validity is strongly related to the portion of the citation rate explained by the examiner's average propensity to grant patents that attract citations. Moreover, the size of this coefficient increases substantially after the inclusion of technology, cohort, and assignee effects, as well as with the inclusion of other characteristics of CAFC-tested patents. If our results were being driven by unobserved variation across examiners in the types of technologies examined, these controls would likely condition out some of this heterogeneity; the fact that our results become stronger after the inclusion of controls

**TABLE 8** Validity Equation

| | Valid | | | |
|---|---|---|---|---|
| Dependent Variable | (8-1) OLS | (8-2) IV[a] | (8-3) IV[a] | (8-4) IV[a] |
| **Patent Characteristics** | | | | |
| Citations Received | −0.0007 | −0.0090 | −0.0228 | −0.0242 |
| | (0.0017) | (0.0042) | (0.0106) | (0.0111) |
| Claims | | | | 0.003 |
| | | | | (0.002) |
| Originality | | | | 0.238 |
| | | | | (0.227) |
| Generality | | | | 0.188 |
| | | | | (0.268) |
| Approval Time | | | | −0.00006 |
| | | | | (0.00009) |
| **Control Variables** | | | | |
| Cohort Fixed Effects | | | Sig. | Sig. |
| Technology Subclass Fixed Effects | | | Sig. | Sig. |
| Assignee Fixed Effects | | | Insig. | Insig. |
| **Regression Statistics** | | | | |
| Adj. R-squared | 0.000 | NA | NA | NA |
| No. of observations | 182.00 | 182.00 | 170.00 | 170.00 |

[a]IV: Endogenous = Citations Received; instrumental variable = Examiner Cites Per Patent.
Note: Standard errors are in parentheses.

makes our findings even more suggestive. In other words, even relying on a test that only allows examiner effects to matter through their impact on the citation rate of the litigated patent, and controlling for differences in the timing and type of litigated technology, we find that the CAFC invalidates patent rights associated with examiners whose degree of specialization or exercise of discretion results in an unusually high level of citations received.

## DISCUSSION AND CONCLUSIONS

We have conducted an empirical investigation, both qualitative and quantitative, of the role that patent examiners play in the allocation of patent rights. In addition to interviewing administrators and patent examiners at the USPTO, we have constructed and analyzed a novel data set on patent examiners and patent litigation outcomes. Starting with a sample of patents for which the CAFC decided on validity between 1997 and 2000, we collected historical data on those who examined these patents at the USPTO. For each of these examiners, we collected data on all of the other patents that they examined during their career, allowing us to compute a number of interesting examiner characteristics. The data set obtained by matching these two sources is, of course, based on a highly selected sample, because very few patents make it to the CAFC. Nonetheless, we view the analysis of these CAFC cases as a very useful first step, largely conditioned by the ease of accessibility to data, in exploring a number of hypotheses about the connection between the patent examination process and the issuance of patent rights. Our results are preliminary, but they suggest a number of interesting findings.

First, patent examiners and the patent examination process are not homogeneous. There is substantial variation in observable characteristics of patent examiners, such as their tenure at the USPTO, the number of patents they have examined, the average approval time per issued patent, and the degree of specialization in technology areas. There is also systematic variation in outcomes of the examination process—such as the volume and pattern of citations made and received by patents—that can be attributed to idiosyncratic differences among examiners. Most interestingly, examiners differ in the number of citations made to "their" issued patents, even after controlling for technology class, issuing cohort, and other factors.

Second, we find no evidence in our sample for the most "naïve" hypotheses about examiner characteristics and quality of examination. In particular, we find no strong statistical association between examiner experience or workload at the time a patent is issued and the probability of the CAFC finding it to be invalid if it is subsequently litigated to the appeals court level. We hesitate to make any policy prescriptions based on this "negative" finding, however, unless it were confirmed in subsequent research using a larger sample.

Third, we find that "examiners matter": Although highly structured, and carefully monitored by USPTO, patent examination is not a mechanical process. Examiners necessarily exercise discretion and are focused in very narrow technology areas, and occasionally the claims allowed under this process are overturned by subsequent judicial review. Our core finding is that the examiners whose patents are cited most often are also more likely to have their patents ruled invalid by the CAFC. Our econometric procedure distinguishes between citations received by a particular patent because of the scope of its claims or the significance of an overall technology area and citations received because of examiner-specific differences in propensity to allow patents that attract citations. It is only the second of these mechanisms that has a statistical relationship with CAFC validity rulings.

The fact that patent examination cannot be mechanistic, and that idiosyncratic aspects of examiner behavior appear to have a significant impact on the nature of the patent rights that they grant, suggests a significant role for the organization, leadership, and management of USPTO. The management literature recognizes the value of corporate culture in the form of informal rules, common values, exemplars of behavior, etc. in providing guidance on how to exercise discretion. Although idiosyncratic behavior of examiners can be controlled to some extent by formal processes such as supervision, selection of examiners, training, and incentives, the institution's cultural norms necessarily play an important role in their exercise of discretion in awarding patent rights. Policy changes that impact the organizational structure and internal culture of the USPTO should be careful to take patent examiner behavior into account.

## REFERENCES

Cohen, W., R.R. Nelson, and J.P Walsh. (2000). "Protecting Their Intellectual Assets: Appropriability Conditions and Why U.S. Manufacturing Firms Patent (or Not)." NBER Working Paper No. 7552.

Griliches, Z., ed. (1984). *R&D, Patents and Productivity.* Chicago, IL: Chicago University Press.

Griliches, Z. (1990). "Patent Statistics as Economic Indicators: A Survey." *Journal of Economic Literature* 92: 630-653.

Hall, B., A. Jaffe, and M. Trajtenberg. (2001). "The NBER Patent Citation Data File: Lessons, Insights and Methodological Tools." NBER Working Paper No. 8498.

Hall, B., and R. Ziedonis. (2001). "The Patent Paradox Revisited: An Empirical Study of Patenting in the US Semiconductor Industry 1979-1995." *RAND Journal of Economics* 32(1): 101-28.

Jaffe, A., R. Henderson, and M. Trajtenberg. (1998). "Universities as a Source of Commercial Technology: A Detailed Analysis of University Patenting, 1965-1988." *Review of Economics and Statistics* 80(1): 119-127

King, J. (2003). "Patent Examination Procedures and Patent Quality." In W. Cohen and S. Merrill, eds., *Patents in the Knowledge-Based Economy.* Washington, D.C.: The National Academies Press.

Lanjouw, J., and M. Schankerman. (2001). "Characteristics of Patent Litigation: A Window on Competition." *RAND Journal of Economics* 32(1): 129-151.

Lanjouw, J., and J. Lerner. (2001). "Tilting the Table? The Use of Preliminary Injunctions." *Journal of Law and Economics* 44(2): 573-603.

Levin, R., A. Klevorick, R. Nelson, and S. Winter. (1987). "Appropriating the Returns from Industrial R&D." *Brookings Papers on Economic Activity*, No. 3: 783-831.

Lerner, J. (2002). "150 Years of Patent Protection." *American Economic Review Papers and Proceedings* 92(2): 221-225.

Waldfogel, J. (1995). "The Selection Hypothesis and the Relationship Between Trial and Plaintiff Victory." *Journal of Political Economy* 103: 224-260.

# Patent Examination Procedures and Patent Quality[1]

John L. King
Economic Research Service
U.S. Department of Agriculture

### ABSTRACT

*This study examines a detailed panel data set of patent examination procedures that affect patent quality. A main conclusion is that the most important of these inputs (examiner hours and examiner actions) have remained largely consistent over time despite an increasing examination workload. Other measures of examination quality (pendency and interference hearings) have declined. Inputs to examination quality are inversely correlated with the rate at which patents are involved in legal complaints, and the expense of increasing examination inputs may be more than offset by the consequent reduction in litigation costs.*

### INTRODUCTION

Patents grant the exclusive rights to use, manufacture, and sell new inventions and are widely sought legal instruments. The numbers of both patent applications and patent awards have more than doubled over the past two decades. Explanations for this increase might include more innovative activity, greater emphasis on intellectual property rights among innovating firms, reduced requirements for patentability, or a variety of other possibilities. Regardless of the cause, the increase in patent activity has created a greater examination workload and placed a greater burden on patent-granting institutions. This chapter examines the

---

[1]Portions of this research first appeared as "An Empirical Investigation of the Economics of Patent Institutions," Ph.D. dissertation, Vanderbilt University, 2000. The views expressed are those of the author alone and do not necessarily reflect views or policies of the U.S. Department of Agriculture. The author wishes to acknowledge suggestions by David Lucking-Reiley, William Lesser, Stephen Merrill, George Elliot, and Wesley Cohen.

increase in empirical detail and analyzes its impact on the quality of patent examination.

Drafting a patent application is a rigorous exercise in technical language that must accommodate the technology underlying the invention, its commercial significance, and relevant statutory and case law. Patent examiners, who read the applications and ultimately decide whether to allow a patent award, must have at least a basic understanding of these and other factors. A skilled staff of patent examiners with adequate training and resources is essential to maintain the validity of patents that issue according to their decisions. The recent rise in patenting activity raises the question of whether patent examiners have adequate resources to fulfill their increased examination responsibilities in a thorough and timely way.

The main contribution of this study is to present data that quantitatively assess the effect of increasing application workloads on the recent examination performance of the U.S. Patent and Trademark Office (USPTO). The study presented in this chapter provides descriptive data about various features of patent examination, largely drawn from detailed Time and Activity Reports of examiners maintained by the USPTO. This data set and information about patent institutions allow inferences about changes in the quality of patent examination. Allison and Lemley (1998) and Jaffe (2000) discuss trends in increasing patenting activity over the past two decades, and Merges (1999) discusses the possible ramifications on patent litigation and incentives to innovate when patent examination standards are unevenly applied. This study extends the empirical analysis of patenting trends in a way that directly addresses issues raised in connection with standards of patent examination.

In addition, this chapter presents results of multivariate regression analysis indicating the correlation between measures of examination quality and areas of policy concern, such as patent litigation and incentives for inventors. Although patent examination is only one element in the complex landscape of intellectual property rights, it is a subject with important implications and one that is amenable to empirical analysis.

Careful patent examination reduces the need for courts to review patent agency decisions in the eventuality of a patent dispute. Hypothetically, a "perfect" patent agency would never issue a patent that was later found invalid in a court of law. In addition, the scope of issued patents would be extremely clear, providing an easy test—and strong deterrent—for infringement should a dispute arise. Approved claims would be broad enough to reward inventors of significant discoveries but sufficiently narrow to allow patents on competing inventions or further improvements. Conversely, factors that constrain the quality of patent examination cause a divergence of patent agency decisions from the determination a court would make if presented with the same facts. This divergence obscures the true strength of a patent in court. A reduction in the quality of patent examination therefore increases the uncertainty of enforcing the intellectual property rights that are inherent in a patent award.

However, the efficiency of the patent examination process requires that examiners make their determinations without the expense and delay that often accompany litigation. Judging patent applications against the relevant technical, legal, and commercial information requires scarce human capital, which the USPTO must acquire and develop. In addition, examiners play their most active role in the examination process in the course of offering an initial rejection, through which they may require amendments to an application such as more citations to prior art, the narrowing or elimination of a specific claim, or one of many corrections that improve the patent application. Doing this in a thoughtful and competent manner requires time, resources, and inputs sufficient to the task. A concern about increasing application workloads is that they might impose constraints on the resources available to examiners, with the possible consequence of jeopardizing examination quality and creating uncertainty in intellectual property rights enforcement.

This is an area of some importance, because uncertain intellectual property rights impose several kinds of costs on the economy. Subsequent legal effort to determine the validity or proper scope of a patent is necessary when examination quality is lower. Legal costs are especially high when patent disputes result in litigation. To the extent that the enforcement costs of patent protection undermine incentives to innovate, low examination quality reduces the amount of innovation in society. Also, when the patent examination process fails to reject patent applications with serious flaws, patent monopolies for inventions with little benefit to society impose additional welfare costs. These costs are more likely to accrue as a greater number of patents are issued each year, underscoring the importance of examination quality in the patent system.

## DATA AND METHODOLOGY

Patent examination quality refers to the ability of patent examiners to make a correct judgment about whether to grant a patent application, meaning that their decisions about validity and scope of protection are consistent with the ruling a court would make after a comprehensive review of the application. Patent examination therefore requires considerable knowledge and skill in the technological area but also knowledge of evolving court rulings.

In the aggregate, it is difficult to assess the quality of patent examination directly. Examination quality is a multifaceted concept covering validity, scope, timeliness, and other attributes. The examination effort necessary to make a determination is likely to vary from application to application, because each patent presumably possesses some unique and novel features. Legal rulings on patents occur too infrequently to provide a comprehensive view of whether courts uphold typical examiner decisions. Moreover, the sample of patents that proceed to court rulings is likely to differ from the population of patents as a whole (Priest and Klein, 1984).

However, inputs to the patent examination process can be observed and quantified. The skill and experience of patent examiners, the time allotted to patent examination, and other factors that contribute to examination can be measured. The principal data used to measure the inputs and outputs of the patent examination process in this study come from Time and Activity Reports obtained from internal USPTO records.[2] The data set contains detailed information about the types of duties performed by examiners, the intensity of examination effort, and some information about examiners. The Time and Activity Reports summarize examination activities covering the period from 1985 to 1997 (Table 1). Some tables and calculations in subsequent sections include aggregate statistics available through 1998, but results that required data at the examination group level extended only through 1997.

Combined with other information about patents, the data facilitate three types of analysis. First, this study provides descriptive data on examination inputs and outputs. Comparison of examination effort to level of examination output measures the intensity of examination, providing some insight into examination quality. Multilinear regressions show which examiner activities are most closely associated with patent awards, establishing a relationship between inputs and outputs. Second, this study analyzes how examination intensity has changed over time, focusing especially on the past two decades of heightened patenting activity. This part of the analysis explores the effects of rising workloads on examination quality. Finally, analysis of a separate data set on patent litigation allows exploration of important consequences outside the patent process itself. In particular, this study relates measurements of patent examination quality with patterns in patent litigation. The combination of these three types of analysis provides an empirical view of examination procedures and allows exploration of their effects on outcomes of policy interest such as patent litigation.

Some of the key variables included in the Time and Activity Reports include:

- **Examination hours.** Examination hours are a primary measure of examiner input. Dividing the total number of hours by 2,000 provides a rough estimate of full-time equivalent (FTE) examiners.[3] The data set distinguishes between regular hours and overtime hours spent examining patents in the various examination groups.
- **GS-12-equivalent examination hours.** An enhanced measure of examiner effort arises from normalizing examination hours by examiner experience and training. The data set includes examination hours adjusted to a GS-12 pay

---

[2]King (2000) discusses FOIA Request 99-118 used to obtain the data.
[3]Assumes a 40-hour work week for 50 weeks per year.

**TABLE 1** Summary of PTO Time and Activity Reports, 1985-1997

| Examination Group | Mean Annual Patent Disposals | Mean GS-12 Examination Hours per Disposal | Mean Actions per Disposal | Mean Rejections per Disposal |
|---|---|---|---|---|
| 1100. General metallurgical, inorganic petroleum, and electrical chemistry and engineering | 16,319 | 18.52 | 2.48 | 0.41 |
| 1200. Organic chemistry | 15,144 | 16.20 | 2.69 | 0.40 |
| 1300. Specialized chemical industries and chemical engineering | 15,734 | 18.66 | 2.54 | 0.41 |
| 1500. High-polymer chemistry, plastics, coating, photography stock materials and compositions | 17,829 | 17.79 | 2.45 | 0.43 |
| 1800. Biotechnology | 15,990 | 22.41 | 3.04 | 0.55 |
| 2100. Industrial electronics, physics, and related elements | 13,676 | 19.44 | 2.27 | 0.26 |
| 2300. Information processing, storage, and retrieval. | 11,888 | 27.52 | 2.51 | 0.40 |
| 2400. Packages, cleaning, textiles, and geometrical instruments | 11,908 | 17.87 | 2.35 | 0.35 |
| 2500. Electronic and optical systems and devices | 17,191 | 20.02 | 2.35 | 0.32 |
| 2600. Communications, measuring, writing, and lamp/discharge | 16,445 | 23.36 | 2.48 | 0.35 |
| 3100. Handling and transportation media | 12,681 | 16.98 | 2.26 | 0.33 |
| 3200. Material shaping, article manufacturing, and tools | 13,677 | 15.78 | 2.23 | 0.31 |
| 3300. Mechanical technologies and husbandry personal treatment information | 16,095 | 16.86 | 2.44 | 0.38 |
| 3400. Solar, heat, power, and fluid engineering devices | 12,259 | 16.27 | 2.17 | 0.24 |
| 3500. General construction, petroleum and mining engineering | 15,235 | 15.67 | 2.29 | 0.30 |

NOTE: Examination Groups 2200 (Special Administrative Unit) and 2900 (Design) were not included in this study.

grade, meaning that examiner hours of more experienced examiners (at higher pay grades) count proportionally more with respect to differences in pay. This measure takes into account differing levels of experience among examiners and is therefore a better measure of effective examiner input.

- **Patent disposals.** Patent disposals are the sum of patent approvals and rejections issued by examination groups, providing a good absolute measure of total examiner output.

- **Examiner actions.** An examiner action occurs when an examiner reviews a patent application and responds to the applicant, which usually happens several times before a patent disposal (i.e., acceptance or rejection of a patent application). Each action provides examiners with an opportunity to require improvements before a patent award or give grounds for a rejection. The number of actions for each patent disposal is another measure of examination intensity.
- **Patent pendency.** The timeliness of patent examination is an important factor because of brisk competition in technology markets; also, time spent in examination counts directly against the 20 years of protection that a patent award allows. The ability of examiners to render accurate, thorough examinations in a brief period enhances the benefits of the patent system to inventions.

The data source used to analyze the effects of patent examination procedures on patent litigation is derived from a U.S. patent law that requires that courts notify the USPTO when a patent becomes involved in a legal dispute.[4] Once the USPTO receives notice of either a complaint or the formal resolution of a case, it is indexed and published commercially by Derwent Publishing in the *LitAlert* database. This notification requirement creates an opportunity to assemble a data source on which patents are involved in litigation, which cases are settled out of court, and how frequently patents are found to be invalid in court decisions. The litigation data include patents issued between 1989 and 1991 that were involved in legal disputes before 2000. Although some of these patents might have been involved in subsequent litigation, the data show that 95 percent of the disputes arose within 3 years of patent issue, indicating that truncation is not a problem for this sample. A more thorough description of the litigation data used here is available in King (2000); it is constructed from the same primary source used by Lanjouw and Schankerman (2001).

The comparison of outputs and workload across the entire population of patents raises a methodological concern about how to treat dissimilar patents. For instance, patents in complex technological areas or areas of relatively rapid innovation such as biotechnology and semiconductors might require additional examination effort. If a complex technology requires more examiner input to provide the same examination quality, regression analysis must account for this difference. Likewise, examiners in more mature technological areas must consider more prior art.

Fortunately, this data set allows the use of panel estimation techniques. These USPTO Time and Activity Reports tracked examination effort in 17 different

---

[4]35 USC 290: "The clerks of the courts of the United States, within one month after the filing of a [legal action involving a patent] shall give notice thereof in writing to the Commissioner.... Within one month after the decision is rendered or a judgment issued the clerk of the court shall give notice thereof to the Commissioner."

technology areas, known as examination groups, over time. To the extent that examination, downstream demand for patents, and other factors affecting patent activity are aligned by technologies, panel estimation techniques allow the isolation of the statistical relationship between examiner inputs and patent quality, holding these other factors constant. Also, examination groups provide a particularly useful unit of analysis for policy makers, suggesting specific areas of patent examination to which resources can be shifted at the margin to improve overall examination performance.

Estimates from the panel data are presented with both random-effects and fixed-effects model specifications. The fixed-effects specification might be preferable if relevant aspects of patent examination are strongly correlated with the broad range of technologies represented in the 17 examination groups. If this is not the case, or because the fixed-effects model is more costly in degrees of freedom, then the random-effects specification might be preferable. Because the purpose of this chapter is not a methodological comparison of these estimation techniques, results from both specifications are reported.

Unfortunately, a reorganization of the USPTO after 1997 limits the time series of the panel data. The 17 examination groups were reorganized into 6 "technology centers" after 1998, and the USPTO did not release the more finely disaggregated numbers in subsequent years. Notwithstanding this temporal limitation, the data set allows detailed analysis of patent examination procedures at the examination group level through a substantial part of the trend of increased patent activity over the past two decades.

## TRENDS UNDER INCREASING PATENT EXAMINER WORKLOADS

This section presents findings from analysis of the data as they pertain to examiner workload and various measures of the thoroughness and timeliness of patent examination. The general finding is that over the period in which examiner workload has increased, several important measures of patent quality have remained consistent, while some others have suffered. The data do not appear to support a general decline in patent examination quality, although a trend for increasing workload might create delays in the examination process if it continues.

Between 1985 and 1998, the USPTO issued approximately 1.4 million patents and 775,000 final rejections, summing to almost 2.2 million patent disposals. On average, a patent disposal received 17.1 hours of examiner time. Adjusting for examiner experience by using USPTO salary calculations, each patent disposal required 19.1 hours of examiner effort paid at the GS-12 level.

Figure 1 shows how the average examination time varied across examination groups. The data reflect more variation across groups than intertemporal variation within groups. For instance, mean examiner hours per patent disposal range between 15.4 and 27.4 across examination groups, a difference of 13.0, whereas most individual examination groups varied within a narrower range of about 3

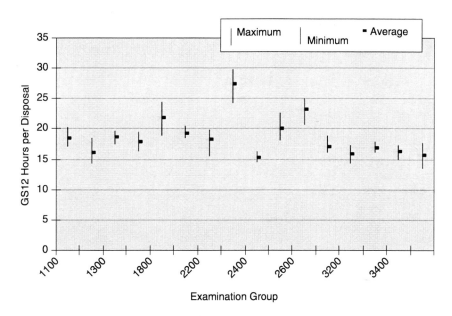

**FIGURE 1** Experience-adjusted examination hours per patent disposal. SOURCE: USPTO Time and Activity Reports, 1985-1997. NOTE: Outliers for Examination Group 2400 in 1995-1997 (28, 24.3, 29.7) omitted.

hours per patent disposal. Although Figure 1 does not show how mean examiner hours per disposal vary over time, the values tend to be evenly distributed around their means and exhibit neither an increasing nor a decreasing trend. Examination Group 2400—Packages, Cleaning, Textiles, and Geometrical Instruments—showed the most dramatic intertemporal variation, but that was mostly confined to an unexplained jump in the final 3 years of the data (which are omitted from the calculations for Figure 1).

An interpretation of Figure 1 that is consistent with the use of fixed-effects modeling is that the time required to examine a patent varies according to examination group. The number of applications and issued patents from each examination group did vary over time, but the amount of examiner effort required to dispose of these applications generally ranged within a fairly narrow band. Panel estimation controls for different examiner input requirements among different technical areas.

With this sense of the level of examiner effort necessary to process patent applications, Figure 2 illustrates how the workload of examiners varied over time. Normalizing examiner activity levels to 100 in 1985, Figure 2 shows the increase in examiner workload from 1985 to 1998. In the first half of the period patent awards slightly outstripped the number of hours devoted to examination, but the

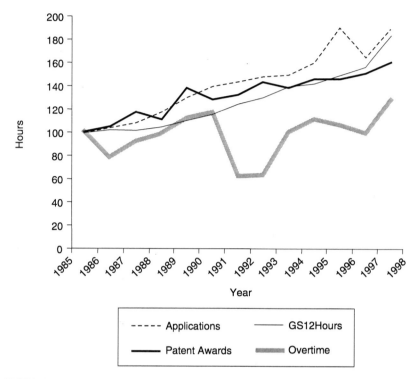

**FIGURE 2** Change in utility patent workload, examination hours in the United States. SOURCE: USPTO Time and Activity Reports, Annual Reports of the Commissioner of the USPTO, 1985-1998.

two variables generally kept pace over the entire span. Although it is not represented in Figure 2, further analysis of the data shows that the same was true within the examination groups as well. Examination groups 2300 (information processing, storage, and retrieval) and 2600 (communications, measuring, writing, and lamp/discharge) experienced temporary decreases in examination hours per patent award, but both reestablished and eventually increased examination intensity.

Figure 2 also shows the use of overtime during this period. The number of overtime hours tended to increase slightly from periodic significant decreases. To the extent that overtime hours are imperfect substitutes for examination by full-time examiners, it would appear that the persistent, rapid increase in patent applications justified employment of more full-time examiners. Indeed, the USPTO hired over 700 additional patent examiners in both 1998 and 1999,[5] although

---

[5]*USPTO Annual Report, 1999.*

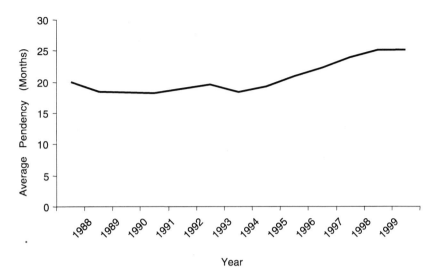

**FIGURE 3** Average pendency for U.S. patents, 1988-2000. SOURCE: Annual Reports of the Commissioner of the USPTO.

some of the additional labor merely replaced attrition. The number of overtime hours might also reflect the difficulty of hiring examiners: The focus of the economic expansion during the late 1990s on high-technology areas may have decreased the availability of suitable examiners.

The number of patent applications, another measure of examiner workload, rose at a faster rate than the number of examination hours (Figure 2). Patent applications outnumbered patent awards by 60 percent in 1980 and grew at a faster average rate (7.3 percent compared to 6.4 percent through 1998). The result was a widening gap between the number of patent applications filed with the USPTO and the time available for examiners to review them. The net effect was an increasing backlog in patent applications. Figure 3 shows the average pendency of patents, i.e., the length of time between filing dates and patent disposals (allowances or rejections). Over a period in which examiner workload was increasing, average pendency increased by 25 percent from 20.0 months in 1988 to 25.0 months in 1999.

To summarize Figures 1 through 3, examination hours per patent disposal varied significantly across groups but were more or less constant within groups over time. Patent awards issued by examination groups varied quite closely with the number of examiner hours each group employed. At the same time, patent applications increased at a faster rate. Pendency, the length of time between application and patent award, increased over time. A possible interpretation of these facts suggests a patent system applying consistent examination intensity to the patents it issues but constrained from examining its increasing workload by the examiner resources available to it.

In 1996 the GAO conducted two studies in patent pendency.[6] Congress requested the studies out of concern for the impact of pendency on effective patent terms, after passage of a law that changed patent terms from 17 years after issue to 20 years after filing.[7] Although the effective patent term for the average patent actually increased under the new law, it is clear from Figure 3 that pendency increased substantially throughout the 1990s. The close relationship between patent awards and the availability of examiner hours employed suggests that greater examiner employment might effectively reduce patent pendency.

Levin et al. (1987) and Cohen et al. (2000) established that lead time is critically important in protecting intellectual property, perhaps more so than patents. So, even apart from the erosion of patent term resulting from longer patent examination delays, the usefulness of patent protection and therefore its incentive effects for innovation suffer. Longer pendency represents one aspect of examination quality that has been compromised in the face of increasing workloads.

Although longer pendency is one effect that can be associated with increasing examination workload, it is important to consider other quantitative relationships in examiner activity. Specifically, data in this study show basic relationships between inputs and outputs of the patent examination process. In the context of regression analysis, the number of hours an examination group devotes to patent examination is strongly correlated with the number of patent awards the group issues. When patent examination output is measured by the natural logarithm of patent disposals (i.e., allowed patents and final rejections) by an examination group, GS-12 examiner hours (Table 2) have the largest and most statistically significant effect on production.[8] Because the coefficients on either independent variable increased in magnitude when the other was excluded, examination hours are probably somewhat collinear with the number of examiner actions per patent disposal. However, the sign and significance of both variables were also robust to this variant on model specification.

The relationship between examiner actions, another input to the examination process, and patent awards is more ambiguous. On its face, a greater number of

---

[6]GAO/RCED-96-190, GAO/RCED-96-152R. The studies provide pendency detail at the art unit level, but only for a 1-year "snapshot" in 1994, so they could not be integrated fully with the data in this study.

[7]Eagerness of inventors to submit applications under the earlier patent term probably explains the 1994 surge and the corresponding drop-off in patent applications the following year, as shown in Figure 1.

[8]The log-log specification is used to maintain consistency with results presented later in this chapter. The positive, significant coefficients are robust to linear and semilog specifications. Also, a specification including linear and quadratic terms returned significant coefficients that were respectively positive and negative, respectively, providing the same goodness of fit ($R^2 = 0.74$) as the log specification within the range of the sample.

**TABLE 2** Contribution of Inputs to Examination Output (Dependent Variable: Logarithm of Patent Disposals)

|  | Random Effects | Fixed Effects |
| --- | --- | --- |
| GS-12 Hours (log) | $0.6670\ t = 16.80^a$ | $0.6995\ t = 17.95^a$ |
| Total Actions (log) | $0.2375\ t = 6.83^a$ | $0.2162\ t = 6.37^a$ |
| Constant | $-1.2842\ t = -4.16^a$ | $-1.4607\ t = -4.86^a$ |
| $R^2$ | 0.75 | 0.74 |

[a]Significant at 1%.

NOTE: 15 examination groups, 1985-1997 (3 missing values).

examiner actions per patent disposal suggests greater examiner scrutiny. Indeed, only 13 percent of patent applications were approved on first action.[9] Each successive action increases the likelihood of a patent award, with 73 percent of third and subsequent actions resulting in approval. This is consistent with examiners interacting with inventors to improve the application and resulting patent award, a manifestation of examination quality. On the other hand, a greater number of examiner actions might indicate a case in which "the squeaky wheel gets the grease": The applications receiving greater scrutiny might be the ones closest to being unpatentable. In this interpretation, more actions per patent disposal suggests a greater number of marginal patents and greater examiner effort implies a higher proportion of patents with weak claims. The positive coefficient on total examiner actions, which was robust to several model specifications (i.e., quadratic, semilog, etc.) is consistent with either interpretation. Additional data are needed to clarify the role of examiner actions in improving examination quality.

### Examination Quality and Patent Litigation

Patent protection is asserted through litigation or through negotiation backed by the threat of litigation. In part because of the quality of the patent examination process, courts presume that patents involved in legal complaints are valid. To the extent that the examination process can rigorously clarify intellectual property rights in advance, examination can help parties avoid legal disputes and can promote efficiency in industries where patents are important. The purpose of this section is to bring data on examination practices to bear on this issue, to see what role the examination process might play in patent litigation outcomes.[10]

---

[9]Practitioner accounts suggest that some patents awarded on first action are actually reworked applications, so that even fewer patents are awarded on their very first attempt.

[10]Other authors to address empirical analysis of patent litigation include Allison and Lemley (1998); Jaffe (2000); and Lanjouw and Schankerman (2001).

Clearly, factors other than patent examination also determine decisions to develop and protect intellectual property. The decision to compete and potentially infringe in markets protected by patents and the decision to litigate against infringers who may have patents of their own involve strategic determinations based on many factors. However, the quality of patent examination affects the validity of issued patents and therefore the incentives for patent infringement. For instance, if issued patents were unlikely to be upheld in court, it is likely that more infringement would be observed. Whether this leads to more patent litigation would depend on the specific technological, legal, and economic issues surrounding each case and whether validity is an issue.

Meurer (1989) models the strategic incentives facing patentholders and potential infringers when the validity of issued patents is uncertain. The model focuses on the question of whether parties will choose to litigate or settle disputes when infringement occurs. However, a theoretical prediction of the model is that greater uncertainty of patent validity leads to a greater incidence of infringement and litigation. Using the data on patent examination quality to stand in for certainty of intellectual property rights, this study performs an empirical test of this hypothesis. To the extent that the variables identified here are correlated with the quality of patent examination, this approach quantifies the effect of examination intensity on patent litigation.

Public records at the USPTO indicate that only 0.22 percent of patents issued between 1989 and 1991 were involved in legal complaints, approximately 200 patents from each year (King, 2000). Patent litigation can be protracted and expensive, which is one reason why so few patents were involved in legal complaints. The cost of litigation also helps explain why many disputes were settled before a final verdict was reached in court: Of complaints involving patents issued between 1989 and 1991 (Table 3), somewhere between 38 and 55 percent of these cases (between 0.084 and 0.121 percent of patents allowed) reached a final verdict.[11]

Table 4 presents estimates of the effects of various measures of examination quality on the rate at which issued patents were involved in legal complaints. The unit of observation is a patent examination group in a sample year. In addition to GS-12 hours per patent disposal and total number of examiner actions per disposal to measure the quality of examination, the regression includes the mean elapsed time between complaint filing and verdict to control for the cost of litigation, which, *ceteris paribus*, should reduce the incidence of litigation. Also, the panel nature of the data holds constant effects that vary with technological area.

The number of GS-12 examiner hours per patent disposal has the greatest effect on the complaint rate of the variables included in the analysis. This effect is

---

[11]USPTO records did not easily allow a precise determination of litigation outcomes; see King (2000) for details.

**TABLE 3** Legal Complaints Involving Patents Issued Between 1989 and 1991

| Examination Group | Annual Rate of Patents Involved in Complaints per Patent Allowed |
|---|---|
| 1100. General metallurgical, inorganic petroleum and electrical chemistry, and engineering | 0.001469 |
| 1200. Organic chemistry | 0.001303 |
| 1300. Specialized chemical industries and chemical engineering | 0.001729 |
| 1500. High-polymer chemistry, plastics, coating, photography stock materials and compositions | 0.001071 |
| 1800. Biotechnology | 0.001371 |
| 2100. Industrial electronics, physics, and related elements | 0.001042 |
| 2300. Information processing, storage, and retrieval. | 0.000535 |
| 2400. Packages, cleaning, textiles, and geometrical instruments | 0.000050 |
| 2500. Electronic and optical systems and devices | 0.000000 |
| 2600. Communications, measuring, writing, and lamp/discharge | 0.001948 |
| 3100. Handling and transportation media | 0.003137 |
| 3200. Material shaping, article manufacturing, and tools | 0.003702 |
| 3300. Mechanical technologies and husbandry personal treatment information | 0.004774 |
| 3400. Solar, heat, power, and fluid engineering devices | 0.004485 |
| 3500. General construction, petroleum and mining engineering | 0.004393 |

SOURCE: Derwent *LitAlert*, USPTO Annual Reports.


**TABLE 4** Panel Regression of Complaint Rate on Measures of Examination Quality (Dependent Variable: Logarithm of Complaint Rate)

|  | Random-Effects Model | Fixed-Effects Model | Random Effects Model | Fixed-Effects Model |
|---|---|---|---|---|
| GS-12 Examiner Hours per Disposal (log) | −1.1749 | −1.0702 | −1.325 | −1.055 |
|  | $t = -3.48^a$ | $t = -2.84^a$ | $t = -2.05^b$ | $t = -3.16^a$ |
| Actions per Disposal (log) | 0.3239 | −0.0814 | — | — |
|  | $t = 1.20$ | $t = -0.09$ |  |  |
| Time to Verdict (log) | −0.0003 | 0.0002 | −0.0003 | −0.0002 |
|  | $t = -0.50$ | $t = 0.36$ | $t = -0.64$ | $t = -0.36$ |
| $R^2$ | 0.40 | 0.27 | 0.36 | 0.30 |

[a]Significant at 1 percent.
[b]Significant at 5 percent.

NOTE: U.S. patents issued by 15 examination groups, observed 1989-1991; 8 missing observations due to absence of verdicts.

statistically significant and robust to model specification.[12] This result implies that examination groups that spent more time examining patent applications issued patents that were less likely to be involved in litigation. Conversely, patent litigation was a greater risk in examination groups in which patent examination hours were low.

Actions per patent disposal did not have a significant effect on the complaint rate. Collinearity with examiner hours is one possible explanation. Model specifications excluding examiner hours made the coefficient on examiner actions significant and positive in a few model specifications, but not robustly. This might indicate that applications receiving more examiner actions were more contentious in some way. In light of the earlier discussion on examiner actions after Table 2, this result lends some support to the hypothesis that marginal patents receive more examiner actions. An alternative interpretation is that examiners have some foresight about which patents are likely to become embroiled in litigation and interact more extensively with those applicants. Because the data do not allow more detailed analysis of this question, Table 4 also presents regression results excluding this variable altogether.

The elapsed time between complaint and verdict, intended as a proxy for litigation costs, had no effect on the dependent variable. The hypothesized negative sign appeared in some model specifications but was extremely weak. Exclusion of this variable did not affect the main result on the effect of examiner hours on the complaint rate.

In practical terms, the significance of the negative coefficient on examiner hours is that a 1 percent increase in examiner hours per patent disposal is associated with a decrease in patent litigation ranging between 1.05 and 1.33 percent. Using the range of elasticity estimates from the regression results and evaluating at the means, this suggests that an additional hour of patent examination would be associated with a decrease in litigation rates from 2.21 to approximately 2.07 complaints per thousand patent awards, i.e., perhaps as many as 24 to 26 litigation complaints annually.[13]

A reduction in the amount of litigation on this order of magnitude would have significant economic impact. Depending on the complexity of a patent, the direct cost of the patent examination process to the applicant is probably on the order of $20,000 per patent, including application fees, attorney expenses, etc. In contrast, an American Intellectual Property Lawyers Association (AIPLA) study (1996) estimates the median cost of patent litigation at $600,000 per side through the discovery phase and $1,200,000 per side if litigation proceeds all the way to a

---

[12]In addition to linear and quadratic specifications, the sign and significance of the examiner hours coefficient were robust to exclusion to one or both of the other explanatory variables.

[13]At 186,000 patent awards per year, and a complaint rate of 0.221 percent, an increase in GS-12 examiner hours from 17.6 to 18.6 translates to 24 fewer cases when elasticity is −1.055, and 26 fewer cases when elasticity is −1.33.

verdict. These figures represent only out-of-pocket litigation costs, ignoring the huge opportunity cost of time spent preparing for litigation by managers and R&D personnel.

The data suggest that an increase in examination quality would reduce the number of patents that courts must review, decreasing the risk of litigation. This reduction in litigation expense benefits the parties involved but also might generate external benefits in the form of greater transparency and less uncertainty regarding intellectual property rights, which contribute to incentives for innovation. Another benefit of decreased litigation from higher examination quality is the reduction in caseloads and associated public expenses for the court system.

Increasing the number of examiner hours by an hour for every patent disposal carries substantial costs, however. A policy of increasing examiner effort across all examination groups by 1 hour per patent disposal would mean an increase in examination costs of roughly 5.5 percent. In 1997, the USPTO employed 4,099,241 GS-12-equivalent examiner hours. By dividing the total number of hours by 2,000 examination hours per full-time equivalent examiner and multiplying the result by an annual salary of $100,000,[14] the total cost of patent examination operations to the USPTO in 1997 is estimated at $205 million. Therefore, an increase in patent examination effort by 5.5 percent would cost roughly $11.3 million.

Although these costs are significant, they point to an interesting result. Assuming that the statistical relationship between increased examination hours and reduced litigation holds, the 1-hour increase in examination would eliminate approximately 25 legal complaints. Reducing litigation to this extent would release significant resources from litigation of patents. This reduction in patent litigation from increased examiner scrutiny could come about either as a result of reducing the number of erroneously granted patents or from improving the ability of granted patents to deter infringement.

Estimating the reduction in patent litigation costs associated with an increase in examiner effort requires certain assumptions. Moore (2001) states that only 5 percent of cases proceed to a litigated verdict, which the 1996 AIPLA estimate of median litigation costs cites at $1.2 million. Another 49 percent of cases were dropped before the start of discovery, which can be assumed to cost much less, perhaps $10,000. The remaining 46 percent of patent cases incur some or all of the costs of discovery. With the 1996 AIPLA estimates of median litigation costs through discovery of $600,000, an expected cost of litigation per case can be calculated as:

$$(49\% \times \$10,000) + (46\% \times \$600,000) + (5\% \times \$1,200,000) = \$340,900$$

---

[14]The U.S. Office of Personnel Management states the full-time salary of a GS-12, Step 5 in the Washington, D.C., area in 2000 as $60,242. Rounding this figure up to $100,000 reflects employee benefits, work space, and additional costs of employment.

By multiplying this figure by 25 cases and two sides per case, an estimate of the reduction in litigation costs achieved by increasing patent examination intensity is equal to $17,045,000. This amount significantly exceeds the $11.3 million cost of increased examination effort, which is based on a $100,000 annual salary for examiners that is probably overstated to begin with. Moreover, the cost of litigation is probably significantly underestimated: Practitioner accounts suggest that the 1996 AIPLA median estimate is low and that skewness in the distribution of legal costs causes the mean litigation costs to be higher than the median. Also, this estimate does not include the public costs of litigation (in the form of courts, judges, etc.) or the opportunity cost of time diverted to litigation by managers, R&D personnel, and other employees of companies involved in litigation.

Table 5 illustrates in greater detail the potential benefits and costs of reducing litigation through greater examination quality. Using the estimated elasticity

**TABLE 5** Ratio of Predicted Benefits and Costs from Increased Examiner Hours

| Examination Group | Estimated Reduction in Litigation Expenses Divided by Cost of Additional Examination Hours |
|---|---|
| 1100. General metallurgical, inorganic petroleum, and electrical chemistry and engineering | 1.27 |
| 1200. Organic chemistry | 1.28 |
| 1300. Specialized chemical industries and chemical engineering | 1.48 |
| 1500. High-polymer chemistry, plastics, coating, photography stock materials and compositions | 0.96 |
| 1800. Biotechnology | 0.98 |
| 2100. Industrial electronics, physics, and related elements | 0.86 |
| 2300. Information processing, storage, and retrieval | 0.31 |
| 2400. Packages, cleaning, textiles, and geometrical instruments | 0.04 |
| 2500. Electronic and optical systems and devices | 0.00 |
| 2600. Communications, measuring, writing, and lamp/discharge | 1.33 |
| 3100. Handling and transportation media | 2.95 |
| 3200. Material shaping, article manufacturing, and tools | 3.74 |
| 3300. Mechanical technologies and husbandry personal treatment information | 4.52 |
| 3400. Solar, heat, power, and fluid engineering devices | 4.40 |
| 3500. General construction, petroleum and mining engineering | 4.47 |
| All | 1.79 |

NOTE: Assumptions: $50 examination cost per hour, $681,800 cost per litigation, and 1.17 percent reduction in complaint rate per 1 percent increase in examiner hours. Examination costs and litigation rates applied to mean patents allowed per year.

of reduced litigation incidence through increased examination hours and controlling for technology and commercial differences with examination group fixed effects (Table 4), Table 5 presents an estimate of potential returns to a policy of dedicating greater resources toward patent examination. The table expresses the predicted benefits of decreasing litigation through greater examination intensity as a ratio of the estimated cost of increased examiner hours.

Because examination groups vary both in the number of examiners they employ and the incidence of litigated patents they issue, not all examination groups have the same expected benefits of increased examination. The examination intensity of groups that employ relatively few examiner hours per patent disposal can be increased relatively cheaply; likewise, the potential reduction in patent litigation incidence varies with the frequency of litigation (from Table 3) and the number of patents awarded (from Table 1).

To fully capitalize on opportunities for reducing patent litigation with greater examination intensity, additional examiner hours could be targeted to the examination groups in which the ratio of benefits to costs is especially high. Groups in which the potential benefits exceed the additional costs of examination are better choices for targeted improvements in examination effort, especially considering that reductions in litigation costs are probably underestimated. Of course, such a policy would have distributional consequences and might also have effects on patents not involved in litigation.[15] The estimates provided here are extremely rough calculations, but they suggest examination groups with the greatest potential benefits from an increase in examination intensity.

## DISCUSSION

To synthesize some results of the previous sections, this study examines inputs to patent examination in the face of an increasing examination workload. Patent examination contributes to the clarity and strength of intellectual property rights and therefore plays an important role in the patent system as a whole. Although the quality of issued patents is difficult to observe in the aggregate, it is possible to quantify inputs necessary for a complete review of patent applications.

A main finding of this study is that inputs into the examination process have remained roughly consistent with the number of patent awards over the past two decades. The most important of these inputs, examiner hours and examiner actions, have kept pace with increases in patent awards and rejections. Although it is possible that the inputs necessary to conduct a thorough examination have increased, examiners appear to devote the same amount of time and effort to each

---

[15]For instance, by changing the strength and clarity of intellectual property rights as a deterrent to infringement.

granted patent despite their increasing application workload. In this sense, empirical analysis suggests that the quality of examination has not declined with the increase in patent activity over the past two decades.

Although examination has kept pace with the number of patent awards, it has fallen somewhat behind with respect to the number of patent applications. The duration of patent pendency has increased, so that applications take longer to go through the examination process. These facts might indicate that standards of patent examination are coming under some pressure from increasing workloads.

Because the main determinant of patent disposals appears to be examiner time and effort, an obvious way to address problems with the examination process is to ensure that the USPTO has sufficient resources. Currently, the USPTO funds examination activities through user fees; however, Congressional appropriations bills from 1996 to 2000 created total budget rescissions approaching $150 million from an annual budget ranging from $621 million to $911 million. These rescissions could have a significant effect if made available to fund examiners. Clearly delineated intellectual property rights could increase incentives to innovate and could have other benefits as well. For instance, inputs to patent examination quality are statistically correlated with lower patent litigation activity, meaning that more examination could lower transaction costs associated with enforcement of intellectual property rights. Implementation of a policy along these lines could be further refined by concentrating additional examiner resources in technologies or markets in which the incidence of litigation is especially high. The estimated benefits in terms of reduced litigation costs are greater than the estimated increase in examination costs on the whole, but the benefit is greater in certain technological areas.

## REFERENCES

AIPLA. (1996). *Report of Economic Survey.* Arlington, VA: American Intellectual Property Lawyers Association.

Allison, J. R., and M. A. Lemley. (1998). "Empirical Evidence on the Validity of Litigated Patents." *American Intellectual Property Lawyers Association Quarterly Journal* 26(3): 185-275.

Cohen, W. M., R. R. Nelson, J.P. Walsh. (2000). "Protecting Their Intellectual Assets: Appropriability Conditions and Why U.S. Manufacturing Firms Patent (or Not)." NBER Working Paper Series No. 7552.

Jaffe, A. B. (2000). "The U.S. Patent System in Transition: Policy Innovation and the Innovation Process." *Research Policy* 29(4-5) (April): 531-557.

King, J. L. (2000). "An Empirical Investigation of the Economics of Patent Institutions." Dissertation, Vanderbilt University, Nashville, Tenn.

Lanjouw, J. O., and M. Schankerman. (2001). "Characteristics of Patent Litigation: A Window on Competition." *RAND Journal of Economics* 32(1) (Spring): 129-151.

Levin, R., A. Klevorick, R. R. Nelson, and S. G. Winter. (1987). "Appropriating the Returns from Industrial R&D." *Brookings Papers on Economic Activity*, 3: 783-820.

Merges, R. (1999). "As Many as Six Impossible Patents Before Breakfast: Property Rights for Business Concepts and Patents System Reform." *Berkeley Technology Law Journal* 14(2) (Spring): 577-615.

Meurer, M. J. (1989). "The Settlement of Patent Litigation." *RAND Journal of Economics* 20(1) (Spring): 77-91.

Moore, K. A. (2001). "Forum Shopping in Patent Cases: Does Geographic Choice Affect Innovation?" *North Carolina Law Review* 79(4) (May): 889-938.

Priest, G., and B. Klein. (1984). "The Selection of Disputes for Litigation." *Journal of Legal Studies* 13(1) (January): 1-55.

# Patent Quality Control: A Comparison of U.S. Patent Re-examinations and European Patent Oppositions[1]

Stuart J. H. Graham
University of California, Berkeley

Bronwyn H. Hall
University of California, Berkeley and
National Bureau of Economic Research

Dietmar Harhoff
Ludwig-Maximilians-Universität München and
Centre for Economic Policy Research

David C. Mowery
University of California, Berkeley and
National Bureau of Economic Research

### ABSTRACT

*We report the results of the first comparative study of the determinants and effects of patent oppositions in Europe and of re-examinations on corresponding patents issued in the United States. The analysis is based on a data set consisting of matched European Patent Office (EPO) and U.S. patents. Our analysis focuses on two broad technology categories—biotechnology and pharmaceuticals and semiconductors and computer software. Within these fields, we collected data on all EPO patents for*

---

[1]We appreciate helpful comments by Robert Blackburn, Wesley Cohen, Markus Herzog, Mark Lemley, Stephen Merrill, Richard Nelson, Cecil Quillen, F. M. Scherer, Rosemarie Ziedonis, an anonymous referee, and seminar audiences in Berkeley, Cambridge (Mass.), Heidelberg, Munich, and Washington, D.C. Thanks to Sophia Kam and Stefan Wagner for excellent research assistance.

*which oppositions were filed at the EPO. We also constructed a random sample of EPO patents with no opposition in these technologies. We matched these EPO patents with the "equivalent" U.S. patents covering the same invention in the United States. Using the matched sample of U.S. Patent and Trademark Office (USPTO) and EPO patents, we compared the determinants of opposition and of re-examination. Our results indicate that valuable patents are more likely to be challenged in both jurisdictions, but the rate of opposition at the EPO is more than thirty times higher than the rate of re-examination at the USPTO. Moreover, opposition leads to a revocation of the patent in about 35 percent of the cases and to a restriction of the patent right in another 33 percent of the cases. Re-examination results in a cancellation of the patent right in only 10 percent of all cases. We also find that re-examination is frequently initiated by the patentholders themselves.*

## INTRODUCTION

Beginning in the 1980s, a series of administrative, judicial, and legislative actions strengthened the economic value of U.S. patents and extended their coverage in such areas as computer software and "business methods." Although many of these actions were undertaken at the behest of the U.S. business community, concerns have been raised since the early 1990s about the potential economic burdens of low-quality patents in an environment of greater deference to the rights of the patentholder (Merges, 1999; Barton, 2000). A number of experts have suggested that the U.S. patent examination system does not impose a sufficiently rigorous review of patent and nonpatent prior art, resulting in the issuing of patents of considerable breadth and insufficient quality. Many of these critics advocate the reform or extension of procedures that would enable interested parties other than U.S. Patent and Trademark Office (USPTO) examiners to bring relevant information to bear on this process either before or shortly after the issue of a patent. However, much of this debate has occurred in an empirical vacuum. Little is known about the characteristics or effectiveness of existing procedures for such post-issue challenges within the U.S. patent system, and virtually no research has compared the characteristics or effects of U.S. post-issue challenge procedures with those available elsewhere in the industrialized world's patent systems.

At present, the primary procedure for such a challenge to the validity of a U.S. patent is the re-examination proceeding, which may be initiated by any party during the life of the patent. A more elaborate and adversarial procedure in the European Patent Office (EPO) is the opposition process. This chapter uses new data in an exploratory comparative analysis of these post-issue challenge proceedings, pursuing two main questions:

1. What are the determinants of post-issue challenges via opposition or reexamination to the validity of patents in the United States and Europe?[2]

2. How do patents pertaining to the same invention fare in the two different administrative systems?

In answering these questions, we use data from both the EPO and the USPTO, including a newly created data set of "twin" patents, that is, patents taken out in both jurisdictions on the same invention.

The institutions that allow for post-grant challenges of patent validity differ considerably between the U.S. and Europe. An important feature of the proceedings at the EPO, the significance of which has been remarked upon widely by practitioners but only minimally analyzed, is the "opposition process."[3] For 9 months after the issue of a patent by the EPO, interested parties can contest its validity by filing an opposition. Typically, opponents argue that an issued patent is invalid because it fails to meet the standard requirements of patentability (novelty, inventive step, industrial application, nonprejudicial disclosures) or it does not disclose the invention with sufficient clarity or completeness.[4] In response to an opposition, the EPO may reject the opposition, amend the patent, or revoke the patent entirely.[5]

Patents issued by the EPO designate the European states in which the applicants wish to patent their inventions. Although the EPO application costs roughly three times more than the typical national application, because an EPO patent grants the applicant a right to patent in any designated state, the EPO process affords significant cost advantages for inventions requiring protection in a number of European markets. However, the centralization of application and examination also allows a centralized legal challenge: Under the European Patent Convention (EPC), any third party can use an opposition proceeding to challenge the granted patent within 9 months after the granting date for all of the designated states, rather than having to pursue legal proceedings in each of the European nations designated in the patent. The EPO opposition process has been cited by Merges (1999) as a more effective means of ensuring "high-quality" patents, especially in novel technological areas, than those available in the United States.

---

[2]We use the terms "European patents" or "opposition in Europe" as shorthand descriptions for patent applications, grants, and challenges administered by/at the EPO. Strictly speaking, a European patent (that is, a patent valid throughout Europe) does not exist, because patent rights are defined within the respective national law. Despite some harmonization, these laws are still heterogeneous.

[3]The opposition process at the EPO resembles the opposition process at the German Patent Office. The frequency of opposition is also quite similar.

[4]Article 100 EPC

[5]Article 102 EPC

U.S. patents are issued on the basis of criteria that are broadly similar to those employed by the EPO. In newly patented or novel technologies a lack of patent-based prior art or the difficulty of accessing nonpatent prior art can result in the issuance of patents of dubious merit or quality. Furthermore, for examiner searches made in the nonpatent prior art, novel technologies create higher search costs that can pose added barriers to effective discovery of prior disclosures. If prior disclosures are missed by the examiner, interested third parties wishing to challenge a U.S. patent after issue have two options: (1) challenge the patent in federal court or (2) request a re-examination of the patent by the USPTO. In absolute terms, patent litigation grew significantly in the United States during the period from 1985 to 2000, although the rate of litigation relative to the number of issued patents remained constant. However, as we suggest below, litigation is a costly and time-consuming means for establishing the validity and/or claims of a patent. In addition, costly patent litigation may contribute to growth in "defensive" patenting, another resource-intensive process with limited social returns (Hall and Ziedonis, 2001).

The patent re-examination procedure was created by federal legislation during the 1980s.[6] The number of annual re-examination requests grew from the mid-1980s through the early 1990s but has scarcely grown since 1994. Unlike litigation or oppositions, the re-examination process is not an adversarial proceeding in which advocates for each side introduce evidence and arguments in support of their position, and there are limits on the types of issues that can be raised within a re-examination. Moreover, Merges (1999) has suggested that the requirement that any opposition be filed within 9 months of the issue of an EPO patent may mean that the validity of EPO patents is determined at a much earlier point in their term than is true of the re-examination or litigation processes.[7] Merges estimates that almost 7 percent of EPO patents trigger opposition proceedings, whereas only 0.3 percent of U.S. patents result in re-examination re-

---

[6]An alternative re-examination procedure, the *inter partes* re-examination, was enacted by the U.S. Congress in 1999 (see the American Inventors Protection Act, codified in 35 USC 311-318). Several commentators have questioned the efficacy of the *inter partes* re-examination on grounds that it allows the third-party requestor limited opportunities of involvement, prevents any adverse findings of the USPTO from being appealed to the courts, and also precludes the raising of any questions of validity on grounds that were, or may have, been raised during the *inter partes* re-examination from being litigated in the courts (Neifeld 2000). The USPTO reports no *inter partes* re-examination requests in 2000 and only one in 2001.

[7]Balanced against this is the fact that EPO patents take longer to issue than U.S. patents, so the median lag between patent application and opposition challenge could be and is in fact longer than the median re-examination lag in our data (see Table 1). Unfortunately, we were unable to obtain data on litigation outcomes in either Europe or the United States that were adequate for addressing the question of the total delay in either system.

quests.[8] In addition, oppositions result in much higher rates of patent revocation than do re-examinations. According to Merges, more than 34 percent of oppositions filed in 1995 resulted in the revocation of the relevant EPO patent, considerably higher than the 12 percent of re-examination requests producing a similar result in U.S. patents during this period.[9]

In this chapter, we report the results of the first comparative study of the determinants and outcomes of patent oppositions in Europe and of re-examinations on corresponding patents issued in the United States. Our analysis focuses on two broad technology categories—biotechnology and pharmaceuticals and semiconductors and computer software.[10] Within these fields, we collected data on all EPO patents for which oppositions were filed at the EPO. We then constructed a random sample of EPO patents in these technology classes that triggered no opposition proceedings. We matched these EPO patents with the "equivalent" USPTO patents covering the same invention in the United States. This approach allows us to compare the post-issue quality control processes for technologically identical patents. Using the sample of matched USPTO and EPO patents, we compared the determinants of either opposition or re-examination.

We explore issues related to the first main topic of the chapter by addressing the following questions:

1. How does the rate of opposition (number of oppositions/all issued patents) vary by patent class within the EPO data and, similarly, for USPTO re-examinations? Which EPO and USPTO patent classes exhibit the highest opposition and re-examination rates, respectively?

2. What are the outcomes of the opposition and the re-examination processes? Do the two procedures consistently lead to a large number of patent revocations or amendments? Do types of outcomes differ significantly with the characteristics of the patent or characteristics of the patent owner or the challenger? For example, is there any evidence suggesting that patents owned by "indepen-

---

[8]Some of this difference in challenge rates may be due to the limited 9-month window available under EPO opposition rules: Because of uncertainty over the competitive threat posed by the new property right, challengers in Europe may be forced to purchase a challenge option by filing within the first 9 months after patent issue. In the United States, conversely, challengers are permitted to observe the development of the competitive landscape and technological trajectory, only filing a challenge when the threat justifies the added costs.

[9]See Merges (1999), pp. 612-614.

[10]The IPC classes included were A61K (except A61K/7), C07G, C12M, C12N, C12P, and C12Q (biotechnology/pharmaceutical) and G01R, G06F, G06K, G11C, H01L, H03F, H03K, H03M, and H04L (semiconductors/computers/software).

dent inventors" are more likely to be challenged than patents owned by corporations?[11]

3. How do the lengths of the average opposition and re-examination processes compare? What is the total time lag between application date and resolution of legal disagreements? Do oppositions, for example, enable a faster resolution of issues of patent quality and/or validity?

Using our matched sample of patents, we address the second main topic (see above) by investigating the following questions:

1. Do EPO oppositions and U.S. re-examinations focus on relatively "important" patents, measured in terms of citations to these patents in subsequent patents? How do the U.S. patents that correspond to opposed EPO patents compare with the U.S. control sample (equivalents to unopposed EPO patents) in terms of the number of post-issue citations?

2. Do we observe significant differences in the probability that a U.S. patent corresponding to an EPO patent for which an opposition is filed will be challenged through a re-examination request, by comparison with patents in the U.S. "control samples"?

More broadly, we wish to use this preliminary analysis as one component of an assessment of the comparative cost and efficiency of the re-examination and opposition processes, including a comparison of the costs, outcomes, and duration of these processes with those of litigation. This more ambitious goal is beyond the scope of this chapter because of the lack of U.S. and European litigation data. Nevertheless, the results reported here provide a useful starting point for the broader analysis.

---

[11]The U.S. re-examination process was altered considerably during congressional consideration in response to pressure from the "independent inventor" community within the United States, and there is some reason to believe that any effort to strengthen the re-examination process or institute an opposition proceeding would encounter considerable opposition from this group. Much of the group's opposition to such changes stems from the belief of many independent inventors that stronger re-examination or opposition proceedings would significantly raise the costs of patenting, because of the added costs of defending patents within these proceedings. Accordingly, information on the incidence of re-examination and opposition proceedings among different classes of patentholders will shed light on the likelihood that a disproportionate share of any such increased costs would be borne by the independent inventor.

## INSTITUTIONAL BACKGROUND

At present, the U.S. and European patent systems have similar aims and requirements for patentability but differ in the allowable subject matter and in their administrative procedures. In this and the next section of the chapter, we provide a brief overview of the operations of the two systems.

In the United States, an invention ("process, machine, manufacture or composition of matter") must satisfy four requirements to be patentable: adequate disclosure, novelty, usefulness, and nonobviousness. In Europe, firms and individuals have been able, since 1978, to submit a single application to the EPO that specifies up to 24 national jurisdictions[12] in which they desire patent protection for an invention. Under the EPO regime, the patentability requirements—adequate disclosure, novelty, industrial application, and inventive step—are broadly similar but not identical to those of the United States. The last two requirements, "industrial application" and "inventive step," map roughly onto the U.S. requirements of "usefulness" and "nonobviousness," respectively.

Figure 1 shows a rough time line covering the period between patent application and grant in the two systems. During the period covered by our data set, the U.S. patent application was kept secret until the patent issued, which meant that the median time between application and publication was 18 months to 2 years, with a long tail. As part of the patent system harmonization legislated in the American Inventors Protection Act of 1999, the United States instituted a policy of publication 18 months after application in November 2000 for many patents with applications pending in jurisdictions outside the United States.[13] In contrast, EPO applications have always been published with an 18-month lag, regardless of whether they have issued.

Both systems have a post-grant procedure through which the validity of the patent can be challenged by other parties, but the two patent systems' post-grant challenge procedures differ significantly. In both systems, interested parties can also bring suit in court over infringement and validity (with some restrictions as to when a suit can be filed). We discuss these administrative processes for post-grant challenges in the following section.

---

[12]Including: Austria, Greece (Hellenic Republic), Belgium, Ireland, Switzerland, Italy, Cyprus, Liechtenstein, Germany, Luxembourg, Denmark, Monaco, Spain, the Netherlands, Finland, Portugal, France, Sweden, Turkey, United Kingdom, Bulgaria, Estonia, Czech Republic, and Slovak Republic.

[13]The American Inventors Protection Act (1999) requires publication of all applications after 18 months but excepts applicants opting to make a declaration that a patent will not be sought in a foreign jurisdiction requiring 18-month publication. 35 USC §122.

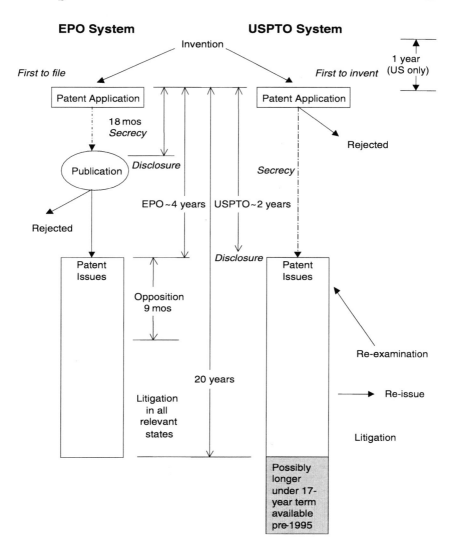

**FIGURE 1** Time line of patent application process in the EPO system and USPTO system.

## THE ADMINISTRATIVE PROCESSES AT THE USPTO AND EPO

### USPTO Examination and Re-examination Procedures

In the United States, inventors may claim a utility patent[14] by making application to the USPTO. Before a patent issues, the USPTO is charged with ensuring that the invention is adequately specified,[15] covers patentable subject matter,[16] and is useful,[17] novel,[18] and nonobvious.[19] Procedurally, the application must be filed within 1 year of the invention's public use or publication,[20] contain an adequate description with one or more claims,[21] and be accompanied by the payment of a fee.[22]

The USPTO patent examiner is the arbiter of the patentability, novelty, usefulness, and nonobviousness requirements cited above, judging these standards against the "prior art," i.e., prior inventions, in the field. Prosecution of the patent has been characterized as a "give-and-take affair," with negotiation and renegotiation between the patentee and the examiner that ordinarily continues for 2–3

---

[14]Although the vast majority of U.S. patents—and the focus of this chapter—are the so-called utility patents authorized by 35 USC §101, patents are also available on plants (35 USC §161) and designs (35 USC §171).

[15]35 USC §112. See *O'Reilly v. Morse*, 56 US 62 (1854) (finding that a claim to all uses of electromagnetic waves did not adequately describe these uses).

[16]35 USC §101. See *Diamond v. Chakrabarty*, 447 US 303 (1980) (determining that man-made living microorganisms are patentable subject matter).

[17]35 USC §101. See *Brenner v. Manson*, 383 US 519 (1966) (upholding examiner's determination that the output of a chemical process was not useful if merely similar to a useful compound).

[18]35 USC §101, 102. See *Jamesbury v. Litton Industrial*, 756 F.2d 1556 (CAFC 1985) (finding that an invention was "novel" when no prior art was precisely equivalent).

[19]35 USC §103. See *Graham v. John Deere Co.*, 383 US 1 (1966) (finding an invention invalid on grounds that the improvement would have been obvious to a person of ordinary skill in the art).

[20]35 USC §102(b).

[21]35 USC §112. Adequate description properly consists of four statutory requirements: enablement, written description, definite claims, and best mode. The "enablement" requirement is intended to allow any person skilled in the art to either make or use the invention. See *Flick-Reedy Corp. v. Hydro-Line Mfg.*, 351 F.2d 546 (7th Cir. 1965) (holding that withholding information from claims failed to adequately describe the invention). The closely-related "written description" requirement ensures that the invention is actually described. See *Permutit v. Graver Corp.*, 284 US 52 (1931) (finding the absence of any writing an insufficient description). The "definite claim" requirement ensures that the boundaries of the patent right are clearly marked. See *Halliburton Oil Well Cementing Co. v. Walker*, 326 US 1 (1946) (finding overbroad and indefinite claims invalid). The "best mode" requirement is intended to ensure that the applicant discloses the most effective method known. See *Chemcast Corp. v. Arco Industries*, 913 F.2d 923 (Fed. Cir. 1990) (finding that a failure to disclose the only known mode violates the best mode requirement).

[22]USPTO regulations set the basic filing fee at $710 for utility patents. 37 CFR §1.16(a). Additional claims may raise the fees payable, and all fees are generally lower for "small entities." 37 CFR §1.16(b),(c),(d).

years (Merges et al., 1997). The costs of prosecuting a patent through the USPTO range from $5,000 to $100,000 (including the USPTO issue fee), depending on the nature of the technology.[23]

Re-examination, originally envisioned as an alternative to expensive and time-consuming litigation, was created by the 1980 Bayh-Dole Act.[24] The legislative history of this act suggests that the re-examination was intended to be a mechanism that would be less expensive and less time-consuming[25] than litigation. During the legislative process, however, the act[26] was purged of its intended adversarial characteristics, reducing the usefulness of the procedure for opponents of a given patent.

Procedurally, the re-examination proceeding permits the patent owner or any other party to notify the USPTO and request that the grounds on which the patent was originally issued be reconsidered by an examiner.[27] Initiation of a re-examination requires that some previously undisclosed "new" and relevant piece of prior art be presented to the agency. Under the statute, a relevant disclosure must be printed in either a prior patent or prior publication—no other source can serve as grounds for the re-examination.

After being initiated by the proponent through a notification and the payment of a fee to the USPTO,[28] the re-examination goes forward only if the USPTO finds a "substantial new question of patentability."[29] Such a determination was intended by lawmakers to prevent the reopening of issues deemed settled in the original examination (Merges, 1997). The USPTO must make this determination within 3 months of the request and, having made the determination, must notify the patent owner.

When the owner is not the re-examination proponent, the patentee is allowed to file a response to the newly discovered prior art within 2 months. If the owner chooses to respond, the requester is afforded an opportunity to reply within 2 months. By choosing not to respond, the owner can limit the requester's partici-

---

[23]Gable and Montague (2001), although it is likely that most patent prosecutions cost less than $10,000. Exclusive of variable costs, e.g., attorney time and search, the USPTO has set utility patent issue fees at $1,240. 37 CFR §1.18(a).

[24]Public Law 96-517 (12/12/80).

[25]Our evidence suggests that the average re-examination takes less than 2 years, slightly shorter than the average duration of a patent lawsuit (31 months). But this difference is not large (especially in view of the high variance of the "average duration" estimate for a trial); some observers have criticized the re-examination system for not having provided a fast and cheap alternative to trial.

[26]"Act to Amend the Patent and Trademark Laws," Pub.L.No. 96-517, 94 Stat. 3015 (1980).

[27]It is also possible for the USPTO Commissioner to request a re-examination. In our sample, approximately 80 of our 4,500 re-examination requests were initiated by the USPTO, a rate of slightly less than 2 percent.

[28]$2,520 in 2001. 37 CFR §1.20(c).

[29]35 USC §303.

pation in the process. The re-examination is thus designed to be an ex parte proceeding between the patent owner and the USPTO, with limited opportunities for third-party involvement.

Any third party, such as a competitor or other opponent of the patent, thus has a limited role in the re-examination process. The requester is entitled to notify the USPTO of the triggering "prior art," to receive a copy of the patentee's reply to the re-examination (if any), and to file a response to that reply. The owner's role in the process is much more involved: The re-examination statute contemplates a second examination, with the same type of "give-and-take" negotiation between owner and patent office that occurs during the initial issuance of a patent. The examiner remains the final arbiter of the process, and it is not uncommon for the original examiner to be assigned the follow-up re-examination, thus putting the question of whether prior art was overlooked in the hands of the same government official who was responsible for ensuring that no prior art was overlooked in the previous search.

Once the re-examination goes forward, however, the statute requires that the Commissioner make a validity determination.[30] The original patent is afforded no statutory presumption of validity in the proceeding, although the practice of assigning re-examinations to the original examiner may produce such a de facto presumption. The re-examination may be neither abandoned nor postponed to await the result of concurrent litigation proceedings.[31] The result of the re-examination may be a cancellation of either all or some of the claims or the confirmation of all or some of the claims. Nothing in the re-examination procedure can expand the scope of the original patent's claims, but claims may be amended or new claims added during the renegotiation between the patent owner and the examiner.

In summary, for parties seeking to invalidate an issued patent, the re-examination procedure involves considerable costs and risks. The filing fee for the re-examination is not insubstantial, and practitioners estimate the average costs of a re-examination at $10,000-$100,000 depending on the complexity of the matter. Although the costs of a re-examination are lower than those of litigation ($1 million—$3 million), the third-party challenger in re-examination is denied a meaningful role in the process, and the patentholder maintains communications with the examining officer, offering amendments or adding new claims during the re-examination. Re-examination may also impose additional costs on challengers seeking redress in court because juries tend to give added weight to re-examined patents and the Court of Appeals for the Federal Circuit (CAFC) has indicated that claims confirmed by the re-examining officer present, in practice, added bar-

---

[30] 35 USC §307. There is no time limit on the duration of a re-examination per se.

[31] However, re-examinations may be stayed during other USPTO proceedings, including re-issue or interferences.

riers to a successful contest.[32] As a result, challengers face powerful incentives to forgo re-examination in favor of litigation, a process that may well be more expensive, more time-consuming, and less expert in testing post-issue validity.

## Patent Litigation in the United States

In the United States, post-issue validity can also be tested in court. The U.S. federal courts obviously are a unified system operating under the same substantive legal requirements, in contrast to the multistate system facing litigants in Europe. Because patent suits are filed at the District Court (trial) level, litigants have considerable control, e.g., through their choice of District Court, over the manner in which litigation unfolds. This opportunity for control is partially mitigated by the existence of the CAFC, which hears all patent appeals. However, only a very small percentage of patent cases are appealed to the CAFC, which means that any differences in judicial philosophy among the many U.S. District Courts may influence the outcomes of litigation.[33]

Procedurally, litigation differs markedly from the re-examination procedure. Unlike the re-examination procedure, litigation is an adversarial appeal to a court-arbiter in which the litigant has a choice over the final arbiter of the dispute and may elect to have the case heard by either a judge or a jury. Because patent suits generally arise from a charge of infringement by the patent owner, the patentee exerts considerable control over the timing of enforcement and litigation in a patent dispute.[34]

Legal standards create a relatively hostile environment in the federal courts for challengers seeking to invalidate an issued patent. Under the statute, patents are "born valid," enjoying a strong presumption of validity during the court proceedings.[35] Furthermore, the evidentiary standard for proving a claim invalid is "clear and convincing" evidence, a standard considerably higher than the mere "preponderance" of proof required in the typical civil suit. Because judges and juries may have limited technical expertise, these presumptions and evidentiary barriers create high costs for challengers. The propatent environment signaled by the creation of the CAFC has compounded these barriers: According to one study,

---

[32]*Kaufman Company v. Lantech, Inc.*, 807 F.2d 970 (CAFC, 1986) (suggesting that evidentiary burdens are likely higher for challengers after re-examination).

[33]However, it is likely true that more valuable patents are more likely to be the subject of an appealed trial verdict.

[34]This owner initiation occurs in many cases in which declaratory validity determinations are being sought by a challenger third party: These suits, which make the patentee the defendant, are often initiated only after a demand by the patentholder for the challenger to stop infringing the patent, thus putting the initial move in the hands of the patentholder.

[35]35 USC §282.

successful challenges to patent validity fell from 50 percent to 33 percent in the years after the creation of the CAFC (Lemley and Allison, 1998).

Direct costs in litigation are also high compared with those of re-examination. Estimates of legal costs in patent litigation run from $1 million to $3 million per suit (AIPLA, 1999) to $500,000 per claim at issue, per side (Barton, 2000). One important driver of these costs is the extensive use of pretrial discovery. The lag between filing a patent suit and reaching a resolution can also be considerable: One study estimates the average length of a District Court patent suit at 31 months (Magrab, 1993). These relatively high costs and long lags have led a number of scholars (e.g., Merges, 1999) to argue that a stronger post-grant challenge system could reduce uncertainty regarding the validity of individual patents and, arguably, contribute to higher patent quality in a less expensive and time-consuming manner. As we noted above, the adversarial elements originally contained in the legislation that established the U.S. re-examination system were largely removed from this procedure during congressional debate of the bill. In contrast, adversarial processes form the basis for the "opposition" procedure adopted by the EPO.

## EPO Examination and Opposition Procedures[36]

Patent protection for European member states can be obtained by filing several national applications at the respective national patent offices or by filing one EPO patent application at the European Patent Office. The EPO application designates the EPC[37] member states for which patent protection is requested. The total cost of a European patent amounts to approximately €29,800, roughly three times as much as a typical national application.[38] Thus, if patent protection is sought for more than three designated states, the application for a European patent is less expensive than independent applications in several jurisdictions. This cost advantage has made the European filing path particularly attractive for applicants selling goods and services in multiple European markets. Increases in the number

---

[36] This section is largely based on the description of the EPO examination and opposition system by Harhoff and Reitzig (2001).

[37] The Convention on the Grant of European Patents, also referred to here as the European Patent Convention (EPC) was enacted in October of 1973. It is the legal foundation for the establishment of the EPO. The full text of the convention is available at http://www3.european-patent-office.org/dwld/epc/epc_2000.pdf.

[38] As in other patent systems, the official patent office fees are a relatively small part of the costs (in this case €4,300). Professional representation before the EPO amounts to €5,500 on average, whereas translation into the languages of eight contracting states requires €11,500. Renewal fees for a patent maintained for 10 years amount to roughly €8,500. See "Cost of an average European patent as at 1.7.99," http://www.european-patent-office.org/epo/new/kosten_e.pdf (Jan. 14, 2002).

of patent applications and grants have given the EPO a level of economic importance that now resembles that of the USPTO.

EPO patent grants are issued for inventions that are novel, mark an inventive step, are commercially applicable, and are not excluded from patentability for other reasons.[39] After the filing of an EPO application, a search report is made available to the applicant by the EPO. The search report is generated by EPO's search office in The Hague and then transferred to the examining staff in the Munich office. The search report describes the state of prior art regarded as relevant according to EPO guidelines for the patentability of the invention, i.e., it contains a list of references to prior patents and/or nonpatent sources.[40] Within 6 months after the announcement of the publication of the search report in the EP Bulletin, applicants can request the examination of their application. This request is a compulsory prerequisite for the patent grant. If examination is not requested, the patent application is deemed to be withdrawn. Eighteen months after the priority date the patent application is published. At this point, the application is normally under examination; thus the patent owner is generally required to reveal some information about his/her invention before the grant of the patent and even if no patent is ever issued.

After examination (if requested) has been performed, the EPO presents an examination report. At this point, the EPO either informs the applicant that the patent will be granted as specified in the original application or requires the applicant to agree to changes in the application that are necessary for the patent grant. In the latter case, a negotiation process similar to that in the U.S. system may ensue. Once the applicant and the EPO have agreed concerning the scope of the allowable subject matter, the patent issues for the designated states and is translated into the relevant national languages. If the EPO declines to grant a patent, the applicant may file an appeal.[41] On average, the issue of a European patent takes about 4.2 years from the date of filing the application (Harhoff and Reitzig, 2001). Within 9 months after the patent has been granted, any third party can oppose the European patent centrally at the EPO by filing an opposition against the granting decision. The outcome of the opposition procedure is binding for all designated states. If opposition is not filed within 9 months after the grant, the patent's validity can only be challenged under the legal rules of the respective designated countries.

The EPO opposition procedure is thus the only centralized challenge process for European patents. An opposition to a European patent is filed with the EPO.

---

[39]See Article 52 EPC.

[40]It is important to note that applicants at the EPO are not required to supply a full list of prior art—as is the case in the U.S. system (see Michel and Bettels, 2001, 191ff).

[41]See Article 106 EPC. Any decisions made by the EPO in receiving, examining, and opposition sections and legal division can be appealed, and the appeal has suspensive effect.

The opponent must substantiate his opposition by presenting evidence that the prerequisites for patentability were not fulfilled, e.g., the opponent must show that the invention lacked novelty and/or an inventive step or that the disclosure was poor or insufficient. At the EPO, an opposition division determines the outcome. The examiner who granted the patent is a member of the three-person opposition chamber but may not be the chairperson. The opposition procedure can have one of three outcomes: The patent may be upheld without amendments, it may be amended,[42] or it may be revoked.[43] As we pointed out above, revocation occurs in about one-third of all opposition cases.[44]

Another interesting aspect of the opposition procedure concerns the restrictions imposed by this process on the opponent's ability to settle "out of court." Once an opposition is filed, the EPO can choose to pursue the case on its own, even if the opposition is withdrawn.[45] Thus the opponent and patentholder may not be free to settle their case outside of the EPO opposition process once the opposition is filed. This provision of the opposition proceeding may discourage its use by opponents seeking to force patentholders to license their patents.

Both the patentholder(s) and the opponent(s) may appeal the outcome of the opposition procedure.[46] The appeal must be filed within 2 months after receipt of the decision of the opposition division, and it must be substantiated within an additional 2 months. The Board of Appeal affords the final opportunity at the EPO to test the validity of the contested European patent. Both parties can bring expert witnesses into the proceedings, and there are various options for having deadlines extended. For the two technical fields considered in this chapter, the median duration of the challenge procedures (opposition and any appeal[47]) is 3.07 years, although there is considerable variation in the duration of individual cases (the interquartile range is 2.8 years).

The official fee for filing an opposition is €613; for filing an appeal against the outcome of opposition, the fee is €1022. However, the total costs to an oppo-

---

[42]See Article 99ff EPC. An amendment normally results in a reduction of the "breadth" of the patent by altering the claims that define the area for which exclusive rights are sought. See Straus (1996) for a discussion of the legal status of the patent under amendment.

[43]On average, the opposition procedure takes around 2.2 years if the patent is revoked and about 4 years if the patent is amended. See Table 2 for similar information on our samples.

[44]See EPO (1999), p. 17 and Merges (1999), pp. 612-614. There are no publicly available data as to the frequency and extent of amendments or the frequency of rejected oppositions. For the technical fields considered in this paper, we compute these figures below.

[45]Rule 60 EPC: "In the event of the death or legal incapacity of an opponent, the opposition proceedings may be continued by the European Patent Office of its own motion, even without the participation of the heirs or legal representatives. The same shall apply when the opposition is withdrawn."

[46]Article 99ff. EPC

[47]For the two technical fields studied in this chapter, an appeal occurs in about one-third of all opposition cases.

nent or the patentholder are much higher. Estimates by patent attorneys of the costs of an opposition range between €15,000 and €25,000 for each party. Patent attorneys we interviewed agreed that there is not much room for the opponent to drive up the patent owner's cost of litigation, because attorney fees are regulated in most European countries, including Germany, where many patent lawyers who have the required EPO registration reside.

## Patent Litigation in Europe

Although the EPO provides a centralized application and examination process, there is no supranational or centralized process of patent litigation in Europe. The attractiveness of the EPO opposition process stems in part from the fragmentation of patent litigation processes in Europe. Unfortunately, there have been very few systematic studies of patent litigation within the various European nations. We therefore confine ourselves to a brief review of the few facts that are known.

After the grant, the EPO patent becomes a bundle of national patent rights that are treated as "normal" national patents, which can be attacked by third parties through legal means allowed for in the respective national legislation. Outcomes in these "local" litigation cases are restricted to the "local" level, e.g., the patent may be invalidated in Spain, but this does not affect its validity in Italy. During the past decade, national patent courts have increasingly taken evidence and decisions from litigation in other European nations into account, but no systematic study has analyzed such legal "spillover" effects (Stauder, 1996; Stauder et al., 1999). Other spillover effects link the outcome of oppositions and those of subsequent litigation. The national authorities involved in the adjudication of these suits can refer to previous proceedings, which may make it more difficult for a plaintiff to win a national validity suit after having lost an EPO opposition proceeding. However, no systematic analysis of these spillovers has yet been undertaken.

The differences among national jurisdictions within Europe are enormous, requiring substantial investments in each national suit and driving up the costs of challenging the national patents emerging from an EPO grant in several of the designated states. The costs of litigation in any national court have been estimated to be between €50,000 and €500,000, depending on the complexity of the case. This cost structure makes an attack at the European level with the opposition procedure particularly attractive for a current or potential competitor of the patentholder. The litigation rate (computed as the number of cases for which a suit is filed divided by the number of patents) in most European countries is roughly 1 percent, slightly lower than the 1.9 percent reported for the United States (Stauder, 1996, 1989; Lanjouw and Schankerman, 2001). However, the quantitative evidence is too sparse to conclude from these figures that the existence of the opposition mechanism leads to a reduction in litigation.

## EXTENT AND DETERMINANTS OF POST-ISSUE CHALLENGES

### Aggregate Statistics

This section presents some aggregate statistics on EPO patent oppositions and USPTO re-examinations during the past two decades. First, we look at the rate at which these post-grant challenges are pursued for all granted patents. We then analyze the length of time until challenge occurs and until it is resolved. Finally, we examine the characteristics of patents that influence the frequency of post-grant challenges in our two technology classes.

Any comparison of opposition and re-examination must begin with a recognition of the fact that there are far more opposition cases (33,599 between 1980 and 1998) than re-examination cases (4,547) during the period of this analysis. This difference reflects the fact that the re-examination proceeding operates very differently from an opposition proceeding. Indeed, one salient difference between the re-examination and opposition procedures concerns the identity of the challengers in these processes. In nearly 40 percent of the re-examination cases during this period, the party initiating the proceeding is identified by USPTO as the patent's "owner." Obviously, there are virtually no circumstances under which the patentholder initiates an opposition proceeding in the EPO. Moreover, because many of the other parties initiating re-examinations are law firms that may be acting on behalf of patentholders, the share of re-examinations initiated by patentholders almost certainly approaches 50 percent.[48] In many cases, patentholders initiate re-examinations to address failures to properly cite prior art, to correct claims, or to repair other flaws in the issued patent. However, this difference between re-examinations and oppositions in the identity of the initiating parties highlights the very different roles of the re-examination and opposition procedures and underscores the need for caution in drawing analogies between these types of post-issue challenges.

Because our technology classes contain relatively few re-examination cases, much of our discussion of re-examinations in this section uses data for all U.S. re-examinations, rather than only those from our two broad technology classes. Figure 2 displays the opposition and re-examination "rates" in all technology sectors

---

[48]Obtaining information on patent re-examinations is difficult. The USPTO makes no effort to supply this information on its website; any amendment of the claims or revocation that results from a re-examination would not be discovered in a search of the patent based on the public data, which seems to us a bit surprising. Therefore, our estimate of owner re-examination requests is based on a somewhat incomplete sample of all re-examination requests, drawn from the incomplete paper records at the USPTO. We identified some additional owner requests by a visual scan of the assignee and requestor for each patent, but for about one-quarter of the requests, the requestor is clearly a law firm or individual and we do not know whom they are representing, if anyone. However, the 40 percent figure is not inconsistent with the current rate of owner requested re-examinations reported by the USPTO, which is in the range of 40 to 50 percent (USPTO, private communication).

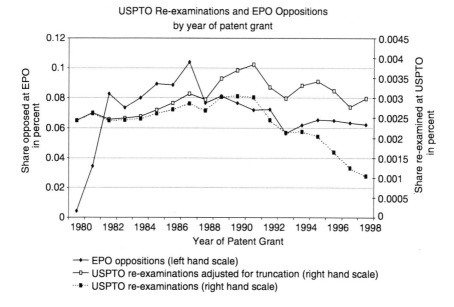

**FIGURE 2** USPTO re-examinations and EPO oppositions by year of patent grant.

during 1980-1998 (based on the year of patent grant), whereas Figure 3 shows the rates for our two technology classes during 1980-1996. The opposition and re-examination "rate" is defined as the share of patents granted in a given year that are ultimately challenged through opposition or re-examination. Our measure of the re-examination rate is truncated because challenges can happen any time during the lifetime of a patent,[49] and we use a simple model of the re-examination lag to compute a minor correction for this truncation. Two facts are immediately apparent from Figures 2 and 3:

1. The opposition rate at the EPO is much higher than the re-examination rate at the USPTO for all technology classes (Figure 2), as has been noted previously by Merges (1999) and Harhoff and Reitzig (2001). The average re-examination rate during the 1981-1998 period was 0.3 percent and the average opposition rate during the period was 8.3 percent. Thus, during 1980-1998, oppositions were about 30 times more likely to be filed than re-examinations.

---

[49]Figure 4 depicts the distribution of the lag between patent *application* and challenge. Roughly three-fourths of the re-examination requests are filed within 8 years of the application date. Because the average pendency period for a U.S. patent application is 2 years, this lag corresponds to approximately 6 years after the grant date.

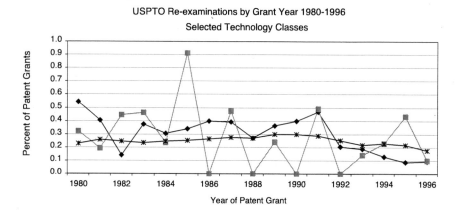

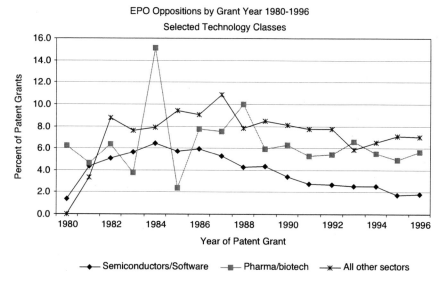

**FIGURE 3** USPTO re-examinations and EPO oppositions by grant year, 1980-1996, by selected technology class.

2. The opposition rate for patents in the semiconductor, computing, and software sector is substantially lower than that for patents in the biotechnology/pharmaceutical sector and for patents in all sectors. Our two technology classes display far smaller differences in their re-examination rates, and their re-examination rates do not differ very much from those for patents in other sectors. The lower opposition rates in semiconductors and software may reflect technological differences, but it is also plausible that firms in the semiconductor and computing industries have developed a pattern of private negotiations (e.g., cross-licensing

negotiations) for resolution of some emerging disputes (Hall and Ziedonis, 2001). The relatively modest interclass differences in re-examination rates reflect the limited utility of this process for use by patent opponents or competitors.

Figure 4 displays the distribution of the average lag between applying for a patent and the filing of a re-examination or opposition request. Because opposi-

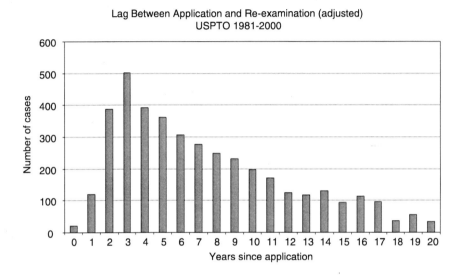

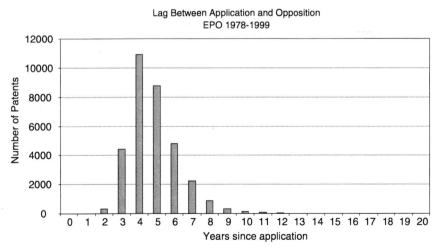

**FIGURE 4** Lag between USPTO application and re-examination (adjusted), 1981-2000, and lag between EPO application and opposition, 1978-1999.

tions must be filed within 9 months of a patent grant, the lag distribution for EPO opposition cases is much tighter than that for re-examinations. However, the grant lag itself in Europe is longer, making the mean lag between application and the filing of an opposition or re-examination action relatively similar for the two proceedings: 4.8 years elapse between the application date and initiation of an EPO opposition, somewhat less than the average lag of 6 years between patent application and a re-examination request in the United States.

Figure 5 depicts the distribution of the time lag between a patent application and final resolution of the post-grant challenge in the two systems. Because prompt resolution of uncertainty over patent validity is one potential source of welfare gain from an efficient system for post-issue challenges, the length of time from patent application to final resolution is an important criterion for evaluating the respective benefits of oppositions and re-examinations. The distributions of the duration of these proceedings differ considerably, and it is clear that the European opposition system takes somewhat longer to resolve patent disputes. The median length of time between application and final outcome of the challenge proceeding is 7.0 years at the EPO and 6.6 years at the USPTO (for re-examinations requested by a nonowner). Confining our analysis to patents applied for before 1991, to minimize the effects of lag truncation, changes the picture slightly: The median lag at the EPO is 7.2 years whereas the resolution of cases takes 7.9 years at the USPTO. The large interquartile ranges for the USPTO cases reflect the more diffuse distribution of these lags at the USPTO. The duration of the period from application to resolution is thus slightly shorter for EPO oppositions than is true of re-examinations.[50]

Table 1 summarizes the characteristics of the lag distributions that were shown in Figure 5 for the two systems. The final panel of the table confines the analysis to USPTO patents with at least one nonowner re-examination request. These cases take one-quarter to one-half year longer to resolve, but the difference is not very significant. For re-examinations requested by a nonowner, the median lag between patent application date and final opposition outcome at the EPO is 0.35 years greater than for re-examination at the USPTO for the overall time period. For pre-1991 applications, this lag is 0.3 years smaller in the EPO system. The interquartile range is 2.7 years smaller within the EPO data for the entire time period and more than 3.3 years smaller for pre-1991 applications. The total durations from application to the outcome of EPO challenge proceedings are slightly shorter on average, but the variance of the lags is greater within the U.S. re-examination proceedings. Because the re-examinations can be initiated at any time during the life of a U.S. patent, this greater variance in the distribution of the "procedural lags" for U.S. re-examinations is hardly surprising.

---

[50]The lag effect may be exacerbated over time in the U.S. system, particularly in certain sectors. In biotechnology, for instance, there is evidence that application pendency increased through the 1990s (Wright, 1997) and that this effect may have been evident before our 1991 cutoff (Rader, 1990).

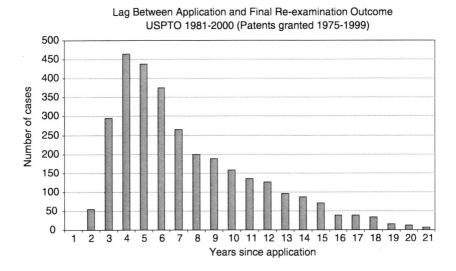
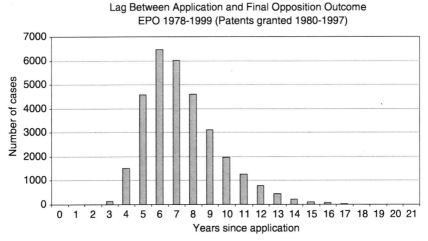

**FIGURE 5** Lag between USPTO application and final re-examination outcome, 1981-2000 (patents granted 1975-1999) and EPO application and final opposition outcome, 1978-1999 (patents granted 1980-1997).

The Appendix to this chapter presents two brief case studies of USPTO and EPO patents covering similar inventions that were opposed in the EPO system (in our terminology, these are "twin" patents). The cases, both of which cover biomedical inventions, indicate that parties opposing patents in the EPO may well pursue litigation simultaneously against the EPO patentholder's U.S. patent. The cases also underscore the point made above about the lengthy duration of the

**TABLE 1** Lags Between Application, Grant, Challenge, and Final Outcome

| | EPO[a] | | | USPTO[b] | | | USPTO (Non-owner Requested) | | |
|---|---|---|---|---|---|---|---|---|---|
| | # Obs. | Median | IQ Range | # Obs. | Median | IQ Range | # Obs. | Median | IQ Range |
| Lag between application & grant | 36,491 | 3.89 | 1.65 | 3117 | 1.73 | 0.90 | 1885 | 1.75 | 0.90 |
| Lag between grant & first challenge | 36,479 | 0.75 | 0.03 | 3117 | 2.68 | 4.62 | 1885 | 2.73 | 4.81 |
| Lag between first challenge & final outcome | 31,389 | 2.10 | 1.91 | 3117 | 1.31 | 0.97 | 1885 | 1.42 | 1.15 |
| Total lag | 31,389 | 6.96 | 2.80 | 3117 | 6.29 | 5.40 | 1885 | 6.61 | 5.71 |
| | | | Pre-1991 Applications Only | | | | | | |
| Lag between application & grant | 24,202 | 3.99 | 1.953 | 2425 | 1.79 | 0.92 | 1506 | 1.80 | 0.90 |
| Lag between grant & first challenge | 24,200 | 0.74 | 0.03 | 2425 | 3.52 | 5.48 | 1506 | 3.45 | 5.68 |
| Lag between first challenge & final outcome | 23,401 | 2.22 | 2.20 | 2425 | 1.30 | 1.04 | 1506 | 1.42 | 1.22 |
| Total lag | 23,401 | 7.23 | 3.11 | 2425 | 7.34 | 6.07 | 1506 | 7.54 | 6.54 |

[a]The numbers in the upper panel are for all oppositions against patents granted from 1980 to 1997.
[b]These numbers are for all re-examined patents (requests 1974-2000; duplicates removed).

NOTE: Neither set has been adjusted for truncation.

EPO opposition system—one lawsuit in the United States over Ortho Pharmaceuticals '799 patent was settled 5 years before the opposition proceeding on the corresponding EPO patent reached a conclusion. The other U.S. lawsuit involving this patent, however, was not settled for 2 years after the conclusion of the opposition process for the EPO "twin." The other case study of the Liposome Corporation's U.S. and EPO patents reveals a similarly complex interaction between the processes of post-grant review or litigation in the U.S. and European systems. In this case, as in the Ortho Pharmaceuticals case, a defendant in an infringement suit filed in the United States by Liposome Corporation was engaged as an opponent to the Liposome Corporation's EPO patent. This case also highlights the strategic use by a patentholder of the U.S. re-examination process to (apparently) strengthen its claims and weaken the position of a competitor. The cases thus indicate considerable interdependence between the EPO opposition process and post-grant challenges in the United States. The dimensions and timing of this interdependence are an important topic for future research.

Summarizing our descriptive findings, the EPO opposition system does not reach a conclusion much more rapidly than the U.S. re-examination procedure when this procedural duration is estimated as the length of time from patent application date to final resolution. The average lag between application date and the initiation of a challenge is substantially greater within the U.S. re-examinations than in the EPO oppositions, but this difference reflects the different time limits on the initiation of such proceedings (the EPO requirement that opposition be filed within 9 months of patent grant). Should we conclude from these comparative data that the longer lags in the EPO opposition system imply a lengthier period of uncertainty, legal expense and, therefore, a higher welfare burden within the innovation systems of these economies? Such a conclusion is unfounded, because it relies on a characterization of the re-examination and opposition proceedings as analogous in their characteristics, rigor, and outcomes. The data presented above on the identity of the parties initiating re-examinations, as well as the abundant evidence discussed above of significant procedural differences between the re-examination and opposition processes, should dispel any such analogies. Any such comparison of challenges also must incorporate data on the next stages of these challenges, which in both Europe and the United States involve litigation. Unfortunately, the analysis of litigation data is beyond the scope of this chapter.

## Analyzing the Determinants of Re-examination at the USPTO

What are the characteristics of the U.S. patents that undergo re-examination? Do they differ from the characteristics that have been identified as determinants of EPO opposition challenges in the study by Harhoff and Reitzig (2001)? To address these questions, we analyzed the characteristics of re-examined patents by analyzing patents issued between 1975 and 1999 in all patent classes for which re-examinations were requested by a nonowner between 1981 and 1999 (a total of 2,462 patents), comparing the characteristics of these patents against those in a 1 percent sample of all U.S. patents issued during this period (yielding a "control sample" of 20,359 patents). To deal with truncation issues, we also analyzed a sample of pre-1991 patents from each system (a sample including 1,513 re-examined patents and 10,099 control patents).

The results for our probit regressions analyzing the determinants of re-examinations of all U.S. patents, which use variables similar to those used by

Harhoff and Reitzig (2001), are shown in Table 2. The first panel shows results for the whole 1975-1999 sample, whereas the second panel restricts the sample to patents granted before 1991, because our measure of forward citations (those during the first 9 years of patent life) is truncated for patents issued in the 1990s. The variables have fairly high predictive value, with a pseudo-R-squared of about 0.13 and an error rate of about 13-17 percent compared with 23 percent for random assignment.

Similar to the findings of Harhoff and Reitzig in their analysis of oppositions, we find that re-examination requests are more likely for patents that are cited more frequently by other patents after their issue. The effects are large and monotonic: A patent cited more than 20 times in the 10 years after its application date is more than 50 percent more likely to be re-examined. Patents owned by individual inventors are about 2 percent more likely to be re-examined than those held by corporations.[51] Patents held by government entities are about 6-9 percent less likely to be re-examined, *ceteris paribus*.

As we noted above, our data include re-examinations in all technology classes, in contrast to the analysis of oppositions by Harhoff and Reitzig. Our analysis of re-examinations of patents in the classes examined by Harhoff and Reitzig indicates that biotechnology/pharmaceutical patents are no more likely to be re-examined and patents in the semiconductor, computer hardware, and software classes are less likely to be re-examined, compared with patents overall.[52] For patents granted before 1991, both biotechnology/pharmaceuticals and semiconductor/computer hardware have re-examination rates that are approximately the same as those for other industries. Only the rate for software is lower, by about 8 percent, although sample sizes are small.

The nationality of the patentholder has little effect on the likelihood of re-examination, although patents held by British, U.S., Canadian, Australian, or Israeli assignees are slightly more likely to be re-examined. Finally, the results for dummy variables indicating the number of claims in the patent suggest that the probability of re-examination rises monotonically with the number of claims; more complex patents are more likely to be re-examined.

---

[51]Unfortunately, it is difficult to be precise about this statement. Removing owner-requested re-examinations, but leaving in those requested by patent law firms, may bias the coefficient of this dummy if independent inventors and individuals are less likely to have their ownership concealed when requesting a re-examination. When the same estimations are performed on the whole sample, whether the patent is assigned to an individual is insignificant.

[52]Looking at the detailed classes, the following are less likely to be re-examined: C12P (fermentation or enzyme-using processes), G06F (electronic digital processing), and H01L (semiconductor devices). More likely to be re-examined are H03K [electronic switching (pulse) devices], G11C (static information storage), and H03F (amplifiers).

**TABLE 2** Probability of a Re-examination Request by Other Than the Patent Owner (Binary Probit Estimation)

| | All Observations (22,821; 2,462 Re-exams) | | | | | Patents Granted Before 1991 (11,612; 1,513 Re-exams) | | | | |
|---|---|---|---|---|---|---|---|---|---|---|
| | No. of re-exams[a] | Dprob/dx[b] | Std. Error | Dprob/dx[b] | Std. Error | No. of re-exams[a] | Dprob/dx[b] | Std. Error | Dprob/dx[b] | Std. Error |
| Bio/pharma | 26 | -0.0063 | 0.0179 | -0.0094 | 0.0167 | 13 | 0.0070 | 0.0358 | -0.0038 | 0.0327 |
| Semiconductor/hardware | 144 | -0.0231 | 0.0065 | -0.0188 | 0.0066 | 87 | -0.0044 | 0.0128 | 0.0002 | 0.0129 |
| Software | 42 | -0.0593 | 0.0061 | -0.0578 | 0.0056 | 19 | -0.0625 | 0.0132 | -0.0612 | 0.0127 |
| Cites (10yr) = 1 or 2[a] | 407 | 0.0505 | 0.0080 | 0.0410 | 0.0077 | 232 | 0.0853 | 0.0153 | 0.0773 | 0.0149 |
| Cites (10yr) = 3 to 10[a] | 982 | 0.1266 | 0.0087 | 0.1074 | 0.0083 | 642 | 0.1641 | 0.0137 | 0.1466 | 0.0134 |
| Cites (10yr) = 10 to 20[a] | 518 | 0.3226 | 0.0174 | 0.2745 | 0.0170 | 361 | 0.4010 | 0.0254 | 0.3591 | 0.0255 |
| Cites (10yr) > 20[a] | 338 | 0.5375 | 0.0222 | 0.4623 | 0.0236 | 223 | 0.6164 | 0.0277 | 0.5486 | 0.0308 |
| Claims = 6 to 9[a] | 481 | | | 0.0165 | 0.0063 | 320 | | | 0.0220 | 0.0091 |
| Claims = 10[a] | 130 | | | 0.0225 | 0.0104 | 92 | | | 0.0369 | 0.0157 |
| Claims = 11 to 15[a] | 455 | | | 0.0332 | 0.0072 | 283 | | | 0.0408 | 0.0106 |
| Claims >15[a] | 1010 | | | 0.0595 | 0.0068 | 579 | | | 0.0880 | 0.0108 |
| Claims missing[a] | 38 | | | -0.0056 | 0.0133 | 5 | | | 0.3658 | 0.1591 |
| Individual assignee | 536 | 0.0310 | 0.0054 | 0.0206 | 0.0051 | 330 | 0.0340 | 0.0080 | 0.0256 | 0.0078 |
| Government assignee[c] | 13 | -0.0566 | 0.0094 | -0.0562 | 0.0084 | 6 | -0.0878 | 0.0101 | -0.0850 | 0.0096 |
| U.S. inventor | 1906 | | | 0.0498 | 0.0039 | 1166 | | | 0.0479 | 0.0061 |
| Inventor from non-US English-speaking country[d] | 87 | | | 0.0532 | 0.0144 | 45 | | | 0.0408 | 0.0218 |
| Pseudo R-squared | | 0.109 | | 0.130 | | | 0.107 | | 0.127 | |
| Log likelihood | | -6954.31 | | -6794.75 | | | -4013.26 | | -3922.83 | |
| Chi-squared (df)[d] | | 1301.07 (9) | | 1620.18 (16) | | | 911.06 (9) | | 1091.93 16 | |

[a]The number of re-exams that have the characteristic described.
[b]This is the increase in probability for a unit change to the dummy (all variables are dummy variables).
[c]UK, Australia, Canada, New Zealand, Israel
[d]Grant year dummies are included in all estimations; the null hypothesis is these dummies only.

NOTE: The left-out category is a corporate-owned patent granted in 1975/76 with fewer than 6 claims and no citations within 10 years after grant.

## Re-examination Outcomes

Table 3 summarizes the results of re-examinations conducted by the USPTO between 1981 and 1998. Where there are two or more re-examination requests and outcomes for a particular patent, we have combined them to produce a final outcome in the following way: Patent revocation replaces any other outcome; claims added, amended, and or cancelled cumulate over re-examinations and replace any earlier no-change result.[53] The first set of columns shows all 3,000 re-examinations for which we have outcome information, and the next two columns show the results for our two main technology classes.[54] The proportions are similar, although claim amendment appears to be more likely than the mere addition or cancellation of claims in our two technology classes, both of which cover relatively new areas of inventive activity. A likelihood ratio test for similarity of the two distributions passes easily.

About 24 percent of the patents are confirmed in full on re-examination, whereas only about 10 percent are revoked in full, a number similar to the 12 percent reported by Merges (1999) for 1995. For the newer technologies, confirmation in full is less likely and revocation more likely. When we confine the sample to re-examinations initiated by nonowners (the third and fourth sets of columns), we find that both revocation in full and no change to the patent are more likely, and that the distribution of outcomes is significantly different for owner-requested re-examinations as compared to the others. The next section compares the results of re-examination to those achieved by the EPO opposition system for these two technology classes.

## Sampling Strategy for U.S.-EPO Equivalents

Thus far, we have examined data on the determinants of re-examinations at the USPTO. We now examine the similarities and differences between the U.S. and European challenge systems, both in terms of the characteristics of patents that trigger challenges and in terms of the outcomes of these challenges. This analysis requires that we control for possible differences between U.S. and EPO patents. To that end, we assembled a data set that includes "twins," i.e., EPO patents that are also issued in the United States or vice versa.

Assembly of this data set of "twin" patents relied on a sampling strategy that could produce a set of U.S.-EPO "twins" and control samples that are similarly

---

[53]Unfortunately, we were unable to find outcomes for a number of re-examinations, because of the problems mentioned above with USPTO paper records. We are missing about 200 outcomes before 1992 (approximately 13 percent) and about 700 from 1992 and later (approximately 30 percent). Some of the later re-examinations are probably not yet completed.

[54]The sample of outcomes is slightly larger than the sample of re-examinations used in Table 2 because a few observations were deleted from the sample used in Table 2 because of problems with missing data.

**TABLE 3** Re-examination outcomes, 1981-1998 (patents granted 1975-1998)

| Final outcome | All Re-exams | | Bio/pharma & Computer Hardware/Software | | Excluding Owner-Only Re-exams | | | |
|---|---|---|---|---|---|---|---|---|
| | | | | | All Re-exams | | Bio/pharma & Computer Hardware/Software | |
| No outcome noted | 902 | 23.1% | 77 | 23.0% | 472 | 20.5% | 37 | 19.0% |
| No change to patent | 716 | 23.9% | 57 | 22.1% | 476 | 25.9% | 41 | 25.9% |
| Patent revoked | 291 | 9.7% | 30 | 11.6% | 209 | 11.4% | 17 | 10.8% |
| Claims added only | 133 | 4.4% | 12 | 4.7% | 86 | 4.7% | 6 | 3.8% |
| Claims cancelled only | 144 | 4.8% | 15 | 5.8% | 93 | 5.1% | 8 | 5.1% |
| Claims amended only | 440 | 14.7% | 47 | 18.2% | 259 | 14.1% | 32 | 20.3% |
| Claims added and amended | 362 | 12.1% | 32 | 12.4% | 213 | 11.6% | 19 | 12.0% |
| Claims added and cancelled | 148 | 4.9% | 5 | 1.9% | 90 | 4.9% | 2 | 1.3% |
| Claims amended and cancelled | 369 | 12.3% | 28 | 10.9% | 208 | 11.3% | 17 | 10.8% |
| Claims added, amended, & cancelled | 397 | 13.2% | 32 | 12.4% | 202 | 11.0% | 16 | 10.1% |
| Total | 3902 | 100.0% | 335 | 100.0% | 2308 | 100.0% | 195 | 100.0% |
| Test for differences in outcomes | | | BioPh/Comp versus other technologies | | Owner-only versus other re-exams | | BioPh/Comp versus other technologies | |
| Chi-squared (degrees of freedom) | | | 11.49 | (9) | 69.57 | (9) | 11.45 | (9) |

Each re-examination appears only once. In the cases where there is more than one re-examination request, the outcomes have been combined. The shares are computed relative to the total for which we have outcomes, except for the first row, which is relative to all re-examinations in the column.

distributed among years and technology classes within the U.S. and EPO patent data. We used the International Patent Classifications (IPC) for our patents, because it is employed by both the USPTO and the EPO. We based our sampling strategy on the IPC classifications done at the EPO, because these assignments are more reliable than the IPC assignments done after the fact at the USPTO.[55] We began by drawing a sample of approximately 2,000 EPO patents that met the following criteria (Figure 6 provides a graphic depiction of this sampling strategy):

- They were granted between 1980 and 1997 (applied for between 1978 and 1995).
- They were classified in one of our two broad technology classes (62 percent in biotechnology/pharmaceuticals and 38 percent in semiconductors/computers/software).
- An opposition was filed against them after grant.

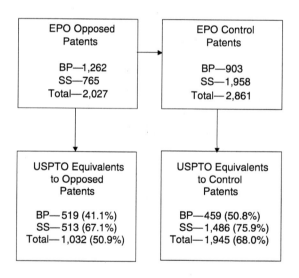

BP = biotechnology/pharmaceuticals
SS = semiconductors/software/computers

**FIGURE 6** EPO-USPTO twin study sampling strategy.

---

[55]This conclusion is based on private communications from more than one U.S. patent examiner. The search system at the USPTO is based on the U.S. patent classification system, and IPCs are assigned only after the fact, based on a rough concordance.

These patents are shown in the upper left corner of Figure 6. Using these 2,027 patents as our sampling frame, we then drew an 8 percent sample of unopposed EPO patents in these technology classes to use as controls in our analysis of oppositions, stratifying on the filing date (month and year) and IPC class, yielding a total of 2,861 patents. These are shown in the upper right corner of Figure 6. Because biotechnology/pharmaceutical patents are opposed at a higher rate, our 8 percent sample of unopposed patents yields a smaller control sample.[56]

U.S. equivalents for these two samples of patents (equivalents are members of the same patent family that have exactly the same priority or priorities in common) were then collected, yielding the patents in the two bottom panels.[57] In about 2-3 percent of the cases, an EPO patent has more than one U.S. equivalent; three patents have more than three US equivalents.[58] The likelihood that an EPO patent has a USPTO equivalent is higher for semiconductor/software than for biotechnology and pharmaceuticals. We have no definitive explanation for this difference at present. It may reflect a greater tendency for patent applicants to pursue national rather than global intellectual property protection strategies in biotechnology and pharmaceuticals, or it may reflect a greater presence of nonindustrial assignees (universities and government laboratories, both of whom are less likely to pursue global filings) in the biotechnology and pharmaceuticals patent databases. However, these possibilities are purely speculative, and additional analysis of our data is needed.

The probability that an EPO patent has a U.S. equivalent is also higher for the controls than for the opposed patents, even when we control for broad technology class. It is possible that this result reflects interdependence between the EPO oppositions and patent filings in the U.S. system. For example, an applicant collecting information—through either patent issue or ex parte discussions with the examiner—that the USPTO patent is likely to be relatively "weak" may be less likely to pursue a "strong" EPO patent, simply because of the nonzero prob-

---

[56]Sixteen patents in the EPO opposed sample and three in the control sample described above had twins that encountered re-examination requests, implying re-examination probabilities of 1.6 percent and 0.15 percent, respectively. This means that the re-examination probability is ten times as high for opposed patents, but still very small overall.

[57]See http://gb.espacenet.com/espacenet/gb/en/help/161.htm for definitions of patent families and equivalents. Equivalents can be identified by using the ESPACENET service of the EPO. This database is available at http://ep.espacenet.com.

[58]This may result because the U.S. and EPO standards for the patenting of embodiments differ, the USPTO permitting a larger number of applications than the EPO's Unity of Invention standard would allow. Article 82 of the European Patent Convention states: "The European patent application shall relate to one invention only or to a group of inventions so linked as to form a single general inventive concept." This international distinction would tend to be exacerbated in the case of pharmaceutical and biotechnology patents, however, the applicants for which have long been recognized to pursue of strategy of "serial patenting" (Merges, 1997).

ability that an opposition to the EPO patent could result in the revocation or significant amendment of the EPO patent. However, this issue requires additional analysis.

### Incidence of Opposition

Table 4 displays the results of a series of probit regressions that relate the probability that a patent is opposed in the EPO system to the characteristics of the patent, its assignee, and the U.S. twin, if there is one. All of the right-hand variables are dummies, and the estimates shown are the change in the probability if the dummy changes from 0 to 1. The first data column of the table gives the number of observations for each variable for which its respective dummy variable was equal to 1.

When we included only grant year dummies, the biotech/pharma dummy, and the U.S. twin dummy in the probit, we obtained the following estimate:

$$\text{Prob(opposition)} = \text{year effects} + 0.290 \, D(\text{biopharm}) - 0.117 \, D(\text{U.S. twin exists})$$

$$(0.015) \quad (0.016)$$

This result essentially summarizes the results of our sampling strategy: Biotechnology/pharmaceutical patents are 30 percent more likely to be opposed, and patents with U.S. twins are approximately 12 percent less likely to be opposed. Including only these variables along with grant year dummies yielded an R-squared of 0.09.

Columns (1) and (2) relate opposition to a number of characteristics of the patent and its holder. In column (2) we replace the biotech/pharma dummy by a full set of dummies for the 15 four-digit IPC classes we are considering. These dummies are clearly significant [$\chi^2(12) = 99.5$], but they have little effect on the estimate of the other coefficients.[59] The other variables in the regression are the following:

- A set of dummies for the number of EPO citations received by the patent between its issue date and 1999. One additional forward cite raises the opposition probability about 3-5 percent, with some diminishing returns, a result that is consistent with the Harhoff-Reitzig results cited above.

---

[59]The degrees of freedom are lowered by the fact that some cells are sparse and therefore not identified in the regression. Those that had much lower opposition probability than average were G06F, G11C, H01L, H03K, H03M, and H04L, which are most of the semiconductor/computing classes. Those that were higher were C07G and C12M. This result essentially confirms the fact that the biopharm dummy captures most of the difference in opposition rates for these technologies.

**TABLE 4**  Probability of an Opposition, Binary Probit Estimation (4868 Observations; 2021 Opposed), Part 1

|  | No. of obs equal to 1 | (1) Dprob/dx[a] | Std. Error | (2) Dprob/dx[a] | Std. Error |
|---|---|---|---|---|---|
| **EPO characteristics** | | | | | |
| Biotech/pharma technology | 2,157 | 0.159 | 0.020 | Full 14 tech dummies[b] | |
| No. of forward EPO cites = 1 | 974 | 0.060 | 0.022 | 0.064 | 0.023 |
| No. of forward EPO cites = 2/5 | 1,311 | 0.163 | 0.021 | 0.173 | 0.021 |
| No. of forward EPO cites = 6/10 | 258 | 0.229 | 0.035 | 0.224 | 0.036 |
| No. of forward EPO cites >10 | 80 | 0.400 | 0.051 | 0.418 | 0.050 |
| No. of designated states 6-10 | 1,082 | 0.137 | 0.022 | 0.128 | 0.023 |
| No. of designated states >10 | 1,733 | 0.175 | 0.024 | 0.169 | 0.025 |
| No. of EPO claims 6-9 | 1,192 | 0.015 | 0.024 | 0.010 | 0.025 |
| No. of EPO claims = 10 | 580 | 0.044 | 0.030 | 0.022 | 0.030 |
| No. of EPO claims 11-15 | 1,068 | 0.051 | 0.026 | 0.033 | 0.026 |
| No. of EPO claims >15 | 1,244 | 0.118 | 0.026 | 0.105 | 0.026 |
| Independent inventor (EPO ass.) | 220 | 0.028 | 0.036 | 0.016 | 0.036 |
| Accelerated search requested | 86 | −0.136 | 0.054 | −0.140 | 0.054 |
| Accelerated exam requested | 140 | 0.243 | 0.045 | 0.240 | 0.046 |
| PCT application | 937 | 0.122 | 0.023 | 0.131 | 0.024 |
| **Nationality of patentholder** | | | | | |
| U.S. | 1,642 | −0.012 | 0.049 | 0.000 | 0.049 |
| Germany | 713 | 0.101 | 0.053 | 0.096 | 0.053 |
| Other West European | 1,240 | 0.075 | 0.050 | 0.072 | 0.050 |
| Japan | 1,154 | 0.042 | 0.051 | 0.045 | 0.051 |
| Chi-squared (4) for region dummies | | 29.50 | | 20.30 | |
| Chi-squared (2) for US,JP | | 5.96 | | 3.85 | |
| **U.S. Twin characteristics** | | | | | |
| U.S. Twin exists | 2,893 | −0.089 | 0.016 | −0.094 | 0.016 |
| More than one U.S. twin | 95 | | | | |
| No. of U.S. forward cites = 1 or 2 | 571 | | | | |
| No. of U.S. forward cites = 3/10 | 1,327 | | | | |
| No. of U.S. forward cites = 10/20 | 512 | | | | |
| No. of U.S. forward cites >20 | 271 | | | | |
| No. of U.S. claims 6-9 | 751 | | | | |
| No. of U.S. claims = 10 | 157 | | | | |
| No. of U.S. claims 11-15 | 555 | | | | |
| No. of U.S. claims >15 | 846 | | | | |
| U.S. app. date prior to EPO | 1,495 | | | | |
| Independent inventor (USPTO ass.) | 124 | | | | |
| Log likelihood | | −2864.89 | | −2810.72 | |
| Pseudo R-squared | | 0.133 | | 0.148 | |
| Chi-squared (df) | | 877.9 (37) | | 977.4 (49) | |

[a]This is the increase in probability for a unit change to the dummy.

[b]One of the dummies predicts opposition perfectly, so the increase in degrees of freedom is only 12 = 13-1.

NOTE: All equations include a complete set of 18 grant year dummies; the left-out category is a corporate patent in semiconductor/software with number of states <6, number of claims <6, zero forward cites, and with holder from a country other than the "triad."

**TABLE 4** Probability of an Opposition, Binary Probit Estimation (4868 Observations; 2021 Opposed), Part 2

|  | (3) Dprob/dx[a] | Std. Error | (4) Dprob/dx[a] | Std. Error |
|---|---|---|---|---|
| **EPO characteristics** | | | | |
| Biotech/pharma technology | 0.165 | 0.020 | 0.181 | 0.020 |
| No. of forward EPO cites = 1 | 0.063 | 0.022 | 0.053 | 0.023 |
| No. of forward EPO cites = 2/5 | 0.168 | 0.021 | 0.144 | 0.021 |
| No. of forward EPO cites = 6/10 | 0.236 | 0.035 | 0.194 | 0.037 |
| No. of forward EPO cites >10 | 0.397 | 0.051 | 0.365 | 0.056 |
| No. of designated states 6-10 | 0.129 | 0.022 | 0.130 | 0.022 |
| No. of designated states >10 | 0.165 | 0.024 | 0.164 | 0.024 |
| No. of EPO claims >15 | 0.084 | 0.018 | 0.074 | 0.019 |
| Accelerated search requested | −0.132 | 0.055 | −0.132 | 0.055 |
| Accelerated exam requested | 0.242 | 0.045 | 0.239 | 0.046 |
| PCT application | 0.119 | 0.023 | 0.100 | 0.023 |
| **Nationality of patentholder** | | | | |
| Germany | 0.091 | 0.023 | 0.091 | 0.024 |
| Other West European | 0.068 | 0.018 | 0.069 | 0.020 |
| **U.S. Twin characteristics** | | | | |
| U.S. Twin exists | −0.088 | 0.016 | −0.141 | 0.038 |
| More than one U.S. twin | | | 0.007 | 0.055 |
| No. of U.S. forward cites = 1 or 2 | | | 0.008 | 0.040 |
| No. of U.S. forward cites = 3/10 | | | 0.090 | 0.036 |
| No. of U.S. forward cites = 10/20 | | | 0.171 | 0.042 |
| No. of U.S. forward cites >20 | | | 0.180 | 0.048 |
| No. of U.S. claims 6-9 | | | −0.025 | 0.028 |
| No. of U.S. claims = 10 | | | 0.000 | 0.027 |
| No. of U.S. claims 11-15 | | | −0.071 | 0.045 |
| No. of U.S. claims >15 | | | −0.037 | 0.029 |
| U.S. app. date prior to EPO | | | −0.041 | 0.021 |
| Independent inventor (USPTO ass.) | | | 0.077 | 0.049 |
| Log likelihood | −2870.21 | | −2848.92 | |
| Pseudo R-squared | 0.131 | | 0.138 | |
| Chi-squared (df) | 867.2 (31) | | 909.8 (42) | |
| Chi-squared for U.S. patent vars. | | | 42.6 (11) | |

[a]This is the increase in probability for a unit change to the dummy.

- A set of dummies for the number of EPO states in which the patent was taken out (1-5, 6-10, and more than 10). Designating more states raises the probability of opposition, which again is consistent with Harhoff and Reitzig (2001).
- A set of dummies for the number of claims (1-5, 6-9, 10, 11-15, more than 15). Having more claims raises the probability of opposition, but only if the number of claims exceeds 10.[60] The meaning of this result is ambiguous, because the number of claims in a patent is itself subject to multiple interpretations. On the one hand, patents with a large number of claims could be seeking protection for a very narrowly defined invention. In other words, these patents are occupying a space in a relatively "crowded" field populated by many similar inventions, raising the probability of an opposition. On the other hand, patents with large numbers of claims may be broader, rather than narrower, and may therefore face a lower probability of opposition (i.e., they could be early occupants of a less crowded invention space). The coefficient implies that the "crowding" effect, which raises the probability of an opposition, becomes significant as the number of claims in the patent exceeds 10.
- Whether the patentholder is an independent inventor,[61] a dummy variable for which the coefficient is insignificant.
- Whether an accelerated search was requested by the patent applicant at the EPO, which lowers the probability of opposition by about 14 percent.[62] Accelerated search is often requested when the applicant is unsure of the state of the art or of whether the invention is patentable. We therefore interpret this result as indicating a relatively "low-quality" (or less important) patent that is less likely to trigger an opposition.[63]
- Whether an accelerated examination was requested by the patent applicant, which raises the probability of opposition by 24 percent. This request indicates that the applicant attaches high value to the patent, e.g., because a patent race is under way. As a result, it is more likely that there will be a competitor that wishes to oppose the patent.
- Filing a Patent Cooperation Treaty (PCT) application, something that enables the applicant to file for protection later in up to 80 countries at the World

---

[60]The focal point at 10 claims is apparently caused by the fact that EPO charges a separate claims fee of €40 for the eleventh and each subsequent claim (Rule 31, 51 and 101 EPC).

[61]This variable disagrees with the U.S. assignment code for the twin in about one-third of the cases, which seems unlikely to us. Some of the differences occur because the EPO records multiple owners for the same patent, both individual and corporate, whereas the USPTO records only the corporation (or university, in many of these cases). We include both the U.S. and the EPO independent inventor variable in the regression to cover all possibilities.

[62]A detailed assessment of this and the next two variables is given by Reitzig (2002).

[63]Accelerated searches may be pursued in the case of commercial necessity, i.e., an applicant's need for a quick patent. In such a circumstance, we can determine no reason why opposition rates would be lower, and indeed these rates might be expected to be higher.

Intellectual Property Organization (WIPO). A PCT application raises the probability of an opposition by about 10-12 percent, which may reflect the higher value of an invention for which broad international patent protection is sought. The PCT also allows strategic delay: A PCT filing gives a patentee up to 31 months in which to make a patent application in a foreign jurisdiction. A PCT filing motivated by such delay would also likely reflect a patent of higher value.

- A set of dummies for the country of the patentholder. Although they are jointly significant, none are significant individually. Because those for Germany and the rest of Western Europe are marginally significant, we retain them.
- A dummy for the presence of one or more U.S. twins. Once we control for other patent characteristics, the relative probability that a patent with a twin is opposed increases slightly, from minus 12 percent to minus 9 percent. This finding is puzzling and requires further analysis. We speculate that this result may once again reflect some interdependence between the information an applicant collects regarding the "weakness" of the USPTO patent and the perceived "strength" or quality of the "twin" EPO patent. The EPO "twins" of patents that are survive USPTO review may be viewed as stronger by potential opponents, and therefore are less likely to trigger an opposition. Obviously, this speculative interpretation requires additional analysis of the timing of filings and oppositions in the U.S. and EPO systems.

In general, the results from the regressions in columns (1) and (2) confirm the findings by Harhoff and Reitzig (2001) that variables positively correlated with the value of a patent increase the probability that the patent will be subject to opposition. It is suggestive that patents held by independent inventors are no more likely to be opposed than other patents, other things being equal. If we do not control for patent characteristics, however, patents held by independent inventors are 11 percent *more* likely to be opposed; the main reason seems to be that they are more likely to be biotechnology/pharmaceutical patents. This result may reflect the greater presence of European university inventions within the biotechnology/pharmaceutical patent class, ownership of many of which remains with the individual faculty member.

Column (3) presents our preferred specification. Eliminating the insignificant variables does not affect the remaining coefficient estimates substantially. Patents held by German and Western European assignees are about 7-9 percent more likely to be opposed than patents held by residents of other countries. We explored the identity of the opposers, finding that they are more likely to come from countries that share a language with Germany or are in close proximity to a country that does. This suggests that the opposition system is more heavily used by those who are familiar with the language and culture of the country in which it is operated. It is natural, therefore, that the opposed patents also come from nearby countries, either because the (potential) opposers are more informed about them or simply because they are more likely to be in the same narrow line of business.

On the other hand, this finding may be caused by the choice of designated states for patent coverage, with Germany being the most favored choice. Inventors and corporations in European countries for which patent protection is not sought will have lower incentives to challenge patents that are not valid in their home country.

Finally, in column (4), we add the following variables concerning any U.S. twins for these patents:

- Whether the patent has more than one U.S. twin, a variable that is insignificant.
- A set of dummies for the number of USPTO citations received by the patent in the first 10 years of its life. One additional forward cite of this type raises opposition probability 1 percent, with some diminishing returns at high citation levels. The slightly lower coefficient for U.S. citations relative to EPO citations may reflect the fact that USPTO patents have many more citations per patent than EPO patents. Although the EPO citation variables fall slightly in the presence of USPTO citation variables, both enter the equation significantly.
- A set of dummies for the number of claims in the U.S. patent (1-5, 6-9, 10, 11-15, more than 15). Unlike citations, these variables are not significant in the presence of the dummy variable for the number of claims in the EPO patent application. When we exclude the dummies for EPO claims, the dummies for the U.S. claims become slightly significant and *negative*. This result may well reflect the difficulty, noted above, of interpreting the meaning of the number of claims in a patent.
- Whether the U.S. application date was before the EPO application date. This reduces the probability of opposition by about 4 percent, possibly reflecting the fact that more of the value of these patents relies on their exploitation in the U.S. market, making opposition in Europe less important. However, the finding also is consistent with the "signaling" interpretation of U.S. patent issue noted above.
- Whether the USPTO coded the inventor as an independent inventor. This increases the probability by about 8 percent, but the coefficient is insignificant. Measuring this more accurately is of some concern, given the reluctance of the U.S. independent inventor community to embrace an opposition system.[64] Controlling for grant year and nothing else, the raw difference in probability is 9.4 percent with a standard error of 4.8 percent.

The set of variables that describe the U.S. twin are jointly significant, with a $\chi^2(11) = 42.6$. Adding them has little effect on the other coefficients beyond a

---

[64] As we indicated above, there are many cases for which the U.S. variable is coded as unassigned that are currently (and perhaps, erroneously) included in the independent inventor class. In EPO applications, the listing of the applicant and of the inventors is compulsory.

reduction in the size of the coefficient for the "U.S. twin" dummy to minus 14 percent.

## Opposition Outcomes

The outcomes of the oppositions for our sample are shown in Table 5. The category "opposition closed" refers to cases in which either the opponent withdraws the opposition and the patent office does not pursue the case on its own behalf, or the patent holder does not renew patent protection, which causes the patent to lapse into the public domain. It is therefore not clear how many of these cases reflect a successful challenge to the patent's validity. Two facts are particularly striking: First, oppositions against patents with U.S. equivalents are more likely to be rejected. This may be due to the fact that patents from non-European applicants are selected carefully for patenting in Europe and are therefore more robust against the opposition challenge. It is also consistent with the argument that USPTO review does have a "quality-enhancing" effect on the issued patent. This result may also be a plausible explanation for the previously discussed negative impact of the "twin status" variable on the likelihood of opposition.

Second, the probability of outright revocation of a patent subjected to opposition is much higher than for re-examination: A total of 35.1 percent of the patents are revoked, and the category of "opposition closed" may contain additional cases which reflect a successful challenge (recall that only 9-11 percent of re-examined U.S. patents are revoked in full). Presumably, these results reflect the wider grounds allowed for opposition and the presence of a third party in the opposition process.

Table 6 explores the relationship between patent characteristics and outcomes with a simple logit model of the following form:

$$P_j = Pr(outcome\ j | X_i) = exp(X_i \beta_j)/\Sigma exp(X_i \beta_k)$$

where $j$ = outcome of the opposition (still pending, rejected, amended, closed, or revoked) and $X_i$ are various characteristics of the $i$th patent. In Table 6 we show the change in probability of each outcome type induced by a one-unit change in the right-hand side dummy variable, holding all other variables constant:

$$\Delta P_j (\Delta X_i^l = 1) = P_j[\beta_j^l - \Sigma \beta_k^l exp(X_i \beta_k)/ \Sigma exp(X_i \beta_k)]$$

where $l$ indexes the right-hand side variables. All effects are measured relative to the opposition pending outcome, so the rows in Table 6 sum to zero.

The results in Table 6 support the following conclusions:

1. Oppositions to patents with more citations or wider European coverage, or where there are multiple oppositions or multiple U.S. twins, tend to take longer to resolve, in the sense that the outcomes are more likely to be pending.

**TABLE 5** Final Outcome of Oppositions

|  | Total | With U.S. Twin | Percent with U.S. Twin | Share of Outcomes | With U.S. Twin |
| --- | --- | --- | --- | --- | --- |
| Opposition rejected | 266 | 173 | 65.0% | | |
| Opposition rejected on appeal | 85 | 47 | 55.3% | | |
| Opposition rejected—total | 351 | 220 | 62.7% | 17.4% | 21.4% |
| Patent amended | 355 | 207 | 58.3% | | |
| Patent amended on appeal | 163 | 81 | 49.7% | | |
| Patent amended—total | 518 | 288 | 55.6% | 25.6% | 28.0% |
| Patent revoked | 366 | 181 | 49.5% | | |
| Patent revoked on appeal | 184 | 92 | 50.0% | | |
| Patent revoked—total | 550 | 273 | 49.6% | 27.2% | 26.6% |
| Opposition closed | 150 | 81 | 54.0% | 7.4% | 7.9% |
| Opposition case pending | 190 | 72 | 37.9% | | |
| Appeals case pending | 262 | 94 | 35.9% | | |
| Case pending—total | 452 | 166 | 36.7% | 22.4% | 16.1% |
| Total | 2021 | 1028 | 50.9% | 100.0% | 100.0% |

Summary

| Outcome | Total | Share of Outcomes | | | |
| --- | --- | --- | --- | --- | --- |
|  |  | Total | Biotech/ Pharma | Computer Hardware/Software | With U.S. Twin |
| Opposition rejected—total | 351 | 22.4% | 19.1% | 26.8% | 25.5% |
| Patent amended—total | 518 | 33.0% | 38.1% | 26.1% | 33.4% |
| Patent revoked—total | 550 | 35.1% | 31.5% | 40.0% | 31.7% |
| Opposition closed | 150 | 9.6% | 11.3% | 7.1% | 9.4% |
| Total with an outcome | 1569 | 100.0% | 100.0% | 100.0% | 100.0% |
| Opposition pending | 452 | 22.4% | 27.7% | 13.6% | 16.1% |
| Total | 2021 | | | | |

2. Oppositions to biotech/pharma patents and/or highly cited patents, or where there are many claims, tend to result in amendment rather than a simple yes or no decision. Amendment is less likely when there are multiple oppositions or the inventor is an individual. More important patents or patents in relatively new, dynamic areas of inventive activity appear on this evidence to be more likely to be amended rather than revoked in an opposition.

**TABLE 6** Multinomial Logit for Opposition Outcomes (2,021 Observations)

| | Change in Probability Going from Dummy=0 to Dummy=1 (#obs.) | | | | | Change in Probability Going from Dummy=0 to Dummy=1 (#obs.) | | | | |
|---|---|---|---|---|---|---|---|---|---|---|
| | Opposition Rejected (351) | Pending (452) | Patent Amended (518) | Opposition Closed (150) | Patent Revoked (550) | Opposition Rejected (351) | Pending (452) | Patent Amended (518) | Opposition Closed (150) | Patent Revoked (550) |
| Biotech/pharma | -2.9% | 1.7% | **10.3%** | 1.4% | -10.5% | -3.9% | 2.8% | **11.7%** | 0.6% | -11.2% |
| EPO Citations 6/10 | -3.2% | 3.0% | 5.5% | -0.6% | -4.6% | | | | | |
| EPO Citations >10 | -15.5% | **14.1%** | **11.3%** | 4.7% | -14.6% | -14.6% | 12.6% | 11.1% | 3.9% | -13.0% |
| Designated states 6/10 | -1.1% | 3.9% | 1.1% | -1.0% | -2.9% | | | | | |
| Designated states >10 | -0.2% | **8.2%** | -3.9% | -0.5% | -3.7% | 0.6% | 5.7% | -5.0% | 0.4% | -1.7% |
| No. of EPO claims 6-9 | -1.7% | -3.9% | **7.4%** | -2.4% | 0.6% | | | | | |
| No. of EPO claims = 10 | -5.1% | -2.4% | **8.2%** | -5.5% | **4.8%** | | | | | |
| No. of EPO claims 11-15 | -2.3% | -5.4% | **13.2%** | -4.4% | -1.0% | -0.2% | -2.7% | *8.1%* | -2.2% | -3.0% |
| No. of EPO claims >15 | -2.5% | -1.0% | **13.5%** | -0.8% | -9.2% | -0.4% | 2.1% | *8.4%* | 1.4% | **-11.5%** |
| Accelerated search requested | 4.5% | 0.8% | 3.7% | 1.8% | -10.8% | | | | | |
| Accelerated exam requested | -5.4% | **5.2%** | 7.3% | -1.0% | -6.2% | -5.3% | 5.3% | 7.3% | -0.7% | -6.6% |
| PCT filing | -2.6% | **2.4%** | -0.2% | 1.3% | -0.9% | | | | | |
| U.S. patentholder | 2.4% | -0.2% | 2.0% | -3.1% | -1.1% | | | | | |
| German patentholder | 7.0% | **-2.5%** | -1.3% | **-4.1%** | 0.9% | 6.5% | -2.3% | -2.6% | -2.7% | 1.1% |
| Japanese patentholder | -0.3% | -1.4% | **10.5%** | **-5.1%** | -3.7% | -0.6% | -2.3% | *9.1%* | -3.5% | -2.7% |
| Same patent & opposer country | -2.2% | -3.2% | 4.1% | **5.0%** | -3.8% | -2.1% | -3.1% | 4.4% | 4.9% | -4.1% |
| Multiple oppositions | -10.7% | **9.6%** | **-6.4%** | -10.3% | **17.9%** | -10.9% | **10.1%** | **-5.8%** | -10.6% | **17.3%** |
| U.S. twin exists | 4.2% | **-1.6%** | 3.2% | 1.8% | **-7.6%** | 4.8% | -1.7% | 3.2% | 1.5% | **-7.8%** |
| Multiple U.S. twins | 1.1% | 5.5% | 2.8% | -1.0% | -8.4% | | | | | |
| USPTO—Indep. Inventor | 7.2% | 0.2% | -3.8% | -0.6% | -3.1% | 0.6% | 5.4% | 2.8% | -0.3% | -8.5% |

NOTES:
Entries in bold are significantly different from the rejection effects at the 5 percent level.
Entries in bold italics are significantly different from the rejection effects at the 10 percent level.
Bi-annual year dummies are included.
Only one (final) outcome per EPO patent included.

3. Amendment is also more likely when an accelerated examination was requested for the patent. Recall that accelerated examinations are associated with a 25 percent higher probability of opposition in the first place. The two facts together suggest that these patents are in relatively new areas that are characterized by higher uncertainty about the technology, prior art, novelty, etc.

4. Revocation is more likely when there are multiple oppositions or few claims and substantially less likely when the patent is in the biotech/pharma area, when the patent is heavily cited by subsequent patents, when an accelerated search was requested, or when there are U.S. twins.

## CONCLUSIONS AND FURTHER QUESTIONS

The determinants and characteristics of patent challenge procedures are an important issue in any assessment of the U.S. or other industrial economies' intellectual property systems. In a "knowledge-based" economy, intellectual property systems are constantly challenged by the advance of technology, a process that among other things creates new artifacts to which the necessarily backward-looking patent system must respond. A "knowledge-based" economy also is one in which the high political salience of national and global intellectual property systems means that they are the focus of political lobbying to strengthen, adapt, or weaken specific features of intellectual property regulation, administration, and law to favor particular interests. Both of these forces have been at work within the U.S. intellectual property system during the past quarter-century; a period of significant strengthening of patentholder rights has triggered a debate over the appropriate level and limits of such rights. Moreover, this debate has important trans-Atlantic and global echoes and analogues.

This chapter has explored one dimension of the operations of the post-issue systems for challenging patent validity in the U.S. and European intellectual property systems. The analysis presented here is preliminary, and many issues remain open for further research. One of the most important gaps in our current data is the lack at present of data on rates of litigation for U.S. patents that are re-examined and EPO patents (and their U.S. "twins") that are opposed. The lack of these data prevents us from examining whether the use of oppositions results in lower rates of litigation, lowering costs and resolving uncertainty more rapidly. Any such conclusion requires that we extend the analysis to incorporate post-challenge litigation, which we hope to do in future research.

Nonetheless, the analysis in this chapter (which itself needs to be extended to cover a broader array of patent classes and to incorporate the length and costs of litigation in the United States and Europe) highlights several interesting features of the patent challenge systems of the U.S. and EPO systems. First, the U.S. re-examination procedure differs dramatically from the EPO opposition procedure in virtually all of its features. Perhaps the most significant of these contrasts is the identity of the party requesting a re-examination, which our data indicate is the patent owner in more than 40 percent of the cases. This characteristic of re-examination hardly qualifies it as the sort of adversarial procedure that EPO oppositions represent. With this fact in mind, comparisons of U.S. re-examinations and EPO opposition proceedings must be treated with great caution.

Keeping in mind the significant differences between the re-examination and opposition processes, our comparative analysis suggests that EPO oppositions do not resolve validity challenges much more quickly than USPTO re-examination proceedings. In other words (and keeping in mind the incomplete nature of our data), for any given patent the EPO opposition process does not resolve uncertainties over the quality and breadth of patents more rapidly than the re-examination process. Indeed, opposition proceedings in some cases (and almost certainly in important, complex cases with numerous opponents, appeals, etc.) may well take as much time to be resolved as litigation in the U.S. system. Nonetheless, the higher frequency of opposition (which is presumably due to the lower cost associated with opposition as compared to the cost of litigation in the United States) within the EPO system is at least consistent with the hypothesis that the opposition process handles many more legal disputes over patent validity than are addressed by the U.S. re-examination process.

Our analysis also indicates that patent amendment, rather than revocation, is more likely for oppositions in relatively new fields of inventive activity, for more "complex" patents, or for oppositions in which numerous opponents participate. Because we lack evidence on the extent to which oppositions are followed by litigation in the European patent system, we cannot determine whether the lack of any "speed advantage" for oppositions in resolving patent disputes quickly is offset by a reduction of litigation rates associated with oppositions. The EPO system may offer few advantages over the U.S. system for post-issue patent challenges, but we cannot address this issue without analyzing litigation data for both the U.S. and European systems. Any comprehensive assessment of the social costs and benefits of the two challenge systems requires that we consider both the "patent office" processes of post-grant challenge (opposition or re-examination) and litigation.

The analysis of EPO oppositions and USPTO re-examinations also indicates that more "valuable" or technologically important patents, based on the usual indicators of such characteristics, are more likely to trigger challenges. This conclusion is consistent with prior research, and if the private and social values of patent rights are correlated, higher levels of scrutiny for more important or valuable patents are welfare-enhancing. Misspecifications of the claims or other characteristics of important patents are likely to produce relatively large welfare losses, e.g., deviations from an optimal trade-off between market power allocated to the patent owner and incentives for R&D (Harhoff and Reitzig, 2001).

Our analysis of "twin patents" also suggests a complex interdependence between the probability of an EPO challenge and the issuance of a U.S. "twin" patent. This interdependence must be explored further, but at least some evidence is consistent with the interpretation that "twin status" reflects selection issues that we have not addressed in this chapter. There also appear to be some interesting issues of the timing of applications and issue of USPTO and EPO patents within these data, and we intend to analyze these issues in greater detail. The existence of such interdependence is hardly surprising in an integrated global economy, but these linkages have received little scrutiny from scholars of intellectual property policy.

The heading for this section thus is used advisedly, because we have raised as many questions as conclusions in this analysis. But this highlights the richness of the agenda for further research.

## REFERENCES

AIPLA (American Intellectual Property Law Association). (1999). "Report of Economic Survey," Washington, D.C.

Barton, J. H. (2000). "Reforming the Patent System." *Science* 287: 1933-1934.

EPO (European Patent Office). (1999). Annual Report. Munich.

Gable, R. L., and M. Montague. (2001). "Strategies to Defer Costs of Patenting: Use Provisional PCT Applications." *New York Law Journal*, March 5: 57, 511-512.

Hall, B. H., and R. H. Ziedonis. (2001). "The Patent Paradox Revisited: An Empirical Study of Patenting in the U.S. Semiconductor Industry, 1979-1995." *Rand Journal of Economics* 32: 101-128.

Harhoff, D., and M. Reitzig. (2001). "Determinants of Opposition against EPO Patent Grants—the Case of Biotechnology and Pharmaceuticals." Muenchen: Ludwig-Maximilians-Universitaet. Photocopied.

Lanjouw, J. O., and M. Schankerman. (2001). "Enforcing Intellectual Property Rights." NBER working paper 8656, December.

Lemley, M. A., and J. R. Allison. (1998). "Empirical Evidence on the Validity of Litigated Patents." *American Intellectual Property Law Association Quarterly Journal*. 26: 185-277.

Levin, R., A. Klevorick, R. Nelson, and S. Winter. (1987). "Appropriating the Returns from Industrial Research and Development." *Brookings Papers on Economic Activity:* 783-820.

Magrab, E. B. (1993). "Patent Validity Determinations of the ITC: Should U.S. District Grant Them Preclusive Effect?" *Journal of the Patent & Trademark Office Society* 75: 125.

Merges, R. P. (1997). *Patent Law and Policy*. Charlottesville, VA: Michie.

Merges, R., P. Menell, M. Lemley, and T. Jorde. (1997). *Intellectual Property in the New Technological Age*. New York: Aspen.

Merges, R. P. (1999). "As Many as Six Impossible Patents Before Breakfast: Property Rights for Business Concepts and Patent System Reform." *Berkeley High Technology Law Journal* 14: 577-615.

Michel, J., and B. Bettels. (2001). "Patent Citation Analysis—A Closer Look at the Basic Input Data from Patent Research Reports." *Scientometrics* 51: 181-201.

Neifeld, R. (2000). "Analysis of the New Patent Laws Enacted November 29, 1999." *Journal of the Patent and Trademark Office Society* 82: 181.

Rader, R. (1990). "Trends in Biotechnology Patenting." Unpublished manuscript presented to Division of Chemistry and Law, American Chemical Society National Meeting, Washington, D.C., Autumn.

Reitzig, M. (2002). "Die Bewertung von Patentrechten—eine Analyse aus betriebswirtschaftlicher Sicht." Muenchen: Ludwig-Maximilians-Universität. Doctoral Thesis.

Stauder, D. (1989). *Patent- und Gebrauchsmusterverletzungsklagen in der Bundesrepublik Deutschland, Großbritannien, Frankreich und Italien*. Schriftenreihe zum Gewerblichen Rechtsschutz (Vol. 89). Köln: Max-Planck-Institut für ausländisches und internationales Patent-, Urheber- und Wettbewerbsrecht.

Stauder, D. (1996). "Aspekte der Durchsetzung gewerblicher Schutzrechte: Fachkundiger Richter, schnelles Verfahren und europaweites Verletzungsgebot." In J. Straus (ed.), *Aktuelle Herausforderungen des geistigen Eigentums*. Köln.

Stauder, D., P. von Rospatt, and M. von Rospatt. (1999). "Protection transfrontalière des brevets europeéns." *Revue Internationals de Droit Economique* 1: 119-133.

Straus, J. (1996). "Die Aufrechterhaltung eines europäischen Patents in geändertem Umfang im Einspruchsverfahren und ihre Folgen." In J. Straus, (ed.), *Aktuelle Herausforderungen des geistigen Eigentums*. Köln, 171-184.

Wright, J. (1997). "Implication of Recent Patent Law Changes on Biotechnology Research and the Biotechnology Industry." *Virginia Journal of Law and Technology* 1(2).

## APPENDIX

### Liposome Corporation—Patent No. 4,880,635 (EPO Publ. No. 190,315)

In July 1985, the Liposome Corporation (LC) submitted an application in the U.S. Patent and Trademark Office (USPTO) for a patent on their "dehydrated liposome" innovation, enabling the use of liposomes—fatty bubbles—that carry drugs to concentrate at the site of an infection. Within a month, the firm submitted an application to the European Patent Office (EPO) to secure patent rights in Europe. The European application was published in August 1986, based on LC's claimed international priority date of August 1984.

After pending in the USPTO for 4 years and 4 months, the U.S. patent issued on November 14, 1989 (patent number 4,880,635—hereafter '635 patent), with nine claims. During the next several years, LC began distributing its drug Abelcet, an antifungal treatment used for AIDS-related infections based on technology disclosed in its '635 patent. Rival Nexstar, Incorporated (formerly known as Vestar) developed a competing liposomal drug, AmBisome, prompting LC to notify Nexstar that the antifungal AmBisome infringed its '635 patent. On May 17, 1993, Nexstar sued LC in the Federal District Court in Delaware, seeking a declaration that the '635 patent was invalid, and LC counterclaimed, charging AmBisome with infringement.

Presented with new prior art that created some likelihood that Nexstar would prevail in court, LC decided on July 13, 1993 to request an "owner-initiated" re-examination on its '635 patent, thus gaining for itself an ex parte proceeding with the USPTO to determine the impact of the new prior art. This re-examination enabled LC to reenter negotiations with the USPTO over the patent's claims. If the USPTO upheld the suspect claims, the presumption of validity of the '635 before the court would be strengthened.

LC was awarded its equivalent European Patent, EP 190315, on October 17, 1993. LC designated Austria, Belgium, Switzerland, Germany, France, Great Britain, Italy, Liechtenstein, Luxembourg, and Sweden as states in which it intended to patent. Nexstar opposed LC's EPO patent on April 6, 1994, and was joined in opposition by Daiichi Pharmaceutical Company on September 21. On December 21, 1994, the Delaware U.S. District Court found that LC's patent was invalid and that Nexstar's product was not infringing. As of this date, no decision has been delivered in the Nextar/Daiichi opposition proceedings, thus suggesting that the cases are essentially closed.

Legal maneuvers kept the U.S. litigation alive through 1995 and on June 7, 1996 LC announced that it had been "upheld" by the USPTO in its re-examination. Company officials declared that the patent's "presumption of validity [was] enhanced" and threatened Nexstar with an injunction to prevent it from selling AmBisome. LC shares were up 3.4 percent on the news that day, whereas

Nexstar's shares dropped 21.5 percent. (Marc Monseau, "Patent Office upholds Liposome patent," *Denver Rocky Mountain News*, June 7, 1996).

The news also appears to have scuttled Nexstar's plans for a $60 million new share offering in June 1996 that would have financed the firm's acquisition of new drugs, marketing its newest product, and research and development. (David Algeo, "Nexstar may kill offering," *The Denver Post*, D:1, June 8, 1996). Nexstar officer said that LC's announcement of the outcome of its patent re-examination had harmed the firm (Jesse Eisinger, "Patent ruling may hamper Nexstar offering," *Denver Rocky Mountain News*, 5B, June 11, 1996).

The USPTO certificate on the re-examination of the '635 patent finally issued on July 2, 1996, and the facts did not entirely support LC's press releases of a month earlier. In reality, B1 Certification 2,937 stated that 3 claims had been cancelled, 6 claims had been amended, and 19 new claims were added to the '635 patent. Nexstar returned to federal court in May of 1997, claiming that LC had purposefully misrepresented the re-examination results to gain advantage and injure Nexstar, and argued that the '635 patent was invalid.

EP 190315 was opposed at the EPO on Feb. 1, 1994 by Nexstar and Daiichi Pharmaceutical. The case is still pending on appeal, and we do not know the preliminary outcome. It is probable, based on the events discussed immediately below, that they are not waiting for the final outcome and the case is essentially closed.

The two competitors ultimately reached a settlement in their U.S. court case on August 11, 1997, jointly stipulating to a dismissal. In the settlement, LC granted Nexstar immunity from future suits in connection with its worldwide manufacture and marketing of AmBisome. The firms agreed to grant reciprocal options to take licenses to the other's patented technologies, whereas Nexstar agreed to unspecified payments to LC. The following day, Nexstar's AmBisome was approved by the Food and Drug Administration for marketing in the United States.

### Ortho Pharmaceuticals—U. S. Patent 4,363,799 (EPO Publ. No. 17,381)

By the early 1980s, monoclonal antibodies had been recognized as a remarkable advance in medical science. The discovery, which allows the identification of so-called T cell subsets of lymphocytes, a type of white blood cell, showed promise for enabling advancements in the treatment of infectious disease, cancer, infertility, autoimmune disorders, heart disease, and other maladies. In 1984, sales of diagnostics and therapies using the technique grossed U.S. $500 million, with projections of annual sales of U.S. $2 billion by 1990 (Lawrence Altman, "A Discovery and Its Impact: Nine Years of Excitement," *New York Times*, C:3, Oct. 16, 1984). The founders of the technique were awarded the 1984 Nobel Prize in "Physiology or Medicine," signaling its path-breaking nature.

On March 20, 1979, the Ortho Pharmaceutical Corporation (Ortho) applied for a U.S. patent on its invention entitled "Monoclonal antibody to human T cells, and methods for preparing same." On March 19, 1980, presumably taking advantage of the 1-year application window allowed in the EPO, Ortho applied for its equivalent European patent, application number EP1980030082, using the U.S. application date as its priority date. On the basis of the application's March 1979 international priority date, the EPO published the application on October 15, 1980, signaling the existence of the pending patent. Ortho designated its European states of interest on that date as Austria, Belgium, Switzerland, Germany, France, Great Britain, the Netherlands, Italy, and Sweden.

On December 14, 1982, after some 2 years and 9 months pending in the USPTO, the U.S. patent issued (number 4,363,799), with 11 claims. Approximately 2 years later, on September 20, 1984, Ortho filed a complaint alleging patent infringement against Becton Dickinson Monoclonal Center, Inc. in the Federal District Court in Wilmington, Delaware. The complaint also covered 12 other patents owned by Ortho. Within 10 months, the European equivalent patent (No. 17381) issued, on July 10, 1985.

During 1986, legal maneuvering on both sides of the Atlantic tested the validity of the Ortho patent. On June 4, 1986, an EPO opposition was filed by Behringwerke AG and Sandoz AG. Within a week, on June 11, a second opposition was filed by Becton, Dickinson & Company and by Boehringer Mannheim Gmbh. On July 24, 1986, Ortho's U.S. infringement action against Becton Dickinson, an opponent to Ortho's EPO patent, was transferred to the U.S. Federal District Court in Northern California. On September 26, Ortho again asserted its patent in an infringement action against Coulter Corporation and Coulter Electronics Corporation in the Southern District of Florida.

By October 3, 1986, Ortho and Becton Dickinson had settled their California litigation. Each party stipulated to a voluntary dismissal of the case and the Court announced that the parties had "resolved their differences." But the EPO opposition proceedings continued, and after the two pending oppositions were consolidated, the EPO patent was revoked on October 17, 1986. Ortho immediately appealed the adverse decision to the EPO, but the appeal was finally rejected on January 8, 1991, 5 years after settlement of the firm's infringement suit against one of the EPO patent opponents.

Ortho's suit against Coulter Corporation and Coulter Electronics Corporation in the Southern District of Florida was finally settled in November 1993, with a consent judgment and a dismissal. Ortho's U.S. patent remains in force but has not been asserted in court since. The patent number is not withdrawn, although the patent is close to expiration.

# Benefits and Costs
# of an Opposition Process[1]

Jonathan Levin
Stanford University

Richard Levin
Yale University

### ABSTRACT

*In recent years, patent protection has extended into new areas, giving rise to serious concern about the lack of clear guidelines for patentability. We analyze the effect of introducing a patent opposition process that would allow patent validity to be challenged directly after a patent is granted. In many cases, such a system would avoid costly litigation at a later date. In other cases, the opposition process would increase the cost of conflict resolution but would also reward holders of valid patents and limit the rewards for invalid patents. Our analysis suggests significant positive welfare gains from the introduction of a patent opposition process.*

### INTRODUCTION

In just over two decades, a succession of legislative and executive actions has served to substantially strengthen the rights of patentholders.[2] At the same

---

[1]Reprinted with permission from Arnott, R., B. Greenwald, R. Kanbur, and B. Nalebuff, eds. (2003). *Economics for an Imperfect World: Essays in Honor of Joseph E. Stiglitz*. Cambridge, MA: The MIT Press. We are indebted to the members and staff of the Committee on Intellectual Property in the Knowledge-Based Economy of the National Academies' Board on Science, Technology, and Economic Policy (STEP) for stimulating our thinking on this topic. We also thank Barry Nalebuff and Brian Wright for helpful comments.

[2]Notable among these actions are the Bayh-Dole Patent and Trademark Amendments Act of 1980, the creation of the Court of Appeals for the Federal Circuit in 1982, the Hatch-Waxman Drug Price Competition and Patent Restoration Act of 1984, the Process Patent Amendments Act of 1988, and the Trade-Related Aspects of Intellectual Property Rights (TRIPS) Agreement of 1994.

time, the number of patents issued in the United States has nearly tripled from 66,290 in 1980 to 184,172 in 2001. Although the surge in patenting has been widely distributed across technologies and industries, decisions by the U.S. Patent and Trademark Office and the courts have expanded patent rights into three important areas of technology in which previously the patentability of innovations was presumed dubious: genetics, software, and business methods.[3] As in other areas of innovation, patents in these fields must meet standards of usefulness, novelty, and nonobviousness. A serious concern, however, in newly emerging areas of technology is that patent examiners may lack the expertise to assess the novelty or nonobviousness of inventions, leading to a large number of patents likely to be invalidated on closer scrutiny by the courts.

Although similar examples could be drawn from the early years of biotechnology and software patenting, economists in particular will appreciate that many recently granted patents on business methods fail to meet a commonsense test for novelty and nonobviousness. Presumably, this occurs because the relevant prior art is unfamiliar to patent examiners trained in science and engineering. Consider U.S. Patent No. 5,822,736, which claims as an invention the act of classifying products in terms of their price sensitivities and charging higher markups for products with low price sensitivity rather than a constant markup for all products. The prior art most relevant to judging the novelty of this application is neither documented in earlier patents nor found in the scientific and technical literature normally consulted by patent examiners. Instead, it is found in textbooks on imperfect competition, public utility pricing, or optimal taxation.

The almost certain unenforceability of this particular business method patent may render it of limited economic value, but other debatable patents have already been employed to exclude potential entrants or extract royalties. A much publicized example is Jay Walker's patent (U.S. Patent No. 5,794,207) covering the price-matching system used by Priceline.com. After several years of legal wrangling, Microsoft Expedia agreed to pay royalties for allegedly infringing on this patent. Many economists, however, would object that Walker's patent covers only a slight variation on procurement mechanisms that have been used for hundreds if not thousands of years. Interestingly, in terms of prior art, Walker's patent application cites several previous patents but not a single book or academic article on auctions, procurement, or market exchange mechanisms.

If challenged in court, a patent on the "inverse elasticity rule" would likely be invalidated for failing to meet the test of novelty or nonobviousness. The Walker patent, a closer call, also might not survive such scrutiny. Current U.S. law, however, permits third-party challenges only under very limited circum-

---

[3]Three landmark cases regarding, respectively, genetics, software, and business methods, are *Diamond v. Chakrabarty,* 447 U.S. 303 (1980); *Diamond v. Diehr,* 450 U.S. 175 (1981); and *State Street Bank & Trust Co. v. Signature Financial Group, Inc.,* 149 F.3d 1368 (Fed Cir 1998).

stances. An administrative procedure, re-examination, is available to third parties who seek to invalidate a patentee's claim by identifying prior art, in the form of an earlier patent or publication, that discloses the precise subject matter of the claimed invention.[4] In practice, however, re-examination is used primarily by patentees to amend their claims after becoming aware of uncited prior art.

Broader objections to a patent's validity can be adjudicated only in response to a patentholder's attempts to enforce rights against an alleged infringer. In response to an infringement suit, the alleged infringer may file a counterclaim of invalidity. In response to a "desist or pay" letter, the alleged infringer may seek a declaratory judgment to invalidate the patent. Generally speaking, such proceedings are very expensive and time consuming. A recent survey estimated the median cost of a litigated patent infringement suit at $1.5 million in cases involving stakes of $1 million to $25 million; when the stakes exceed $25 million, the median cost of a suit was estimated to be $3 million (American Intellectual Property Law Association, 2001). A typical infringement suit might take 2 to 5 years from initial filing to final resolution.

What are the costs of uncertainty surrounding patent validity in areas of emerging technology? First, uncertainty may induce a considerable volume of costly litigation. Second, in the absence of litigation, the holders of dubious patents may be unjustly enriched and the entry of competitive products and services that would enhance consumer welfare may be deterred. Third, uncertainty about what is patentable in an emerging technology may discourage investment in innovation and product development until the courts clarify the law, or, in the alternative, inventors may choose to incur the cost of product development only to abandon the market years later when their technology is deemed to infringe. In sum, one suspects a timelier and more efficient method of establishing ground rules for patent validity could benefit innovators, followers, and consumers alike.

One recently suggested remedy is to expand the rights of third parties to challenge the validity of a patent in a low-cost administrative procedure before sinking costly investments in the development of a potentially infringing product, process, or service (see Merges, 1999 and Levin, 2002). Instead of the current re-examination procedure, which allows post-grant challenges only on very narrow grounds, the United States might adopt an opposition procedure more akin to that practiced in Europe, where patents may be challenged on grounds of failing to meet any of the relevant standards: novelty, nonobviousness, utility, written description, or enablement. The European system requires only minimal expenditure by the parties. When interviewed, senior representatives of the European Patent Office estimated expenditures by each party at less than $100,000. The

---

[4]Prior art invalidating the inverse elasticity patent could probably be found. On the other hand, patents such as Walker's that are close but not identical to past published ideas typically cannot be overturned on re-examination.

time required for adjudication, however, is extremely long, nearly 3 years, owing to very generous deadlines for filing of claims, counterclaims, and rebuttals.[5]

The idea of a streamlined, efficient U.S. administrative procedure for challenges to patent validity is clearly gaining momentum in the response to mounting concern about the quality of patents in new technology areas. In its recently released *21st Century Strategic Plan*, the U.S. Patent and Trademark Office stated as one of its intended actions: "Make patents more reliable by proposing amendments to patent laws to improve a [sic] post-grant review of patents" (U.S. Patent and Trademark Office, 2002).[6]

This chapter makes a modest attempt to evaluate the potential costs and benefits of introducing such a post-grant opposition process. In the next two sections, we develop a simple model of patent enforcement and patent oppositions. We model patent oppositions as essentially a cheaper and earlier way to obtain a ruling on patent validity. The one further difference between patent opposition and litigation captured by the model is that patent oppositions can be generated by potential infringers, whereas litigation must be initiated or triggered by the patentholder. The analysis divides naturally into two cases: one case in which the potentially infringing use of the patent is rivalrous (i.e., it competes directly with the patentee's product) and one case in which the uses are nonrivalrous (i.e., independent or complementary). The key difference between these cases is that in the former the patentholder wants to deter entry whereas in the latter the patentholder simply wants to negotiate for a large licensing fee.

We identify several effects of introducing an opposition process. First, if the parties foresee costly litigation in the absence of an opposition, they have a clear incentive to use the cheaper opposition process to resolve their dispute. This lowers legal costs and potentially prevents wasteful expenditure on product development. At the same time, giving the parties a lower-cost method of resolving disputes can lead to oppositions in cases when the entering firm might either have refrained from development or been able to negotiate a license without litigation. These new oppositions have a welfare cost in that the firms incur deadweight costs from preparing their opposition suits. Nevertheless, these oppositions generate potential benefits. They can prevent unwarranted patents from resulting in monopoly profits, and, more broadly, if decisions under the opposition process are more informed than those made directly by the patent examiners, the rewards to patentholders end up more closely aligned with the true novelty and non-

---

[5]See Graham, Hall, Harhoff, and Mowery, this volume, for this and other details of the European Patent Office's opposition procedure.

[6]Of course, an alternative way to reduce uncertainty about patent validity would be to intensify the U.S. Patent Office's screening of applications. Lemley (2001) argues that this approach is unlikely to be cost-effective because resources would not necessarily be focused on economically meaningful patents. In contrast, an opposition process encourages early scrutiny of patents that are both debatable and economically significant.

obviousness of their invention. From a dynamic welfare standpoint, this has the favorable effect of providing more accurate rewards for innovation.

The model suggests that in some cases, introducing an opposition process will have an unambiguous welfare benefit, whereas in other cases there will be a trade-off between static welfare costs and static and dynamic welfare benefits. In the fourth section of this chapter, we use available information on the cost of litigation and plausible parameters for market size and the cost of development to provide a rough quantitative sense of the welfare effects. Our general conclusion is that the costs of introducing an opposition system are likely to be small in relation to the potential benefits.

The fifth section concludes with a discussion of some aspects of the opposition process not captured in our simple modeling approach. The model provides a reasonable assessment of how an opposition system affects the gains and losses realized by a single inventor, a single potential infringer, and their respective customers. It ignores, however, substantial positive externalities from greater certainty and more timely information about the likely validity of patents that would flow to other parties contemplating innovation and entry in a new technology area. In this respect, our analytic and quantitative findings probably understate the full social benefit of introducing a low-cost, timely system for challenging patent validity.

## A MODEL OF PATENT ENFORCEMENT

We start by developing a simple benchmark model from which we can investigate the effect of an opposition process. There are two firms. Firm $A$ has a newly patented innovation, and Firm $B$ would like to develop a product that appears to infringe on $A$'s patent. The dilemma is that the legitimacy of $A$'s patent is uncertain. In the event of litigation, $B$ may be able to argue convincingly that part or all of the patent should be voided.

The interaction between the firms unfolds as follows. Initially, Firm $B$ must decide whether to develop its technology into a viable product. Let $k$ denote the costs of development. If $B$ does not develop, $A$ will be the monopoly user of its technology. If $B$ does develop, it can enter negotiations to license $A$'s technology. If negotiations are successful, $B$ pays a licensing fee (the precise amount will be determined by bargaining) and both parties use the technology. If $B$ does not obtain a license, it may still introduce its product. In this event, $A$ can either allow $B$ to market its product unhindered or file suit to enforce its intellectual property rights. If $A$ files suit, the parties enter litigation.

We adopt a simple formulation for thinking about litigation. In litigation, each party incurs a cost $L$ to prepare its case. At trial, the court assesses the validity of $A$'s patent and whether $B$'s patent infringes upon it. We focus on the determination of validity, because this is the aspect of patent disputes for which an opposition process has relevance. Let $p_A$ and $p_B$ denote the subjective probabilities that Firms $A$ and $B$ assign to the court upholding the patent and let $p$ denote the true objective probability of validity. We assume that the firms' subjective probabilities (but not the true objective probability) are commonly known, although not necessarily equal.[7]

If the court invalidates the relevant parts of Firm $A$'s patent, $B$ is free to market its product. In contrast, if the patent is upheld, $A$ has the option of excluding $B$ from the market. Firm $B$ may try again to negotiate a license, but if it fails $A$ proceeds to market alone.

The firms' profits depend on whether $B$'s product reaches the market and whether they incur litigation costs. Let $\pi_{A|B}$ and $\pi_A$ denote the gross profits that $A$ will realize if $B$'s product does or does not reach the market, respectively. Let $\pi_B$ denote the gross profits that $B$'s product will generate. In making decisions, the firms must factor in these eventual profits as well as development costs, litigation costs, and licensing fees in the event of a licensing agreement.

We model licensing negotiations, both before and after litigation, by using the Nash bargaining solution. This means that if there are perceived gains to licensing, each party captures its perceived payoff in the absence of a license and the additional surplus generated by the agreement is divided equally.

The timing of the benchmark model is displayed in Figure 1. After development, the firms can negotiate a license. If this fails, $B$ must make a decision about whether to enter and $A$ can respond by litigating. If there is litigation, the court rules on the patent's validity, at which point the parties have another opportunity to negotiate a license.

In thinking about this benchmark situation and the effects of an opposition process, we have found it useful to distinguish two prototypical situations. In the first, which we refer to as the case of *nonrivalrous innovation*, the firms have a joint interest in bringing Firm $B$'s product to market. This is the situation, for instance, when Firm $A$'s patent covers a research tool or perhaps a component of a product that $B$ can produce at lower cost than $A$. In the second case, *rivalrous innovation*, Firm $B$'s product will compete directly with $A$'s product and the introduction of $B$'s product will decrease joint profits through intensified competition. Think, for instance, of $A$ as a drug company and $B$ as a rival with a closely related therapeutic.

We analyze these situations separately for a simple reason. When innovations are nonrivalrous, litigation and opposition hearings will not bar entry. They

---

[7] The assumption that $p_A$ and $p_B$ are commonly known but not necessarily equal means that firms will not update beliefs when they negotiate as in standard asymmetric information models. Rather, they "agree to disagree" about patent validity. This is a simple way to capture the fact that parties may sometimes end up in court rather than settling. Note that the uncertainty about patent validity is the only uncertainty in the model—for instance, there is no uncertainty or learning about whether $B$'s development will succeed or about the size of the product market. Accounting for these realistic forms of uncertainty would change the quantitative, but not the qualitative, conclusions of our model.

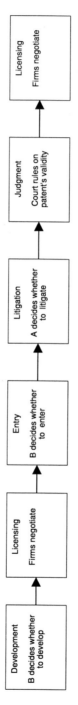

**FIGURE 1** Timing in the benchmark model.

serve only to affect the terms of licensing agreements. In contrast, with rivalrous innovation, litigation is an instrument for Firm A to defend its monopoly status. In this regard, we assume that antitrust law precludes A from paying B not to enter or from designing a licensing agreement that manipulates future competition.[8] Thus if A's patent rights are upheld, it denies its rival access to the market. Changing the method for resolving disputes from litigation to an opposition may substantively affect what products eventually reach market.

## Nonrivalrous Innovation

We start by considering nonrivalrous innovation. To focus attention on this case, we make the following parametric assumption, which is sufficient to ensure that introducing Firm B's product generates a joint gain for the two firms.

**Assumption NR** $\pi_{A|B} + \pi_B - 2k \geq \pi_A$.

In fact, this assumption is slightly stronger than is needed to ensure nonrivalry. A weaker condition would be that $\pi_{A|B} + \pi_B - k \geq \pi_A$. The stronger condition has the benefit of guaranteeing that Firm B will have a sufficient incentive to develop its product before negotiating a license, rather than needing to seek a license before development. Because the effect of an opposition proceeding turns out to be essentially the same in this latter case, we omit it for the sake of clarity.[9]

To analyze the model, we work backward. First, we describe what happens if the parties wind up in litigation. We then consider whether litigation will occur or whether B will negotiate a license or simply enter with impunity. Finally, we consider B's incentives to develop its product.

### Outcomes of Litigation

Suppose that Firm B introduces its product without a license and Firm A pursues litigation. Two outcomes can result. If the court voids the relevant sections of A's patent, B can enter without paying for a license. If the court upholds A's patent, B must seek a license. Because the products are nonrivalrous, there is a gain $\pi_{A|B} + \pi_B - \pi_A > 0$ to be realized from an agreement. Development costs do not appear in the calculation of the gain from introducing B's product because

---

[8]See Meurer (1989) for a model in which the patentholder may use the terms of a licensing agreement to restrict future competition.

[9]Note that our definition of nonrivalry does allow Firm A's profits to decrease if B enters. A more traditional notion of nonrivalry might require that $\pi_{A|B} \geq \pi_A$. Our more encompassing definition focuses on joint profitability, which is natural once one realizes that Firm A will be capture some of Firm B's profits through licensing fees.

they have already been sunk. Nash bargaining means that this gain is split equally through a licensing fee $F_V$:

$$F_V = \frac{1}{2}(\pi_A - \pi_{A|B}) + \frac{1}{2}\pi_B$$

Here we use the subscript $V$ to refer to bargaining under the presumption that $A$'s patent is valid.

Factoring in these two possible outcomes of litigation, we can calculate the (subjective) expected payoffs to the two firms upon entering litigation. These are $\pi_{A|B} - L + p_A F_V$ for Firm $A$ and $\pi_B - k - L - p_B F_V$ for Firm $B$.

### Determinants of Litigation

We now back up and ask what will happen if Firm $B$ develops its technology.

The first question is whether $A$ has a credible threat to litigate if $B$ attempts to market its product without a license. Because $A$'s subjective gains from litigation are $p_A F_V - L$, it will want to pursue litigation only if

$$p_A F_V - L \geq 0. \tag{A}$$

If this inequality fails, Firm $A$ has a *weak patent*—the benefit of enforcing it is smaller than the litigation costs. If $A$'s patent is weak, Firm $B$ can simply ignore it and enter without fear of reprisal. Indeed, even if an opposition system is in place, $B$ would never want to use it because $A$'s patent is already of no meaningful consequence. This makes the weak patent case relatively uninteresting from our perspective. For this reason, we assume from here on that $A$'s patent is not weak.

Given that Firm $A$ has a credible threat to litigate, we now ask whether litigation will actually occur. The parties will end up in court if and only if the following two conditions are met:

$$\pi_B - p_B F_V - L \geq 0 \tag{B}$$

and

$$(p_A - p_B)F_V - 2L \geq 0. \tag{L}$$

The first condition says that Firm $B$ would prefer to endure litigation than to withdraw its product. The second condition says that the two firms have a joint incentive to resolve the patent's validity in court rather than reaching a licensing agreement with validity unresolved. Note that this can only occur if the parties disagree about the probable outcome in court (i.e., if $p_A > p_B$). Moreover, it is more likely to occur if litigation costs are small relative to the value generated by $B$'s product.

If either condition $(B)$ or condition $(L)$ fails, litigation will not occur. Rather, the parties will negotiate a license without resolving the patent's validity. The specific license fee is determined by Nash bargaining with the parties splitting the surplus above their threat points should negotiations fail. If $(B)$ fails, Firm $B$ does not have a credible threat to litigate so Nash bargaining results in a licensing fee $F_V$—in essence, the parties treat the patent *as if* it were valid. In contrast, if Firm $B$ has a credible threat to litigate but there is no joint gain to licensing after litigating (i.e., $(L)$ fails), the alternative to licensing is litigation. In this case, $B$ will pay a somewhat lower fee, $F_U$:

$$F_U = \frac{1}{2}(p_A + p_B)F_V.$$

Here, the subscript $U$ refers to bargaining under uncertainty about the validity of the patent. Intuitively, the licensing fee is lower when there is uncertainty about the patent's validity.

## *Development*

The last piece of the model is to show that Firm $B$ has an incentive to develop its product regardless of whether it anticipates licensing or litigation. The worst outcome for $B$ is that $(B)$ fails and it is forced to pay a licensing cost $F_V$. Even in this case, however,

$$\pi_B - k - F_V = \frac{1}{2}(\pi_{A|B} + \pi_B - \pi_A) - k \geq 0.$$

So $B$ still has an incentive to develop its product, a conclusion that follows directly from Assumption NR.

We can now summarize the benchmark outcomes when innovation is nonrivalrous.

**Proposition 1** Suppose the innovation is nonrivalrous and that Firm $A$'s patent is not weak. The possible outcomes are:

• *(Litigation)* If both $(B)$ and $(L)$ hold, Firm $B$ develops its product and there is litigation to determine patent validity. If the patent is upheld, Firm $B$ pays $F_V$ for a license.

• *(Licensing without Litigation)* If either $(B)$ or $(L)$ fails, Firm $B$ develops and negotiates a license. The fee is either $F_U$ if $(B)$ holds or $F_V$ if not.

The table below summarizes the (objective) payoffs to the two firms in each scenario.

|  | $A$'s profit | $B$'s profit |
|---|---|---|
| Litigation | $\pi_{A|B} + pF_V - L$ | $\pi_B - k - pF_V - L$ |
| Licensing | $\pi_{A|B} + \{F_U, F_V\}$ | $\pi_B - k - \{F_U, F_V\}$ |

### Rivalrous Innovation

Next we consider the case of rivalrous innovation. To do this, we assume that introducing Firm $B$'s product reduces joint profits. The following assumption is sufficient to imply this.

**Assumption R** $\pi_{A|B} + \pi_B/p_A + 2L/p_A < \pi_A$.

As in the previous section, this is slightly stronger than is needed. A weaker condition that would guarantee rivalry is that $\pi_{A|B} + \pi_B - k < \pi_A$. The stronger condition implies that if Firm $B$ chooses to enter, then not only will Firm $A$ have an incentive to litigate (ruling out the weak patent case), it will not want to license just to avoid costly litigation. We rule out this latter situation in an effort to keep the model as simple as possible. Nevertheless, it can be worked out, and in such a circumstance the effect of an opposition process corresponds closely to the nonrivalrous environment described above.[10]

To analyze the possible outcomes, we again work backward. We first consider what would happen in the event of litigation, then ask whether litigation will occur if $B$ develops, and finally consider the incentive to develop.

*Outcomes of Litigation*

If Firm $B$ introduces its product and there is litigation, there are two possible outcomes. If the court voids the patent, $B$ can market its product without paying any licensing fee. If the court upholds $A$'s patent, the rivalry of the products means that $A$ will deny $B$ a license. Thus the firms' (subjective) profit expectations entering litigation are $p_A\pi_A + (1 - p_A)\pi_{A|B} - L$ for Firm $A$ and $(1 - p_B)\pi_B - k - L$ for Firm $B$.

---

[10]There is also another reason why the firms might want to avoid litigation, which is that if there are other potential entrants, Firm $A$ may incur a larger cost from having its patent invalidated than from just allowing $B$'s entry. We discuss the case of multiple entrants in the fifth section of this chapter.

## Determinants of Litigation

Now consider what would happen should $B$ develop its product. If $B$ attempts to introduce its product, Assumption R implies that $A$ will certainly want to initiate litigation because:

$$p_A(\pi_A - \pi_{A|B}) - L > 0. \tag{A}$$

That is, Assumption R rules out the weak patent case where Firm $A$ is not willing to defend its intellectual property rights.

At the same time, Firm $B$ is willing to introduce its product and face litigation if and only if

$$(1 - p_B)\pi_B - L \geq 0. \tag{B}$$

If this inequality fails, the litigation cost outweighs $B$'s expected benefit from a product introduction. If it holds, $B$ will introduce its product and the parties will end up in court. To see this, we note that under Assumption R, the sum of the perceived gains from litigation necessarily outweigh the litigation costs so long as $(B)$ is satisfied. In particular, combining $(B)$ and Assumption R shows that

$$p_A(\pi_A - \pi_{A|B}) - p_B\pi_B - 2L \geq 0, \tag{L}$$

so there is a joint gain to litigation versus a licensing agreement.

## Development

Finally, we consider Firm $B$'s incentive to develop. If $B$ would not introduce a product it developed, it should certainly not develop the product. On the other hand, $B$'s subjective expected profits from litigation are greater than zero if

$$(1 - p_B)\pi_B - L - k \geq 0. \tag{E}$$

Importantly, whenever $(E)$ holds, so will $(B)$. That is, if $B$ is willing to develop in expectation of litigation, it certainly wants to litigate having sunk the development costs. Intuitively, $B$ is more likely to develop and endure litigation if litigation costs are relatively low, if $A$'s patent does not seem certain to be upheld, or if the potential profits from entry are large.

It is now easy to summarize the equilibrium outcomes.

**Proposition 2** Suppose that $B$'s product is rivalrous. The possible outcomes are:

- *(Litigation)* If (E) holds, Firm B will develop its product and there will be litigation. Firm B will enter if and only if Firm A's patent is voided.
- *(Deterrence)* If (E) fails, Firm B is deterred from developing by the threat of litigation.

The following table summarizes the firm's expected payoffs in the two cases.

|  | A's profit | B's profit |
| --- | --- | --- |
| Litigation | $p\pi_A + (1-p)\pi_{A|B} - L$ | $(1-p)\pi_B - k - L$ |
| No Entry | $\pi_A$ | 0 |

## AN OPPOSITION PROCESS

In this section, we introduce an opposition process that allows for the validity of Firm A's patent to be assessed immediately after the granting of the patent. Then, starting with the benchmark outcomes derived in the previous section, we examine the effect of allowing for opposition hearings.

With an opposition process, the timing proceeds as follows. After the grant of the patent, Firm B is given the opportunity to challenge Firm A's patent. Before initiating a challenge, B can approach A and attempt to license its technology. If B does not obtain a license, it must decide whether to challenge. If B declines to challenge, everything unfolds exactly as in the earlier case—that is, B retains the option of developing and either licensing or facing litigation. On the other hand, if B initiates a challenge, the parties enter a formal opposition hearing.

We model the opposition proceeding essentially as a less expensive way of verifying patent validity than litigation. In an opposition proceeding, each firm incurs a cost $C \leq L$ to prepare its case. There are several reasons to believe that the costs of an opposition would be lower than litigation should the United States adopt an opposition process. First, an opposition hearing would be a relatively streamlined administrative procedure rather than a judicial process with all the associated costs of extensive discovery. Second, as noted above, the cost of an opposition in Europe is estimated by European Patent Office officials to be less than 10% of the cost of litigation. Although the crossover to the United States is imperfect, it suggests that an opposition procedure could be made relatively inexpensive if that were a desired goal.

Once the parties present their cases in an opposition hearing, an administrator rules on the patent's validity. We assume that the firms assign the same subjective probabilities ($p_A$ and $p_B$) to A's patent being upheld in the opposition process as in litigation and also that the objective probability $p$ is the same. Similarly, if A's patent is upheld in the opposition, Firm B must obtain a license to market its product. (In particular, Firm A need not endure another round of costly litigation to enforce its property rights against $B$.) Conversely, if the relevant parts of $A$'s patent are voided, $B$ can develop and market its product without fear of reprisal.

## Nonrivalrous Innovation

We now derive the equilibrium outcomes with an opposition process and contrast these to the benchmark outcomes without an opposition.

The first question is whether Firm $B$ has any incentive to use the opposition process. If not, the change will have no effect. Assume as before that Firm $A$'s patent is not weak (in which case the patent could simply be ignored). Then Firm $B$ has an incentive to use the opposition process if and only if

$$\pi_B - k - p_B F_V - C \geq \Pi_B. \tag{BC}$$

Here $\Pi_B$ denotes Firm $B$'s subjective expected payoff should it decline to challenge. That is, $\Pi_B$ is the payoff derived for $B$ in the previous section.

If Firm $B$ has a credible threat to use the opposition process, an opposition proceeding will still only occur if the parties do not have a joint gain from negotiating a settlement. The sum of their subjective expected payoffs from an opposition hearing exceeds their joint payoff from licensing if and only if:

$$(p_A - p_B)F_V - 2C \geq 0. \tag{C}$$

Note that this condition is precisely the same as that which characterizes whether there is a joint gain from litigation, except that the litigation cost $L$ is replaced by the opposition cost $C$.

If both $(BC)$ and $(C)$ hold, the result is an opposition proceeding. If the patent is upheld, $B$ will be forced to pay a fee $F_V$ for a license. On the other hand if $(BC)$ holds but $(C)$ does not, there will be licensing under uncertainty at a fee $F_U$.

From here, it is easy to see that the effect of introducing the opposition process depends on the relevant no-opposition benchmark. If the result without an opposition process was litigation, then because the incentives to enter an opposition process are at least as strong as the incentives to enter litigation (because $C \leq L$), the new outcome will be an opposition hearing. Importantly, because an opposition is less expensive than litigation, both firms benefit from the introduction of the opposition process.

In contrast, suppose that the result without an opposition process would be licensing at a fee of either $F_V$ or $F_U$. In this case, simple calculations show that both $(BC)$ and $(C)$ may or may not hold. The new outcome depends on the exact parameters. One possibility with the opposition system in place is that there is no change. Another possibility is that Firm $B$ goes from not having a credible threat to fight the patent's validity in litigation to having a credible threat to launch an opposition. In this event, the licensing fee drops from $F_V$ to $F_U$. The last possibility is that an opposition proceeding occurs.

What is certain in all these cases is that Firm $B$'s expected payoff with the opposition proceeding is at least as high as without it. This should be intuitive. Introducing the opposition process gives Firm $B$ an option—it can always decline to challenge and still get its old payoff. On the other hand, $A$'s expected payoff may increase or decrease. The case in which litigation costs decrease benefits $A$; the case in which licensing fees decrease hurts $A$. The case in which an opposition proceeding replaces licensing certainly hurts $A$ if the earlier licensing fee would have been $F_V$ but could potentially benefit $A$ if the licensing fee would have been $F_U$.

In the simple static model we are looking at, the direct welfare effects are limited to the cost of conflict resolution and the change in licensing fees. An important point, however, is that the impact on $A$ depends on whether its patent is valid. In particular, the opposition process tends to help $A$ if its patent is valid and hurt it if its patent is invalid. Because the opposition process tends to more closely align the rewards to innovation with truly novel inventions, it seems clear that in a richer dynamic model in which $A$ was to make decisions about R&D expenditures and patent filing, the opposition process would have an additional positive incentive effect. We argue in the next section that this effect might be fairly large in practice relative to the costs of oppositions.

The next result summarizes oppositions in the nonrivalrous case.

**Proposition 3** Suppose that the products are nonrivalrous and that $A$'s patent is not weak. The introduction of a opposition process will have the following effects depending on the outcome in the benchmark case of no oppositions:

- *(Litigation)* If the benchmark outcome was litigation, the outcome with an opposition process will be an opposition. Legal costs are reduced, and both firms benefit.

- *(Licensing)* If the benchmark outcome was licensing, the outcome with an opposition process may be the same, licensing before development, or an opposition. Legal costs may be higher, but license fees will tend to go down for invalid patents and up for valid patents. The social welfare effects are ambiguous, because the deadweight loss from the opposition process is offset by the increased incentive to file valid patents.

### Rivalrous Innovation

We now turn to the case of rivalrous innovation and again consider the effects of introducing the opposition process.

The first question again is whether Firm $B$ has an incentive to make use of the opposition procedure. Firm $B$ is willing to initiate an opposition if and only if:

$$(1 - p_B)(\pi_B - k) - C \geq \Pi_B. \tag{BC}$$

Again $\Pi_B$ denotes Firm $B$'s subjective expected payoff in the absence of oppositions.

Unlike in the nonrivalrous case, $(BC)$ is not just a necessary condition for an opposition proceeding to occur but also a sufficient condition. If $(BC)$ holds, then Assumption R implies that the joint benefit from the opposition proceeding exceeds the costs. In particular, combining $(BC)$ and Assumption R shows that:

$$p_A(\pi_B - \pi_{A|B}) - p_B(\pi_B - k) - 2C \geq 0,$$

so there is no gain from licensing rather than facing the opposition process. Thus if $(BC)$ holds the new outcome is an opposition, whereas if it fails the outcome is unchanged from the no-opposition benchmark.

To see how the opposition process affects previous outcomes, imagine that the result without an opposition process was litigation. In this case, $B$ was willing to face litigation for an opportunity to market its product so it will certainly be willing to ante up the opposition costs. By taking the opposition route rather than the litigation route, $B$ can also avoid sinking the development cost $k$ in the event that $A$'s patent is upheld rather than voided. It follows that the previous litigation over the validity of $A$'s patent will be replaced by opposition hearings.

In contrast, suppose the result without an opposition process was that Firm $B$ chose not to enter. Now the introduction of oppositions may encourage $B$ to initiate a challenge. $B$ can enter if the challenge succeeds. From a welfare standpoint, this potential change has a cost, which is that both firms will have to spend opposition cost $C$ on the challenge. It also has the benefit of increased competition. Although $B$'s entry will decrease industry profits, the increase in consumer surplus typically will exceed this loss. Thus the net welfare gain depends on whether the potential increase in market surplus is greater than $2C$.

As in the nonrivalrous case, Firm $B$ always gains from the introduction of the opposition process. Because it need not use the opposition option, it can certainly do no worse. Firm $A$'s situation is more complex. If it previously would have had to litigate, it benefits from the cheaper opposition process. If it previously would have been able to deter entry without litigation, it loses from having to pay the opposition costs and loses substantially if its patent, which would not have been litigated, is held invalid and its monopoly profits disappear.

**Proposition 4** Suppose the products are rivalrous. Depending on the benchmark outcome, an opposition system has the following effects:

- *(Litigation)* If the outcome without oppositions was litigation, the new outcome is an opposition hearing. This reduces dispute costs and saves on wasted development costs in the event of a valid patent.
- *(Deterrence)* If the outcome without oppositions was deterred entry, the new outcome may be an opposition. If it is, dispute costs increase but Firm B is able to enter if the patent is invalid.

As in the nonrivalrous case, there is a potential dynamic welfare effect in addition to the static effects. The static welfare effects are limited to the cost of conflict resolution, the possible reduction in monopoly power, and the potential savings on wasted development. Dynamically, the opposition process also serves to reward valid patents and punish invalid patents. So, again, the better alignment of rewards with true innovation should tend to provide better incentives for R&D and patent filing decisions.

## WELFARE EFFECTS OF AN OPPOSITION PROCESS

Figure 2 summarizes the welfare effects of introducing an opposition system. The first column distinguishes cases in which Firms A and B are nonrivalrous and rivalrous. The second column classifies the possible behaviors under a regime comparable to the current status quo. As the figure illustrates, there are four pos-

| Type of Innovation | Behavior w/o Oppositions | Behavior w/ Oppositions | Static Wefare Effect | Dynamic Wefare Effect |
|---|---|---|---|---|
| Nonrivalrous | Litigation<br>- license if valid<br>- free entry if invalid | Opposition (1)<br>- license if valid<br>- free entry if invalid | Gain = 2(L-C) | Positive |
| | Licensing w/o Litigation | No Change (2) | None | None |
| | | License at $F_U$ not $F_V$ (3) | None | Ambiguous |
| | | Opposition (4)<br>- license if valid<br>- free entry if invalid | Loss = 2C | Positive due to sorting of valid/invalid patents. |
| Rivalrous | Litigation<br>- monopoly if valid<br>- free entry if invalid | Opposition (5)<br>- monopoly if valid<br>- free entry if invalid | Gain = 2(L-C)<br>+ k if valid | Positive |
| | Deterrence w/o Litigation | No Change (6) | None | None |
| | | Opposition (7)<br>- monopoly if valid<br>- free entry if invalid | Loss = 2C; Gain from eliminating monopoly if invalid. | Positive due to sorting of valid/invalid patents. |

**FIGURE 2** Welfare economics of patent oppositions.

sible outcomes: litigation and licensing without litigation in the nonrivalrous case and litigation and deterrence without litigation in the rivalrous case.

The third column of the figure indicates how behavior changes when Firm $A$'s patent is subject to challenge in an opposition proceeding. There are now seven possible outcomes, as described in the third section of this chapter, and columns four and five indicate the static welfare and dynamic incentive effects of each outcome.

One striking implication of our model, which is apparent from inspection of Figure 2, is that once a challenge procedure is available, full-scale litigation never occurs. This conclusion depends on several of the model's assumptions concerning full information that are unlikely to represent every empirical situation with accuracy. For example, some patents are (allegedly) infringed and thus may become the subject of lawsuits, without the knowledge of the (alleged) infringer, who may be ignorant that his product, process, or service is potentially covered by the patent. Or suppose that both Firms $A$ and $B$ initially agree that the probability of a patent's validity is very low. This is the weak patent case that we noted but did not analyze, in which $B$'s entry is accommodated by $A$. In such a circumstance $B$ would not file a challenge, but if, subsequent to $B$'s entry, $A$ revised its estimate of validity significantly upward, it might sue for infringement. Finally, an opposition system would rule only on the validity of $A$'s patent or specific claims within the patent. It would not pass judgment on whether a particular aspect of $B$'s product infringed on $A$'s patent. For all these reasons, we clearly would not expect an opposition system to supplant litigation entirely.

To get a sense of the likely magnitude of the welfare effects displayed in Figure 2, we constructed a simple simulation model, which we calibrated with empirically plausible parameter estimates. The theoretical model contains nine parameters ($\pi_A$, $\pi_{A|B}$, $\pi_B$, $p_A$, $p_B$, $p$, $L$, $C$, and $k$). We add three more in order to make welfare calculations. The first of these additional parameters is the consumer surplus generated by the entry of Firm $B$. The other two parameters represent an attempt to capture the dynamic incentive effects implicit in an otherwise static model. Thus we assume not only that Firm $A$'s profits enter directly into a social welfare function that sums consumer and producer surpluses but that extra weight is given to $A$'s profits when it has a valid patent and some weight is subtracted when it licenses or exclusively exploits an invalid patent.

With so many parameters to vary, a comprehensive presentation of simulation results would be tedious. Therefore, we limit ourselves to describing just two plausible cases, one nonrivalrous and the other rivalrous. In both cases we assume that the present value of Firm $A$'s monopoly profit from its patent is $100 million and that Firm $B$ must spend $20 million to develop its innovation. We also assume that patent litigation costs each party $2.5 million, which, given the size of the market, is consistent with the estimates reported by the American Intellectual Property Law Association. We assume, given the U.S. propensity to spend on lawyers, that the cost of an opposition proceeding would be 20 percent of the cost of litigation, or $500,000 for each party. This is a conservative assumption in light of the report of the European Patent Office that oppositions cost less than $100,000. Finally, in both nonrivalrous and rivalrous examples, we assume that the objective probability of the validity of Firm $A$'s patent is 0.55, corresponding to the empirical frequency of validity calculated by Allison and Lemley (1998) on all litigated patent cases from 1989 through 1996.

In the nonrivalrous case we assume that Firm $B$'s entry would yield it a gross profit of $60 million and generate an equivalent amount of consumer surplus. We also assume no decline in Firm $A$'s gross profit given $B$'s entry. This leaves us free to examine what happens as we vary first the subjective probabilities of validity and then the dynamic welfare parameters. For simplicity, we assume that the subjective probabilities of $A$ and $B$ are symmetric around the objective probability of 0.55.

Under these circumstances, if the firms have very similar expectations about the validity of the patent, there will be no litigation before the introduction of a challenge system and no use of the opposition procedure thereafter. This situation is represented as Case (2) in Figure 2. The introduction of an opposition system has no effect on either static or dynamic welfare.

If the expectations of the firms diverge by more than 0.032 but less than 0.166 (i.e., as Firm $A$'s subjective probability of validity increases from 0.566 to 0.633), there would be no litigation before the introduction of a challenge system but Firm $B$ would initiate an opposition proceeding. This situation is represented as Case (4) in Figure 2. There is a net static welfare loss equal to the total cost of an opposition proceeding, or $1 million. Still, the opposition process has advantages because it sorts out valid from invalid patents. If, when the patent is valid, we give an additional positive weight of only 14 percent to Firm $A$'s profit as a proxy for the incentive effect, then the welfare benefits of an opposition system outweigh the cost of a proceeding. If we subtract an equal percentage from $A$'s profit when its patent is ruled invalid, we need subtract only an 8 percent weight to offset the cost of the opposition proceeding. If we give substantial weight to these incentive effects, such as counting as a component of social welfare 150 percent of $A$'s profit in the case of a valid patent and only 50 percent if the patent is invalid, then introducing an opposition system increases social welfare by $6.4 million.

The final possibility arises when the divergence in subjective probabilities exceeds 0.166 (i.e., Firm $A$'s subjective probability exceeds 0.633). In this instance, there is an unambiguous social benefit of the difference between the total cost of litigation and the total cost of opposition, as represented in Case (1) in Figure 2. Given our assumptions, this produces a gain of $4 million. Because our model implies that half the gain is realized by Firm $A$, there is a small (favorable) dynamic incentive effect. In this case, however, the gain comes not from sorting valid from invalid patents but from Firm $A$'s capture of a portion of the social saving.

To explore the rivalrous case, we vary only two parameters and assume that the present value of post-entry gross profits of Firms $A$ and $B$ are now $45 million. Again, if the subjective probabilities of validity are close together, litigation will not occur, because Firm $B$'s entry can be deterred without it. In this instance, if the difference in subjective probabilities does not exceed 0.1 (i.e., Firm $A$'s subjective probability does not exceed 0.6), there will be no litigation but $B$ will challenge $A$ if oppositions are permitted. As shown in Case (7) in Figure 2, there is a static welfare loss equal to the total cost of the challenge ($1 million) if the patent is valid. If the patent is not valid, there is a substantial net gain of $29 million, representing the incremental producer plus consumer surplus ($50 million) created by $B$'s entry minus the development cost ($20 million) minus the cost of the challenge ($1 million).

Finally, if Firms $A$ and $B$ have subjective probabilities that differ by more than 0.1, litigation will occur when oppositions are not permitted. If oppositions are allowed, a challenge will be lodged and, as in Case (5) in Figure 2, there will be an unambiguous gain in static welfare, amounting to $4 million if the patent is invalid and $24 million if the patent is valid, because $B$ will not sink the cost of development if it loses a challenge.

In all, it would appear that the cost of introducing an opposition procedure is quite small relative to the potential static welfare gains and dynamic incentive effects. A static welfare loss arises only when a challenge is lodged under circumstances that would not have given rise to litigation, such as when the parties do not differ greatly in their subjective expectations of the patent's validity. In such instances the loss is never greater than the cost of both parties participating in the administrative proceeding, which, if European experience is any guide, is likely to be modest. By contrast, both the potential static and dynamic welfare gains that arise under other circumstances will be considerably larger. The low-cost opposition procedure will often supplant higher-cost litigation; larger profits to the innovator will provide a favorable dynamic incentive; and wasteful development expenses may sometimes be avoided. All of these effects are likely to be larger in magnitude than the cost of an opposition proceeding.

## DISCUSSION

The analysis of our two-firm model of a patentee and a potential entrant makes clear that in this simple framework an opportunity to contest the validity of an issued patent is likely to yield net social benefits. In the model, however, benefits and costs are evaluated strictly by the standard welfare metrics in the product markets occupied by Firms $A$ and $B$, assuming that there are no additional firms that might potentially infringe on $A$'s patent. As a result, the model fails to capture several additional effects and likely benefits of introducing an opposition system.

First, opposition proceedings should speed the education of patent examiners in emerging technologies. Third parties will tend to have far greater knowledge of the prior art in fields that are new to the U.S. Patent and Trademark Office. Allowing the testimony of outside experts to inform the opposition proceedings should have substantial spillovers in pointing patent examiners to relevant bodies of prior art, thus making them more likely to recognize non-novel or obvious inventions when they first encounter them.

Second, in an emerging area of technology, a speedy clarification of what is patentable and what is not confers substantial external benefit on those who wish to employ the new technologies. Because precedent matters in litigation and would presumably matter in opposition proceedings, a decision in one case, to the extent that it articulates principles and gives reasons, has implications for many others. Clarifying the standard of patentability in an area could have significant effects on firms developing related technologies, even if these technologies are unlikely to infringe on the patent being examined. Early decisions making clear the standard of patentability would encourage prospective inventors to invest in technology that is appropriable and to shun costly investments in technology that might later prove to be unprotected.

More narrowly, clarifying the validity of a patent has an obvious effect on future users of the technology.[11] In fact, it is not difficult to broaden our two-firm model to allow for future infringers on $A$'s patent. One important change then is that $A$'s future profits are likely to depend on whether or not a definitive decision is handed down concerning the validity of its patent. In principle, this future patent value effect might make $A$ either more or less inclined to grant an early infringer a license. To the extent that $A$ becomes more inclined to grant a license, this can lead to one new outcome not captured in the model—the firms may negotiate a license even if $B$'s product is rivalrous. In this case, the introduction of oppositions can result in a hearing when without the opposition process the result would have been licensing, with consequent positive and negative welfare effects.

In closing, we note that we have offered little guidance about the specific design of a system permitting post-grant review of patent validity. To be effective, such a system should have a broader mandate than the current re-examination process, which is not an adversary proceeding and which allows third-party

---

[11] See Choi (1998) for a model in which there is a single patentholder and several potential infringers. Choi points out that a free rider problem may arise in this environment, whereby a potential infringer on a patent may hesitate to introduce its product in hopes that another infringer on the same patent will enter first and the ensuing litigation will clarify the patent's validity. This kind of free rider problem could also arise with an opposition process, although it would be mitigated to the extent that the cost of oppositions can be kept low.

intervention on only very limited grounds. Presumably, a more thoroughgoing U.S. system would allow challenges to validity on any of the familiar grounds now available to litigants in a court proceeding. The testimony of experts and the opportunity for cross-examination would seem desirable as means of probing questions of novelty and nonobviousness. Still, it would be important to avoid extensive prehearing discovery, unlimited prehearing motions, and protracted hearings. The costs of using a challenge system should be kept substantially below those of full-scale infringement litigation or its benefits will be negligible. In designing an opposition system, we would do well to examine the diverse experience with administrative proceedings in various federal agencies and imitate the best practices.

## REFERENCES

Allison, J., and M. Lemley. (1998). "Empirical Evidence on the Validity of Litigated Patents." *American Intellectual Property Law Association Quarterly Journal* 26: 185-277.

American Intellectual Property Law Association. (2001). *Report of the Economic Survey*. Arlington, VA: American Intellectual Property Law Association.

Choi, J. P. (1998). "Patent Litigation as an Information Transmission Mechanism." *American Economic Review* 88(December): 1249-1263.

Graham, S., B. Hall, D. Harhoff, and D. Mowery. (2003). "Patent Quality Control: A Comparison of U.S. Patent Re-examinations and European Patent Oppositions." In W. Cohen and S. Merrill, eds., *Patents in the Knowledge-Based Economy*. Washington, D.C.: The National Academies Press.

Lemley, M. (2001). "Rational Ignorance at the Patent Office." *Northwestern University Law Review* 26: 1495-1532.

Levin, R. (2002). Testimony before the FTC-DOJ Joint Hearings on Competition and Intellectual Property Law, Washington, D.C., February 6.

Merges, R. (1999). "As Many as Six Impossible Patents before Breakfast: Property Rights for Business Methods and Patent System Reform." *Berkeley Technology Law Journal*, 14(Spring): 577-616.

Meurer, M. (1989). "The Settlement of Patent Litigation." *Rand Journal of Economics* 20(Spring): 77-91.

U.S. Patent and Trademark Office. (2002). *The 21st Century Strategic Plan*, June 3. http://www.uspto.gov/web/offices/com/strat21/index.htm

# Patent Litigation

# Enforcement of Patent Rights in the United States[1]

Jean O. Lanjouw
Department of Agricultural and Natural Resource Economics
University of California, Berkeley
and the Brookings Institution

Mark Schankerman
Department of Economics
London School of Economics and Political Science

### ABSTRACT

*We study the determinants of patent suits and their outcomes over the period 1978-1999 by linking detailed information from the U.S. Patent and Trademark Office, the federal court system, and industry sources. The probability of being involved in a suit is very heterogeneous, being much higher for valuable patents and for patents owned by individuals or by firms with small patent portfolios. Thus the patent system generates incentives, net of expected enforcement costs, which differ across inventors. Patentees with a large portfolio of patents to trade, or having other characteristics that encourage "cooperative" interaction with disputants, more successfully avoid court actions. At the same time, key post-suit outcomes do not depend on observed characteristics. This is good news: Advantages in settlement are exercised quickly, before*

---

[1] We thank the National Academy of Sciences and the Brookings Institution for financial support and Derwent for generously providing access to the detailed patent information from their *LitAlert* database, which was critical to making this project feasible. We also thank Bronwyn Hall and Adam Jaffe for their input and provision of data, Joe Cecil from the Federal Judicial Center and Jim Hirabayashi of the U.S. Patent and Trademark Office for helpful discussions about the court and patent data, as well as Marty Adelman, Wesley Cohen, Kimberly Moore, and seminar participants at the National Academy of Sciences, the University of Maryland, Wharton, and Berkeley for useful comments. Maria Fitzpatrick provided excellent research assistance.

*extensive legal proceedings consume both court and firm resources. But it is bad news in that the more frequent involvement of smaller patentees in court actions is not offset by a more rapid resolution of their suits. However, our estimates of the heterogeneity in litigation risk can facilitate development of private patent litigation insurance to mitigate this adverse affect of high enforcement costs.*

## INTRODUCTION

Although the central purpose of the patent system is to encourage R&D investment, there is increasing concern among scholars and the business community that "patent thickets" are beginning to impede the ability of firms to conduct R&D activity effectively (Eisenberg, 1999; Shapiro, 2001). The perception is that patenting strategies have increasingly made disputes over rights unavoidable and that, as a result, research firms are burdened by growing enforcement costs. The fact that patent litigation grew rapidly during the period 1978-1999 encourages this view. The number of patent suits rose by almost tenfold, with much of this increase occurring during the 1990s. We show here, however, that a focus on the level of litigation gives a misleading picture. The growth in patenting has been comparable to the growth in litigation, with the consequence that the rate of suit filings has been roughly constant over these two decades. Nonetheless, although our data indicate that the likelihood of litigation has not increased, survey evidence suggests that involvement in a patent suit has become substantially more costly over the past decade (American Intellectual Property Law Association, 2001). Thus the overall burden of enforcement may well be on the rise.

Perhaps of greater importance, we show that the exposure to litigation varies widely across technology fields and patent profiles. Although the *average* rate is relatively low, 19.0 suits per thousand patents, rates vary from a low of 11.8 per thousand chemical patents to 25-35 per thousand computer, biotechnology, and nondrug health patents. Moreover, within any given technology field, probabilities of litigation differ very substantially and are systematically related to patent characteristics associated with their economic value and to characteristics of their owners.

This variation in litigation risk across patents and their owners is a central issue for the enforcement of intellectual property rights and its economic consequences. Lerner (1995), for example, provides evidence that small firms avoid R&D areas where the threat of litigation from larger firms is high. Lanjouw and Lerner (2001) argue that the use of preliminary injunctions by large firms can discourage R&D by small firms, and this may apply to other legal mechanisms. Even if parties can settle their patent disputes without resorting to suits, the threat of litigation will influence settlement terms and thus, ultimately, the incentives to undertake R&D. Using a comprehensive new data set covering all recorded patent litigation in the United States over the period 1978-1999, we determine the characteristics that affect the decision to begin a suit and the decision of whether to end with a settlement or to proceed to adjudication at trial.[2]

One of our key empirical findings is that the observed characteristics of both patents and their owners only affect the decision to file suits. The key post-suit outcomes—the probability of settlement and the plaintiff win rates at trial—are almost completely independent of these characteristics. This implies that advantages in resolving disputes come into play quickly, before a suit is filed. This helps to mitigate legal costs and reduce the private (and social) costs of enforcement. Two additional findings are encouraging: First, post-suit settlement rates are high (about 95 percent), and second, most settlement occurs soon after the suit is filed, often before the pretrial hearing is held.

Patentees have a number of mechanisms for settling disputes without resorting to litigation. They may "trade" intellectual property. Trading takes various forms, including cross-licensing agreements and patent exchanges, sometimes with balancing cash payments (Grindley and Teece, 1997). One motivation for accumulating patents may be to facilitate such trading (Hall and Ziedonis, 2001). From this perspective, extensive patenting may be beneficial by lowering costs once a dispute arises. Settlement may also be promoted if patentees interact with each other often and expect to continue doing so in the future. Theoretical models suggest that repeated interaction increases both the ability and the incentive to settle disputes "cooperatively"—that is, without filing suits (Tirole, 1994, Chapter 6). However, there is very little econometric evidence to support this prediction.[3]

The role of patent trading and the role of repeated interaction over time both imply that there may be economies of scale in resolving patent disputes. Greater research and patenting experience may speed settlement as parties become better able to anticipate the result should a dispute go to court. Experienced firms may also make higher-quality patent applications that give rise to fewer disputes in the first place (Graham et al., 2003). Three key findings in this chapter support the importance of scale. First, we find strong evidence of a patent portfolio effect: Having a larger portfolio of patents reduces the probability of filing a suit on any individual patent, conditional on its observed characteristics. The quantitative effect is large. For a (small) domestic unlisted company with a small portfolio of 100 patents, the average probability of litigating a given patent is 2 percent. For a similar company but with a moderate portfolio of 500 patents, the figure drops to only 0.5 percent. Second, we find that the (marginal) effect of patent portfolio size is stronger for *smaller* companies, as measured by employment. This is con-

---

[2] P'ng (1983), Bebchuk (1984), Priest and Klein (1984), and Spier (1992) provide theoretical models of this decision process.

[3] A notable exception is Siegelman and Waldfogel (1999), who construct measures of repeat play and find evidence that reputation matters in various areas of litigation.

sistent with the idea that having a portfolio of patents to "trade" is the key mechanism for avoiding litigation for small firms, whereas larger firms can also rely on repeated interaction in intellectual property and product markets to discipline behavior. Third, firms operating in technology areas that are more concentrated (in which patenting is dominated by fewer companies) are much less likely to be involved in patent infringement suits. Such firms are likely to have more interaction with one another. Together these results are consistent with the view that having either a portfolio of intellectual property to trade or other dimensions of interaction that promote "cooperative" behavior confers important advantages in avoiding litigation. We also find that asymmetry of firm size affects litigation risk. Patent owners who are large *relative* to the disputants they are likely to encounter less frequently resort to the courts to settle disputes.

The characteristics of a given patent also strongly affect litigation risk in ways that are consistent with existing hypotheses in the economics literature (as in Lanjouw and Schankerman, 2001). We illustrate this with two examples. First, more valuable patents, as measured by the number of claims and citations per claim, are much more likely to be involved in suits. Second, patents that are related to subsequent technological activity by the firm (cumulative innovation), as measured by the extent of self-citation in patents, are more likely to be litigated. This supports the idea that when there are interlinkages between inventions owners are more willing to protect each of them, especially the key (early) innovations (Scotchmer, 1991). We show here that differences in these, and other, patent characteristics lead to wide variations in the probability of litigation within any given technology field.

The chapter is organized as follows. The second section summarizes the analytical framework, including the litigation stages and outcomes that we study. The third section describes the construction of the data set and the main characteristics of the patents and their owners on which we focus and discusses how they relate to economic hypotheses about the causes of litigation. The fourth section presents and discusses evidence on the relationship between these characteristics and the filing of suits and their outcomes. The fifth section presents econometric analyses of the determinants of litigation for infringement suits and declaratory judgment suits and the determinants of post-suit settlement. Concluding remarks summarize directions for future research.

## ANALYTICAL FRAMEWORK

For analytical purposes, we break down the litigation process into four stages:

1. suit filing,
2. the pretrial hearing,
3. commencement of the trial, and
4. adjudication at the conclusion of trial.

According to our discussions with patent lawyers, legal costs are more closely related to how many stages the case reaches than to the actual length of the case, which is strongly affected by the availability of court resources and other external factors.

There are three possible outcomes to a suit:

1. settlement,
2. win for the plaintiff, or
3. win for the defendant (the identity of the patentee depends on whether it is an infringement or invalidity suit).[4]

If a patent dispute is settled before a suit is filed, we do not observe the dispute in the data. Thus low filing rates can either reflect low rates of infringement (disputes) or high probability of pre-suit settlement. After a suit is filed, settlement can occur before the pretrial hearing, after the hearing but before the trial begins, or during the trial. Otherwise, the trial concludes with a court judgment in favor of one of the parties.[5]

Lanjouw and Schankerman (2001) analyzed the determinants of the probability of litigation (case filings). For this chapter, we have constructed a larger data set that allows us to study both case filings and post-suit outcomes. In particular, we analyze:

1. The probability of a suit being filed
2. The probability of settlement, conditional on a suit being filed
3. The timing of settlement, i.e., the conditional probability that the suit is resolved before the pretrial hearing or after
4. The plaintiff win rates, conditional on adjudication at trial

Information on win rates is relevant for assessing overall litigation risk (e.g., in pricing patent insurance). Such information is also useful in testing competing economic models of litigation because the models generate different predictions about plaintiff win rates at trial (Waldfogel, 1998; Siegelman and Waldfogel, 1999). There are two main models, divergent expectations (Priest and Klein, 1984) and asymmetric information (Bebchuk, 1984). In the divergent expectations model, each party estimates the quality of his or her case (equivalently, the rel-

---

[4] A win for both parties can arise, e.g., infringement suits when there is a counterclaim for invalidity by the defendant. The court may rule that infringement occurred but strike down the validity of some of the patent claims. When a win for both parties is recorded, we count it for both the plaintiff and the defendant rather than as a separate category.

[5] Apart from settlement, the court may dismiss the case before trial without the request of one of the parties. We have dropped these cases from the sample. In this chapter we do not distinguish different forms of adjudication, such as court verdicts, jury verdicts, and directed verdicts.

evant legal standard) with error and cases go to trial when the plaintiff is sufficiently more optimistic than the defendant. This is most likely to occur when true case quality is near the court's decision standard. This selection mechanism drives the plaintiff win rate at trial toward 50 percent.[6] In the asymmetric information model, one party knows the probability that the plaintiff will win at trial, whereas the other party knows only the distribution of plaintiff win rates. The uninformed party makes a settlement offer (or a sequence of offers, in dynamic versions of the model; Spier, 1992), which will be accepted only by informed defendants who face a relatively low probability of winning at trial. Trials can arise in equilibrium because settlement offers have some probability of failing when one of the parties has private information. Because of this one-sided selection mechanism, the asymmetric information model predicts that the win rate for the party with private information should tend toward 100 percent. As we discuss in the fourth section of this chapter, the empirical evidence for patent litigation strongly favors the divergent expectations model.

Litigation models explain why cases reaching trial are a selected sample of filed cases. Similar selection will be at work on filed cases, to the extent that potential plaintiffs may not file suits on certain types of patents (or defendants may settle before suit). Lanjouw and Schankerman (2001) show that the observed characteristics of patents and their owners strongly affect the probability of filing a suit. We confirm, and extend, those findings in this chapter. At the same time, we find that post-suit outcomes—for example, whether parties settle or who wins if the case reaches trial—are unrelated to these same characteristics.

## DESCRIPTION OF DATA

The data source used to identify litigated patents is the *LitAlert* database produced by Derwent, a private vendor. This database is primarily constructed from information collected by the U.S. Patent and Trademark Office (USPTO). The data used include 13,625 patent cases filed during the period 1978-1999. Each case filing identifies the main patent in dispute, although other patents may also be listed. We use only the main listed patent in our analysis, for reasons explained below. There are 9,345 patents involved in our sample of suits.

We also obtained information on all U.S. patent-related cases (those coded 830) from the court database organized by the Federal Judicial Center (FJC). This information runs through the end of 1997 and includes the progress or resolution of suits—for example, whether the case is settled and at which stage of the proceedings this occurs, whether the case proceeds to trial, and the outcome of the

---

[6]If parties have differential stakes (e.g., one firm also gets reputation gains from winning), the divergent expectations model predicts higher win rates for the party with higher stakes.

trial.[7] The form of docket numbering was made (by hand) consistent across the two data sets, so they could be merged.

To create a control group, we generated a "matched" set of patents from the population of all U.S. patents (both litigated and unlitigated) from the USPTO. For each litigated patent, a patent was chosen at random from the set of all U.S. patents with the same application year and primary three-digit U.S. Patent Classification (USPC) class assignment. With a population sample constructed in this way, the comparisons we present between litigated patents and matched patents largely control for technology and cohort effects. The control is not perfect, however, because we have 12,771 matched patents. This is more than the number of litigated patents for two reasons. First, the more recent part of our sample includes matches for both main and other patents in each suit, whereas we only use the main litigated patents in the analysis. Second, in combining our old (1978-1991) and new (1990-1999) data, we dropped duplicate cases in the overlapping years when counting litigated patents. We do not have identifiers in either round of subsetting the litigated data that would allow us to easily delete the corresponding matched patents. We do not expect this to create any systematic bias.

Although the U.S. federal courts are required to report to the USPTO every case filing that involves a U.S. patent, underreporting occurs in practice. Thus the USPTO (and Derwent) data comprise a subset of all patent cases. To estimate the reporting rates, we take the number of cases filed according to Derwent divided by the number in the same year that are coded as a patent case by the FJC. We can compute the reporting rates through 1998 (we use the last value for 1999). They stabilize in the 1990s at about 55 percent (see Appendix 1). We found no evidence of selection bias in the underreporting by the courts to the USPTO: There are no significant differences between reported and unreported cases for a range of variables in the federal database.

A truncation issue arises because we observe suit filings only through 1999, so later cohorts of patents look like they are less litigated by construction. We use the lag structure for case filings for cohorts 1982-1986 to adjust for this truncation. The estimates are based on the pooled sample and are applied to each technology field. The truncation rate is about 50 percent for the 1992 cohort (i.e., lag of 7 years), and it jumps sharply to 75 percent for the 1995 cohort. Appendix 1 presents the estimated truncation rates.

---

[7]Discussions with the FJC indicated that the data probably do not cover all cases involving patents, because some may be coded under other categories by the court (e.g., the patent issue may be part of a broader contractual dispute). This is also evident in the data where a small percentage of cases identified in Derwent are not in the FJC database (see Somaya, 2003, for a breakdown between typos and coding differences). However, there is no reason to expect any selection bias from the perspective of the issues we analyze.

From the main USPTO database we obtained information on the following characteristics for each litigated and matched patent:

*Number of Claims:* A patent is composed of a set of claims that delineates the boundaries of the property rights provided by the patent. The principal claims define the essential novel features of the invention in their broadest form, and the subordinate claims are more restricted and may describe detailed features of the innovation claimed. The patentee has an incentive to claim as much as possible in the application, but the patent examiner may require that the claims be narrowed before granting.

*Technology Field:* Each patent is assigned by the patent examiner to three-digit classes of the USPC system, of which there are 421 in total. The USPC is a hierarchical, technology-based classification system, and patents may be assigned to more than one class. In the empirical analysis, we use the set of all three-digit classes to which a patent was assigned. We use the categorization developed by Adam Jaffe to aggregate these classes to a two-digit level (used for some purposes explained later) and then to the eight broad technology groups used in most of this paper: Drugs, Other Health, Chemical, Electronics (excluding computers), Mechanical, Computers, Biotechnology, and Miscellaneous. Assignments to the biotechnology group are based on the categorization used by the USPTO when determining who examines a patent. The technology field composition of cases is given in Table 1.

*Citations:* An inventor must cite all related prior U.S. patents in the patent application. A patent examiner who is an expert in the field is responsible for ensuring that all appropriate patents have been cited. Like claims, the citations in the patent document help to define the property rights of the patentee. For each patent in the litigated and matched data, we obtained the number of prior patents cited in the application (backward citations) and their USPC subclass assignments. We obtained the same information on all of the subsequent patents that had cited a given patent in their own applications, as of 1998 (forward citations).

**TABLE 1** Composition of Sample: All Filed Cases, Cohorts 1978-1995

| Technology | Number | Percent |
|---|---|---|
| Drugs | 573 | 5.6 |
| Other Health | 825 | 8.0 |
| Chemical | 1,378 | 13.4 |
| Electronics | 1,924 | 18.7 |
| Mechanical | 2,848 | 27.7 |
| Computers | 183 | 1.8 |
| Biotechnology | 92 | 0.9 |
| Miscellaneous | 2,456 | 23.9 |
| TOTAL | 10,279 | 100.0 |

For recent patents there is substantial truncation in the number of forward citations, because citation lags can be long (Jaffe and Trajtenberg, 1999). To minimize truncation bias, we limit parts of the analysis to cohorts before 1993. For older patents there is considerable missing information on the USPC subclass assignments of backward citations, because comprehensive data are only available from about 1970, but the number of backward citations is complete for all patents.

*Ownership:* We identify each patent owner as an individual, an unlisted company, or a listed company.[8] Individual and firm owners are indicated as such in the USPTO data. Bronwyn Hall and Adam Jaffe were generous in providing us with their link between USPTO company codes and Standard and Poor's CUSIP identification code, based on the 1989 industry structure. We call a patent-owning company "listed" if we are able to identify it as having a Standard and Poor's CUSIP code at that time.[9] Unlisted companies are typically smaller than listed ones, but there is wide variation in both categories. Individuals and listed companies are more predominantly domestic (81.0 and 95.6 percent, respectively) than unlisted companies (60.4 percent). We also break down listed firms into "large" firms (those with employment above the median of 5,425) and "small" firms with employment below the median. Unless otherwise noted, we classify the nearly 40 percent of firms without employment data as large firms because they have similar litigation and settlement patterns.

*Nationality:* We use the USPTO designation of companies as domestic or foreign if there is an assignee and the address of the first listed inventor if there is no assignee. Domestic patents account for 73.4 percent of the total.

*Case Type:* We manually matched the owner of each litigated patent to the appropriate party in the suit (plaintiff, defendant, neither). We identify a filed case as an infringement suit if the patent owner is a plaintiff and as a suit for a declaratory judgment if the patent owner is a defendant. This could be done for about 65 percent of the suits. For those cases, infringement suits account for about 85 percent of the total. In most of the analysis we treat those suits in which the patentee is not one of the litigants as infringement suits, because they are likely to be suits brought either by an exclusive licensee or by a subsidiary or head office of the patent-owning entity.

*Patent Portfolio Size:* The USPTO gives a company code to each company that is assigned a patent by the inventor. This allows us to construct a measure of the size of an owner's patent portfolio, as it looks around the application date of

---

[8] A small share of patents are assigned to institutions, such as universities, hospitals, or governments. We treat these as unlisted companies.

[9] Two points are worth noting here. First, companies that merged after 1989 stop accumulating patent portfolios because their subsequent patenting is listed under a different (merged company) code. Second, any listed company that started after 1989 will not have a CUSIP in our data and thus will be coded as an unlisted company.

each of our sample patents. The relevant portfolio variable (Portsize) is defined as the number of patents owned by a company that have an application date within 10 years in either direction of the patent in question. It should be noted that this portfolio size variable may differ across patents for a given company. As expected, domestic listed companies tend to have larger portfolios—roughly one-third of patents owned by domestic listed companies are in portfolios in each of size groups 1-100, 100-900, and >900 patents. By contrast, about 90 percent of patents owned by domestic unlisted companies, and two-thirds of patents owned by foreign companies, are in portfolios with fewer than 100 patents.

*Relative Size:* We construct a measure of the asymmetry in portfolio size between a patentee and the "representative" disputant he/she can expect to face on each patent. Disputes will often occur between firms engaged in similar research. Firms pursuing similar research programs will also be in the position of citing each other's patents as prior art. Thus we identify firms patenting in the same technology areas as a given patent's forward citations as the likely potential disputants for the patent. This identification is supported by an analysis of the three-digit classifications of patents owned by actual defendants. We compare these to the technology classes of the forward citations to the patent in a suit. The share of classes that overlap ranges from 0.16 to 0.47 depending on the type of innovation. By contrast, the overlap for a random set of patents from the same cohorts is about one-tenth the size, ranging from 0.016 to 0.059. On the basis of this result, relative portfolio size is defined as the firm's total portfolio size (including all patents since 1978) divided by a weighted average of the portfolio size of firms in classes from which its forward citations come.[10]

For a patentee who is the plaintiff (infringement suits) being relatively large confers greater threat power (e.g., holding cross-licensing of other patents hostage to this dispute), and this should facilitate settlement with the infringer. This is less clear-cut when the patentee is the defendant. A stronger defendant may be less willing to settle (or be able to extract more favorable settlement terms from the plaintiff). Thus we expect the probability of litigation to decline with relative size in infringement suits, but the prediction for declaratory judgment suits is ambiguous.

*Technology Concentration:* We construct a measure of firm concentration in the technology area of each patent. To do this, we first construct, for each two-digit USPC class, a four-firm concentration index, measured as the patenting share of the top four firms. A firm's share is the size of its patent portfolio in that class divided by the sum of all firms' patents in that class. For each patent we then

---

[10]Formally, let $Z_f = \Sigma_c Z_{cf}$ be the portfolio size for firm $f$ and $Z^*_c = Z_c/n_c$ be the average portfolio size of the $n_c$ firms with patents in class $c$. The relative portfolio size of firm $f$ for patent $i$ is $R_{if} = Z_f / \Sigma_c w_{ci} Z^*_c$, where $w_{ci} = F_{ci}/F_{.i}$ is the fraction of the forward citations to patent $i$ that fall into technology class $c$.

construct a weighted average of the concentration indices for the different classes, where the weights are the shares of the forward citations to the patent that fall in that technology class.[11] If a company operates in more concentrated technological areas, it faces a greater chance of encountering other firms in patent disputes more than once. This expectation of repeated interaction should lower the litigation rate (i.e., promote pre-suit settlement).

*Other Information:* From Standard and Poor's information on listed companies, we downloaded financial and other company information for the listed firms either owning patents involved in litigation or in our matched sample.

The preceding variables are designed to capture the main determinants of patent suits:[12]

1. the number of potential disputes—measured by the number of claims, the diversity of technology classes into which the patent falls, and the technological similarity of future patents that cite the original patents;

2. the size of the stakes—measured by the number of future citations the patent receives and the extent of self-citation (as an indicator of the firm's cumulative investment in that technology);

3. the degree of certainty about outcomes—measured by patent portfolio size and ownership type (as indicators of experience); and

4. relative costs of settlement and prosecuting a suit—again measured by patent portfolio size and ownership type and, in addition, technology concentration, relative size, and nationality of the patentee.

## NONPARAMETRIC EVIDENCE

Although the number of patent infringement suits has risen by almost tenfold since 1978, the increase has not been uniform across technology fields—it was particularly high in Drugs, Biotechnology, Computers, and Other Electronics. Closer examination of the data shows that the increase in the aggregate number of suits has been driven both by the sharp increases in the number of patent applications in each technology field and by the shift of patenting toward technology fields with higher litigation rates. The total number of patent applications grew by 71 percent over the period, but in Drugs, Biotechnology, and Medical Instruments patenting nearly tripled, and in Computers it grew by fourfold. Once the

---

[11]Formally, let $Z_{cf}$ be the portfolio size for firm $f$ in technology class $c$ (including all patents since 1978) and $Z_c = \Sigma_f Z_{cf}$. The concentration index for the class is $C4_c = \Sigma_f Z_{cf}/Z_c$, where the sum is over the top four firms in terms of shares in that class. The weighted technology concentration index for patent $i$ is $C4_i = \Sigma_c w_{ci} C4_c$, where $w_{ci}$ is defined as above.

[12]For a good, general discussion of the economic determinants of litigation, see Cooter and Rubinfeld (1989).

growth in patenting is taken into account, we find that there has been *no trend increase in the filing rates of suits* in any technology field over this period. (Note again, however, that with increasing expenditures per suit, legal enforcement costs may well have grown over the period.)

Table 2 presents estimates of average filing rates for three subperiods: 1978-1984, 1985-1990, and 1991-1995. We measure filing rates as the number of suits filed per thousand patents from a given *cohort*.[13] These include all of the suits filed in connection with these patents through 1999 (that is, we count multiple cases for the same patent), and they are adjusted both for underreporting in the Derwent data and truncation associated with time lags in case filings.[14]

**TABLE 2** Filing Rates by Technology Fields and Cohort Groups

| Technology Field | Filing Rate (cases per thousand) | | | |
|---|---|---|---|---|
| | Total: 1978-1995 | 1978-1984 | 1985-1990 | 1991-1995 |
| Aggregate | **19.0** | **19.3** | **16.6** | **21.1** |
| | (0.21) | (0.31) | (0.28) | (0.44) |
| Drugs | **22.2** | **22.5** | **18.9** | **24.3** |
| | (1.05) | (1.62) | (1.34) | (1.97) |
| Other Health | **34.6** | **48.2** | **35.2** | **27.3** |
| | (1.33) | (2.67) | (1.98) | (2.23) |
| Chemicals | **11.8** | **11.6** | **10.9** | **13.0** |
| | (0.35) | (0.50) | (0.49) | (0.80) |
| Electronics | **15.4** | **16.2** | **13.1** | **16.8** |
| | (0.40) | (0.61) | (0.51) | (0.79) |
| Mechanical | **16.9** | **17.7** | **14.5** | **18.7** |
| | (0.2) | (0.53) | (0.46) | (0.79) |
| Computers | **25.6** | **32.6** | **21.2** | **25.9** |
| | (2.25) | (4.24) | (2.80) | (3.78) |
| Biotechnology | **27.9** | **33.3** | **27.6** | **25.5** |
| | (3.36) | (6.13) | (5.16) | (5.52) |
| Miscellaneous | **34.2** | **32.4** | **28.9** | **40.7** |
| | (0.76) | (1.10) | (0.98) | (1.66) |

NOTE: The filing rates cover all patent suits filed through 1999, including multiple suits on the same patent. Figures are adjusted both for underreporting and for truncation (based on the filing rate structure for cohorts 1982-1986). Numbers in bold are statistically significant at the 0.01 level.

---

[13]We do not compute rates based on filing year for two reasons: 1. the population of patents alive at any date (the denominator of the filing rate) is unknown because it depends on the pattern of patent renewals for the preceding 20 cohorts, and 2. the age structure of the population changes over time as patenting rates increase, and age and filing rates are related.

[14]Given the acceleration of patenting activity, the stock of patents grew more slowly than the flow during this period, so that the number of filed cases relative to the stock of patents did rise (not reported).

The table also shows that mean filing rates vary substantially across technology fields. A formal test that the filing rates are the same across fields is strongly rejected [$\chi^2(7) = 1,103$; $P$-value $< 0.001$]. For the aggregate (pooled technology field) data, there are 19.0 case filings per thousand patents. The lowest rates are found in Chemicals (11.8), Electronics (15.4), and Mechanical (16.9). Interestingly, filing rates for pharmaceutical patents are only modestly higher than the average. The filing rates are much higher for patents in Other Health, Computers, Biotechnology, and Miscellaneous. Computers and Biotechnology are both newer areas in which one might expect there to be greater uncertainty about legal outcomes.

Although we observe little evidence of trends in filing rates, the *level* of filing rates may be understated by Table 2. These are calculated by using only the main patents in each suit, whereas there may in fact be several patents per suit. We present these calculations because, for filing years before 1990, we only have information about the main patents (mixing the subsidiary patents for later years would distort litigation trends). The filing rates we compute are underestimates of the "true" rates if one views being a subsidiary patent in a case as equivalent to being the main litigated patent. To estimate the difference, one could scale up the filing rate by dividing by the ratio of subsidiary to main patents. This ratio is 0.24 percent overall, but it varies across technology fields.[15]

It is important to look beyond average filing rates for given technology fields, because they conceal huge heterogeneity. Lanjouw and Schankerman (2001) showed that litigated patents have more claims and more forward citations per claim. Table 3 confirms this finding on the larger data set. The table presents the mean number of claims, and citations per claim, for litigated and matched patents, broken down by ownership type. Litigated patents have far more claims than matched patents, and this holds for each ownership type. They also have *more* forward citations per claim and *fewer* backward cites per claim (i.e., more backward citations are an indication that the technology area is well-developed and the innovation is more likely to be derivative and less valuable). Both of these findings indicate that valuable patents are more likely to be involved in litigation.

There are also large differences across different types of patent owners. Table 4 summarizes the mean filing and settlement rates for four ownership categories: individuals, domestic unlisted and listed companies, and foreign companies. Domestic listed companies are far less likely to file suits on their patents than unlisted companies and individuals: Their mean filing rate is 10.4 suits per thousand patents, compared to 35-45 suits for the smaller owners. Moreover, filing rates

---

[15] The percentages for the individual technology fields are Drugs 0.25, Other Health 0.36, Chemicals 0.20, Electronics 0.37, Mechanical 0.20, Computers 0.34, Biotechnology 0.46, and Miscellaneous 0.15.

**TABLE 3** Mean Citations and Claims per Patent, by Ownership Type

| Mean | Domestic Listed | | Domestic Unlisted | |
|---|---|---|---|---|
| | Filed Cases | Matched | Filed Cases | Matched |
| Claims | **18.8** | **13.1** | **18.6** | **14.0** |
| | (0.60) | (0.25) | (0.21) | (0.23) |
| Forward Cites/Claim | **2.17** | **0.98** | **1.25** | **0.85** |
| | (0.10) | (0.02) | (0.03) | (0.04) |
| Backward Cites/Claim | **1.00** | 1.18 | 1.11 | 1.20 |
| | (0.04) | (0.02) | (0.03) | (0.04) |

| Mean | Foreign Firms | | Individuals | |
|---|---|---|---|---|
| | Filed Cases | Matched | Filed Cases | Matched |
| Claims | **14.5** | **10.6** | **14.2** | **11.0** |
| | (0.38) | (0.13) | (0.19) | (0.17) |
| Forward Cites/Claim | **1.58** | **0.76** | **1.57** | **0.84** |
| | (0.07) | (0.02) | (0.04) | (0.03) |
| Backward Cites/Claim | **0.95** | **0.99** | **1.09** | **1.34** |
| | (0.03) | (0.02) | (0.02) | (0.03) |

NOTE: Citations include self-citations. Estimated standard errors are in parentheses. Numbers in bold are statistically significant at the 0.01 level.

for foreign patentees (mostly unlisted firms) are much lower than for their domestic counterparts. These differences in mean filing rates are statistically significant, and the joint null hypothesis that they are the same is decisively rejected [$\chi^2 (3) = 11{,}853$; $P$-value $< 0.001$].

Although filing rates differ sharply across ownership types, we find that ownership does not affect the probability that a suit is settled before it reaches the end

**TABLE 4** Filing and Settlement Rates, by Ownership Type

| | Individuals | Domestic Unlisted | Domestic Listed | Foreign Firms |
|---|---|---|---|---|
| Filing Rate | **35.2** | **46.0** | **10.4** | **4.2** |
| (cases/thousand) | (0.65) | (0.78) | (0.27) | (0.16) |
| Settlement Rate | **94.7** | **94.0** | **94.1** | **94.5** |
| (percent) | (1.4) | (0.4) | (0.7) | (0.9) |

NOTE: Foreign firms include both listed and unlisted companies. The filing rate is the number of suits filed per thousand patents, including multiple suits on the same patent, from cohorts 1978-1995 (as in Table 2). The settlement rate is the fraction of filed cases reported to have been settled at some time before court judgment, according to the FJC. Settlement rates are computed for suits filed during the period 1978-1992 to minimize truncation bias and include only infringement suits. Estimated standard errors are in parentheses. Numbers in bold are statistically significant at the 0.01 level.

of trial—which we call post-suit settlement. The formal $\chi^2(3)$ test statistic is 4.55 ($P$-value $\approx 0.2$). Overall, about 95 percent of all patent suits filed are settled by the parties before the conclusion of trial (and most of those before the trial begins). However, which suits these are is not related to observed characteristics.

One explanation for why listed and unlisted firms have such different filing rates may be that the listed firms are typically larger and there may be advantages to size. As discussed above, there are several distinct aspects to such advantages. First, firms with larger patent portfolios may be more experienced or better able to settle disputes through trading intellectual property, without resorting to suits (the *portfolio size effect*). Second, if imperfect capital markets constrain the ability of smaller firms to finance litigation, relatively large firms may be better able to settle because they pose greater litigation threats when confronting smaller firms. And when large firms have disputes with each other, they are likely to have many points of interaction other than trading intellectual property, especially through competition in product markets. This expectation of repeated interaction in other dimensions should promote settlement. Large firms are also likely to be relatively experienced. We call these latter aspects *firm size effects*. The detailed patent data enable us to discriminate between the portfolio size and firm size effects on litigation.

We begin by examining how the probability of litigation (i.e., of being involved in at least one suit over the life of the patent) and the probability of post-suit settlement vary with different portfolio sizes. To compute these probabilities, we adjust for the fact that patents from large portfolios are disproportionately represented in the matched data (because the matching was not stratified by portfolio size—see Appendix 2 for details). Table 5 shows that the probability of litigation sharply declines with portfolio size. A formal test confirms this finding [$\chi^2(6) = 2{,}610$; $P$-value $< 0.001$]. The probability of filing a suit involving a patent in a portfolio with a small number of other patents (0-10) is 1.7 percent, compared to about 0.5 percent for a patent in a portfolio with 100-300 other

**TABLE 5** Probability of Litigation and Settlement, by Patent Portfolio Size

| Portfolio Size | Probability of Litigation (percent) | | Settlement Rate (percent) | |
| --- | --- | --- | --- | --- |
| 0-10 | 1.71 | (0.05) | 95.0 | (0.5) |
| 11-100 | 1.20 | (0.05) | 93.3 | (0.7) |
| 101-200 | 0.52 | (0.05) | 93.0 | (1.7) |
| 201-300 | 0.43 | (0.06) | 97.0 | (1.3) |
| 301-600 | 0.39 | (0.04) | 90.9 | (1.9) |
| 601-900 | 0.34 | (0.04) | 93.3 | (2.5) |
| >900 | 0.26 | (0.01) | 93.2 | (1.1) |

NOTE: The probability of litigation is adjusted for underreporting and truncation and for the overrepresentation of patents from large portfolios (Appendix 2). Estimated standard errors are in parentheses. See also notes to Table 4.

patents and only 0.25 percent for those in large portfolios (>900 patents). These are large differences, and they show that having bigger portfolios confers substantial advantages in settling patent disputes without filing suits, but again, we observe only small differences in the post-suit settlement rates across portfolio size. The differences in point estimates are marginally statistically significant $[\chi^2(6) = 14.2; P\text{-value} \approx 0.05]$.

To distinguish between the advantages of portfolio size and firm size, we divide domestic listed firms into two groups—those with employment around 1989 above the median level of 5,463 ("large") and those below the median ("small").[16] Panel A in Table 6 presents the litigation probability broken down by

**TABLE 6** Probability of Litigation and Settlement, by Patent Portfolio Size and Ownership Type

| | Panel A. Probability of Litigation (percent) | | | |
|---|---|---|---|---|
| Portfolio | Large Domestic Listed | Small Domestic Listed | Domestic Unlisted | Foreign Firms |
| 0-10 | 0.55 (0.26) | 1.09 (0.49) | 2.63 (0.09) | 0.48 (0.03) |
| 11-100 | 1.16 (0.25) | 1.78 (0.32) | 2.00 (0.09) | 0.37 (0.03) |
| 101-200 | 0.70 (0.14) | 0.77 (0.28) | 0.67 (0.12) | 0.23 (0.03) |
| 201-300 | 0.49 (0.17) | 0.82 (0.32) | 0.84 (0.27) | 0.18 (0.04) |
| 301-600 | 0.54 (0.10) | 0.70 (0.31) | 0.56 (0.10) | 0.19 (0.03) |
| 601-900 | 0.62 (0.10) | 0.44 (0.25) | 0.34 (0.12) | 0.18 (0.04) |
| >900 | 0.39 (0.02) | nc | 0.37 (0.06) | 0.12 (0.01) |

| | Panel B. Settlement Rates (percent) | | |
|---|---|---|---|
| Portfolio | Domestic Listed | Domestic Unlisted | Foreign Firms |
| 0-10 | 90.0 (3.1) | 95.0 (0.5) | 95.9 (1.3) |
| 11-100 | 95.0 (1.3) | 93.0 (0.9) | 91.2 (2.3) |
| 101-200 | 92.9 (2.4) | 92.1 (3.4) | 95.0 (3.4) |
| 201-300 | 98.8 (1.2) | 97.9 (2.1) | 90.3 (5.3) |
| 301-600 | 92.0 (2.4) | 85.2 (3.9) | 100.0 (0.0) |
| 601-900 | 96.3 (2.6) | 87.5 (5.8) | 94.4 (5.4) |
| >900 | 94.1 (1.3) | 88.8 (3.0) | 95.3 (2.6) |

NOTE: The probability of litigation is the number of patents involved in suits (multiple suits not counted) per hundred patents, adjusted for underreporting and truncation and for the overrepresentation of patents from large portfolios (Appendix 2). "nc" denotes an empty cell. Estimated standard errors are in parentheses. See also notes to Table 4.

---

[16]Employment data are missing for 38 percent of our listed firms, either because their 1989 Standard & Poors CUSIP code does not match to a 2000 CUSIP code or because their employment is not recorded. This group is not included for this test.

both portfolio size and this measure of company size. First, we see a fall in litigation probability with portfolio size within each ownership type, at least in terms of the point estimates. However, the fall is by far more precipitous for domestic unlisted companies. For a patent owned by such a company and in a portfolio of 0-10 other patents, the *average* probability of being involved in litigation is 2.6 percent, whereas for patents in the same-sized portfolio but owned by listed domestic companies it is closer to 1 percent. At the same time, there is little evidence that size—either in terms of public listing or employment—matters once more than about 100 patents are held. For any given portfolio size, foreign companies are much less likely to file suits than other types of firms. The relationship between probability of litigation and portfolio size holds in each of the technology fields (not reported).

Consistent with the results in Table 2, we find that the probability of litigation differs substantially across technology areas for any given ownership type (see Table 7). However, here we also see that the pattern of differences across technology fields depends on the type of owner.

One explanation for these differences in litigation probabilities is that firms with larger portfolios may have a higher propensity to patent their innovations (harvesting) and thus more often have patents that are not worth fighting over. However, the evidence contradicts this hypothesis. Portfolio size is *positively*, and significantly, correlated with forward citations and forward citations per claim. The correlation coefficients are 0.10 and 0.06, respectively (these are computed with the matched sample and cohorts 1978-1988 to avoid spurious correlation due to both portfolio size and citations being truncated). Even within electronics, where firms have often been described as following a patent harvesting strategy, there is no evidence that the average quality of patents falls in larger

**TABLE 7** Probability of Litigation, by Technology and Firm Ownership (in percent)

| Technology | Domestic Unlisted | | Small Domestic Listed | | Large Domestic Listed | | Foreign Firms | |
|---|---|---|---|---|---|---|---|---|
| Drugs | 9.1 | (0.2) | 2.9 | (0.2) | 4.2 | (0.2) | 3.3 | (0.1) |
| Other Health | 10.5 | (0.2) | 9.1 | (0.4) | 4.1 | (0.3) | 2.2 | (0.1) |
| Chemicals | 3.8 | (0.1) | 3.9 | (0.1) | 1.2 | (0.05) | 0.5 | (0.02) |
| Electronics | 6.6 | (0.1) | 12.3 | (0.1) | 11.2 | (0.1) | 0.8 | (0.02) |
| Mechanical | 6.8 | (0.1) | 3.9 | (0.1) | 11.2 | (0.1) | 0.7 | (0.02) |
| Computers | 14.9 | (0.6) | nc | | 1.3 | (0.2) | 0.3 | (0.06) |
| Biotechnology | 20.1 | (0.7) | 3.9 | (0.6) | 3.4 | (0.6) | 7.2 | (0.5) |
| Miscellaneous | 11.2 | (0.2) | 4.2 | (0.1) | 2.6 | (0.1) | 1.3 | (0.04) |

NOTE: Estimated standard errors are in parentheses. See notes to Table 5.

portfolios (Hall and Ziedonis, 2001). The same positive and significant relationships are found, and the same is true for all other technology areas. Thus it appears that the link between litigation probability and portfolio size does actually reflect the advantages that large portfolios give to firms in settling disputes.

However, this is only half the story. Panel B in Table 6 presents the average probability of settlement for different portfolio sizes and ownership categories, conditional on a suit being filed. Here we see that post-suit settlement rates do *not* vary significantly with portfolio size or with ownership type controlling for portfolio size.

In short, the likelihood of filing a suit (i.e., of *not* settling beforehand) is much higher for patents owned by individuals and unlisted companies and for patentees with smaller patent portfolios to trade. However, these differences do not appear in post-suit settlement rates. Thus almost all of the settlement of disputes, as determined by observed characteristics of patents and patentees, occurs before suits are filed, not afterwards in the courts.

To this point we have focused on the probability of litigation and of post-suit settlement. We now turn to the timing of such settlements and the win rates for cases that reach the trial adjudication stage. Table 8 summarizes this information broken down by ownership type—domestic listed, domestic unlisted, and foreign firms and all individuals. About 80 percent of all suits that are ever settled (without third-party adjudication) are settled before a pretrial hearing is held. This suggests that the filing of a suit sends a strong signal about the seriousness of the plaintiff to use legal means and quickly triggers resolution before substantial le-

**TABLE 8** Timing of Settlement and Trial Win Rates, by Ownership Type

|  | Domestic Listed | Domestic Unlisted | Foreign Firms | Individuals |
|---|---|---|---|---|
| Timing of Settlement (%): | | | | |
| Before Pretrial Hearing | **81.2** (1.2) | **83.0** (0.7) | **78.8** (1.7) | **84.7** (0.8) |
| Before Trial | **18.0** (1.2) | **15.5** (0.7) | **19.9** (1.7) | **14.2** (1.8) |
| During Trial | 0.8 (0.3) | 1.5 (0.2) | 1.3 (0.5) | 1.1 (1.9) |
| Plaintiff Win Rate at Trial | **51.2** (3.8) | **49.1** (2.3) | **42.7** (4.9) | **46.5** (2.3) |

NOTE: The timing of settlements is computed on the basis of all infringement cases filed during the period 1978-1992 and terminated by settlement before or during trial, according to the FJC. Cases that proceed beyond trial (e.g., on appeal or remand, which are about 5 percent) are not included. The plaintiff win rate is the number of infringement cases in which the court judgment favors the patentee divided by the total number of cases. When the FJC reports a judgment in favor of both parties, we treat it as a win for each party and adjust the total appropriately. Estimated standard errors in parentheses are based on the binomial formula. Numbers in bold are statistically significant at the 0.01 level.

gal costs are incurred.[17] Nearly all of the remaining settlement occurs before the trial commences. However, the table shows that the timing of settlements differs little by ownership type.

The table also shows the trial win rates (for infringement suits). For domestic listed and unlisted firms, the win rates are very close to 50 percent, as predicted by the divergent expectations model of litigation. They are sharply inconsistent with the win rates of either zero or 100 percent predicted by the asymmetric information models. The point estimate of the win rate for foreign corporate patentees is only 42.7 percent, but the standard error is relatively large.

## ECONOMETRIC ANALYSIS

In this section we present estimates of probit regressions on the determinants of the probabilities of infringement suits and post-suit settlement for the pooled data. These endogenous variables are related to the following regressors: the number of claims, forward citations per claim, backward citations per claim, and the percentage of backward and forward citations that are self-citations (as measures of cumulative technology), the number of three-digit USPCs as a measure of patent breadth, the size of the patent portfolio, the relative size of the patent portfolio (as a measure of asymmetry between a patent owner and likely disputants), the technology concentration index, and ownership dummy variables that distinguish between patentees who are foreign or domestic individuals and unlisted or listed firms. The effects of technology and cohort on litigation probabilities are largely controlled by the matching, but because the litigated and matched data contain somewhat different numbers of patents, we also include technology group dummies.

We use the Derwent data as the basis for the sample, because it contains the link to patent numbers, and then include only those cases that can also be linked into the FJC database, which contains the outcomes information. This procedure yields 6,538 litigated main patents. In analyzing the determinants of the *litigation probability* (filing of suits), we do not count multiple cases involving the same patent. We do this to avoid undue influence by a few patentees suing many infringers in separate but related cases. We include multiple cases in the econometric analysis of the suit outcomes for two main reasons: first, it is unclear how one would choose the "representative" suit when there are multiple cases and, second, the sample size for outcomes (especially trials) is relatively small even when we include multiple cases.[18]

---

[17]Pooling all cases, the median number of months that pass before settlement occurs are 8, 16, and 25 for those settling before pretrial hearing, after a hearing but before trial, and after trial, respectively.

[18]In addition, it is appropriate to include multiple cases if one wants to use the empirical analysis to assess litigation risk in order to set actuarial prices for patent insurance.

Panel A in Table 9 summarizes the parameter estimates and the *sample* marginal effect of each variable on the probability of litigation for a randomly drawn patent in the matched sample (i.e., at matched sample means). This is done separately for patent infringement and declaratory judgment suits. Because the sample litigation rate is close to 40 percent by construction, we must multiply the reported marginal effects by a conversion factor to obtain the marginal effects for a randomly drawn patent in the population (the conversion factors are given at the bottom of Table 9; see Appendix 3 for computational details). The statistical significance of variables and the relative size of their effects are preserved through this conversion, although magnitudes will depend on the specific population of interest. We focus the discussion on the results for patent infringement cases. Because the pattern of results is similar for declaratory judgment suits (Panel B of Table 9), we do not discuss them in detail. Variable definitions are listed in Table 10.

The probability of litigation increases with the number of claims and forward citations per claim at a declining rate, and the effects are substantial. Evaluated at population means (litigation probability of 1.35 percent), a 10 percent increase in the number of claims (1.2 claims at the mean) implies an increase of 3.1 percent in the *population* probability of litigation. We also find that a 10 percent increase in the number of forward citations per claim raises the probability of an infringement suit by 1.8 percent. These findings confirm the importance of the value of a patent in determining infringement suits. In related work on the determinants of re-examinations at the USPTO and opposition proceedings at the European Patent Office—both events suggesting that the use of a patent is subject to dispute—Graham et al. (2003) and Harhoff and Reitzig (2000) find similar positive relationships.

The likelihood of an infringement suit falls with the number of backward citations per claim (at a declining rate). At mean values, a 10 percent increase in the number of backward citations per claim reduces the litigation probability by 0.7 percent. Although the effect is small, this finding is consistent with the view that backward citations are an indication that the patent is in a relatively well-developed technology area, in which many related patents have been taken out and where uncertainty about property rights is less likely to cause frequent patent disputes (Lanjouw and Schankerman, 2001).

We have also argued that forward *self-citation* to a patent (given its total number of forward citations) indicates the presence of "cumulative innovation" by the patentee. That is, the patent owner is engaged in subsequent inventions that build on this earlier patent and, as a result, he has a greater incentive to protect his property rights in this area. This hypothesis is supported by the positive and significant coefficient on the variable FWDSELF, the percentage of citations that is self-citation. At the mean (FWDSELF = 0.065), increasing the percentage of forward self-citations by 10 percent would raise the probability of an infringement suit by 0.4 percent (the estimate is proportionately higher for larger

**TABLE 9** Probit Estimation of Litigation Probability: Case Filings

| Variable | Panel A Infringements | | Panel B Declaratory Judgments | |
|---|---|---|---|---|
| | Parameter | Marginal | Parameter | Marginal |
| Claims | **0.023** | 0.007 | **0.029** | 0.002 |
| | (0.001) | | (0.003) | |
| Claims$^2$ ($\times 10^3$) | **−0.024** | | **−0.15** | |
| | (0.002) | | (0.038) | |
| FWD cites/claim | **0.19** | 0.059 | **0.20** | 0.0008 |
| | (0.008) | | (0.017) | |
| [FWD cites/claim]$^2$ ($\times 10^3$) | **−4.38** | | **−5.65** | |
| | (.32) | | (.83) | |
| BWD cites/claim | **−0.056** | −0.017 | **−0.072** | −0.005 |
| | (0.010) | | (.019) | |
| [BWD cites/claim]$^2$ ($\times 10^3$) | **0.89** | | **1.47** | |
| | (0.43) | | (.62) | |
| FWDSELF | **0.51** | 0.17 | **0.63** | 0.05 |
| | (0.058) | | (.10) | |
| BWDSELF | **−0.31** | −0.10 | −0.16 | −0.01 |
| | (0.08) | | (.15) | |
| NO3USPC | **−0.068** | −0.022 | −0.014 | −0.003 |
| | (.008) | | (.015) | |
| Portsize ($\times 10^3$) | **−0.104** | −0.025 | −0.21 | −0.015 |
| | (.037) | | (.13) | |
| Portsize$^2$ ($\times 10^6$) | **0.009** | | 0.005 | |
| | (0.001) | | (.0003) | |
| PortNondrug ($\times 10^3$) | **−0.061** | −0.021 | 0.056 | 0.004 |
| | (0.033) | | (.12) | |
| PortUNLIST ($\times 10^3$) | **−0.027** | −0.009 | −0.07 | −0.005 |
| | (0.013) | | (.04) | |
| PortFLIST ($\times 10^3$) | 0.001 | 0.0003 | 0.05 | 0.004 |
| | (0.020) | | (.05) | |
| PortDLIST-S ($\times 10^3$) | **−0.6** | −0.20 | −0.36 | 0.028 |
| | (.27) | | (.41) | |
| Tech. Concentration ($C4$) | **−4.17** | −1.36 | **−6.15** | −0.48 |
| | (.23) | | (.46) | |
| Relsize ($\times 10^3$) | **−3.1** | −1.01 | −0.91 | −0.07 |
| | (1.12) | | (2.67) | |
| FIND | **−0.54** | −0.12 | **−1.84** | −0.036 |
| | (0.09) | | (.17) | |
| DIND | 0.13 | 0.14 | **−1.30** | 0.012 |
| | (0.08) | | (.15) | |
| FUNLIST | **−0.69** | −0.22 | **−1.81** | −0.045 |
| | (.08) | | (.15) | |
| DUNLIST | **0.21** | 0.19 | **−1.06** | 0.058 |
| | (.08) | | (.15) | |
| FLIST | −0.15 | 0.007 | **−1.77** | −0.030 |
| | (.19) | | (.42) | |

*continues*

**TABLE 9** Continued

| Variable | Panel A Infringements | | Panel B Declaratory Judgments | |
|---|---|---|---|---|
| | Parameter | Marginal | Parameter | Marginal |
| DLIST-S | **0.27** | **0.17** | **0.46** | **0.060** |
| | (.11) | | (.15) | |
| D-LIST-B | **−0.23** | −0.03 | **−0.98** | −0.006 |
| | (.08) | | (.20) | |
| No. Observations | 17,443 | | 11,061 | |
| Pseudo-$R^2$ | 0.162 | | 0.164 | |
| $\chi^2$ | 3858.3 | | 1098.9 | |

| Conversion Factors to Estimate Population Marginal Effects | | |
|---|---|---|
| Technology Field | Infringements | Declaratory Judgments |
| Aggregate | .048 | .021 |
| Drugs | .050 | .018 |
| Other Health | .089 | .039 |
| Chemicals | .031 | .014 |
| Electronics | .038 | .020 |
| Mechanical | .045 | .021 |
| Computers | .063 | .034 |
| Biotechnology | .076 | .030 |
| Miscellaneous | .084 | .031 |

NOTE: Estimated standard errors are in parentheses. Numbers in bold are significant at the 0.01 level. The conversion factors are computed as described in Appendix 3.

values of self-citing). At the same time, we find that greater backward self-citation (BWDSELF) significantly reduces the likelihood of litigation, but the effect is again small at the mean: Raising the percentage of backward self-citations by 10 percent lowers the litigation probability by about 0.25 percent. Greater backward self-citation in a patent indicates that an invention builds more extensively on one's own past research and is thus more likely to be a "derivative" invention. This evidence supports the idea that there is complementarity among technologically related inventions in a firm's R&D portfolio and that this raises the willingness to protect the property rights of the key, early inventions in the chain.

In our earlier work (Lanjouw and Schankerman, 2001), we found that greater technological similarity of forward citations increased the probability of litigation.[19] The similarity measure was used as an index of whether the technology

---

[19]Similarity measures whether subsequent citing patents fall in similar technology fields as the patent in question. It is calculated by finding the percentage of three-digit USPC assignments of each citing patent that overlap with those of the patent itself and averaging over all citing patents.

**TABLE 10** Variable Definitions

| | |
|---|---|
| Claims | Number of claims in the patent specification |
| FWD cites/claim | Number of citations to the patent by subsequent patents, divided by claims. |
| BWD cites/claim | Number of citations to prior patents in the patent specification, divided by claims. |
| FWDSELF | Percentage of forward citations that are from patents owned by the same company code. For individuals it is set to zero. |
| BWDSELF | Percentage of backward citations that are to patents owned by the same company. For individuals it is set to zero. |
| NO3USPC | Number of unique three-digit technology classes to which the patent is assigned by the patent office examiner. |
| Portsize | Number of other patents owned by the same assignee that have an application year within a ten-year window of the application year of the patent in question. For individuals it is set to one. |
| PortNondrug | Portsize times an indicator variable that is one if the patent is not a drug innovation, zero if it is a drug innovation. |
| PortUNLIST | Portsize times UNLIST (see below) |
| PortFLIST | Portsize times FLIST (see below) |
| PortDLIST-S | Portsize times DLIST-S (see below) |
| Tech. Concentration (C4) | Firm C4 concentration measures – weighted average over the technology areas of the patent's forward citations. |
| Relsize | Total portfolio size of the patent owner divided by a weighted average of portfolio sizes of firms in the technology areas of the patent's forward citations. |
| FIND | Foreign (non-U.S.) individual |
| DIND | Domestic (U.S.) individual |
| FUNLIST | Foreign company assignee without a Standard & Poor's (S&P) CUSIP code |
| DUNLIST | Domestic company assignee without an S&P CUSIP code |
| FLIST | Foreign publicly listed company with an S&P CUSIP code |
| DLIST-S | Domestic publicly listed company with fewer than the median number of employees for such firms (5,425) |
| DLIST-B | Domestic publicly listed company with more than the median number of employees. |

area was "crowded" and thus more likely to generate potential disputes. However, we do not find any evidence of that link in the current expanded data set.

Lerner (1994) suggests that patents with uses in many technological areas—"broad" patents—are more likely to be litigated because they face more potential infringers. Using the number of technology class assignments as a measure of patent breadth, he confirmed the hypothesis on a sample of biotechnology patents. Using more comprehensive data for various technology fields, Lanjouw and Schankerman (2001) found that broader patents are *less* likely to be involved in suits, but the evidence was weak. We test this hypothesis on our expanded and more recent data set, using the number of three-digit USPC classes as the measure of breadth (NO3USPC). The estimated coefficient is similar to the earlier estimate by Lanjouw and Schankerman and highly significant. A 10 percent increase in NO3USPC (the mean number of technology field assignments is 2.2) reduces the litigation probability by about 1.7 percent.[20] This finding suggests that it is harder to detect infringements when the patented innovation is used in more technology areas and that this effect dominates any increase in the number of potential infringers associated with greater patent breadth.

An important finding is that the probability of litigation is negatively related to the size of the patent portfolio, with an elasticity (at the mean) of –0.13. The marginal effect of portfolio size declines with larger portfolios (positive quadratic term), but the point estimate of the portfolio effect is negative over most of the sample range. This means that having a larger portfolio of patents reduces the probability of being involved in a suit on any individual patent owned by the firm, e.g., there are beneficial "enforcement spillovers" across patents within a given firm. We can compute by how much increasing portfolio size reduces the litigation probability of any constituent patent. For example, raising the portfolio from 100 to 500 patents lowers the litigation probability on an "average" patent (with characteristics at their mean values) by 0.13 percentage points, or about 10 percent of the mean probability. Going from a portfolio of 500 to 2,500 reduces the probability by 0.21 percentage points, or by about 15 percent. Harhoff and Reitzig (2000) find that larger portfolios also tend to keep owners out of European opposition proceedings.

The impact of portfolio size on the probability of litigation is smaller for drug patents than for patents in other technology fields. Estimation at the technology field level (not reported) suggested this hypothesis (the other differences in the estimated portfolio coefficients across technology fields were not statistically significant). To test the hypothesis, we include a portfolio dummy variable for nondrug technology fields (PortNondrug). The estimated coefficient is negative and large relative to the baseline portfolio effect. Using the estimated coefficients on Portsize and PortNondrug, we find that the marginal effect of portfolio size on the litigation probability is nearly twice as large for nondrug patents as compared to drug patents. This finding is consistent with the idea that trading intellectual property is especially important in areas in which innovation is "complex" in the sense that it may rely on multiple components or research tools that may be patented by other firms (see Cohen et al., 2000). This feature has been less important in drugs. Somaya (2003) finds a similar difference, using somewhat overlapping technology definitions and a related variable for portfolio size. He finds that the size of a patentee's portfolio has an insignificant effect on the litigation of patents for research medicine, whereas it has a negative effect for computer patents.

---

[20]The point estimates in the separate technology fields (not reported) are negative and statistically significant in five cases, negative but insignificant in two, and positive but insignificant in one.

The portfolio effect captures the ability of firms to trade patents as a means of settling disputes. Smaller companies may have few alternative mechanisms to facilitate settlement, so we expect portfolio size to be more important for smaller firms. To test this hypothesis, we include interaction effects between portfolio size and ownership type (unlisted and small domestic and foreign listed, with large domestic listed firms being the reference category). The point estimates strongly support the hypothesis that company size affects the importance of having larger patent portfolios. For a small domestic listed company with the mean portfolio size (1,420 patents), the *marginal effect* of portfolio size on the probability of litigation is about eight times larger than for a large listed company with the same portfolio (compare marginal effects for Portsize and PortDLIST-S). The marginal effect of portfolio size for small listed firms is even greater than that for unlisted firms.

Additional evidence that the expectation of repeated interaction promotes settlement is provided by the technology concentration variable (*C4*), defined in the third section of this chapter. If a company operates in concentrated technology areas (i.e., where the top four firms account for a larger share of patenting), there is a greater chance that the company will be involved in repeated patent disputes with the same firms. This should increase the likelihood of settlement and thus reduce the probability of litigation. As predicted, the estimated coefficient on the technology concentration index is negative and highly significant and the quantitative effect on the litigation probability is large. A 10 percent increase in the four-firm technology concentration index reduces the probability of a suit by 4.6 percent.

The portfolio size, company size, and technology concentration variables capture the ability to trade and the role of repeated interaction. We also find that the litigation probability is influenced by the *asymmetry in portfolio size* between the patent owner and likely disputants, which we interpret as reflecting relative threat power of the parties. The coefficient on the relative size variable (RelSize) is significantly negative for infringement suits, as expected.[21] If a patent owner is large relative to typical disputants, the probability of litigation is lower (settlement is more likely). However, the effect is not very large—a 10 percent increase in relative size lowers the litigation probability by 0.5 percent. Interestingly, relative size does not matter in declaratory judgment suits, those in which the patent owner is the defendant (Panel B of Table 9). The prediction was that larger relative size (of the patentee) would make settlement more difficult or have no effect for declaratory judgment suits, and we find the latter.

---

[21]Two points should be noted. For patents without any forward citations, the denominator in the RelSize variable is set equal to the average portfolio size for other patents in the same two-digit USPC class as the patent in question. For all individuals, and for about 900 cases in which company patentees had only one patent, we set RelSize equal to zero.

We easily reject the hypothesis that there are no ownership differences when we control for other factors [$\chi^2(6) = 978.8$; $P$-value $< 0.001$]. The pattern of marginal effects on the ownership dummies points to five main findings about the conditional effects of ownership type on the propensity to litigate. First, foreign individuals and unlisted (smaller) companies are much less likely to engage in infringement suits than their domestic counterparts. Comparing the marginal effects of *FIND* and *DIND*, we see that the probability of litigation is much lower—by about 1.2 percentage points—for foreign individual owners than for their domestic counterparts. Comparing foreign and domestic unlisted companies (*FUNLIST* and *DUNLIST*), the difference is even larger, about 2.0 percentage points. Second, larger domestic and foreign listed companies are equally likely to file suits. Third, domestic individuals and unlisted and small listed companies are equally likely to litigate (the differences in point estimates are not statistically significant). Fourth, domestic individuals and unlisted companies are more likely to litigate than large domestic listed firms, by about 0.9 percentage points. And finally, small listed companies are far more likely to file suits than larger ones, the difference being about 1.0 percentage points on average.

To summarize, we find the following ranking of the propensity to litigate, in descending order: Small domestic listed companies, domestic unlisted companies and domestic individuals have the highest propensity to sue (given the characteristics of a patent), and there are no significant differences among them. Large domestic listed companies and foreign listed companies have the next highest propensity to litigate. Foreign individuals and foreign unlisted companies are least likely to be involved in patent infringement suits.[22] Because these effects are conditional on portfolio and company size (both of which relate to the cost of settling), this ranking should reflect two main factors, the cost of litigation and access to information about potential infringements. We expect that the cost of litigating for domestic patentees is less than (or equal to) that for foreign patentees and that it is harder for foreign owners to detect infringements in the United States. Given the cost of settling disputes, these hypotheses predict that domestic owners should litigate more often than their foreign counterparts. That is what we find, except for listed companies. This exception is not surprising, because foreign firms that are listed, and have a presence, in the United States are less likely to be at much disadvantage in terms of litigation costs and access to information.

Table 11 highlights the enormous variation in litigation risk implied by these estimation results. We calculate the population probability of involvement in an infringement suit for each patent in the matched sample, given the patent's full set of characteristics. The 50th-99th percentile cutoffs for the distribution of these

---

[22] In terms of the variable names in Table 9, this ranking is: *DLISTS* = *DUNLIST* = *DIND* > *DLISTB* = *FLIST* > *FIND* = *FUNLIST*, where *DLISTS* and *DLISTB* are small and large (or unclassified) listed domestic firms, respectively.

**TABLE 11** Predicted Probabilities of Infringement Suits

| Percentile of Distribution | 99th | 95th | 90th | 50th |
|---|---|---|---|---|
| Aggregate | 7.9% | 3.8% | 2.8% | 0.8% |
| Technology Field | | | | |
| Drugs | 9.4% | 3.9% | 2.8% | 0.9% |
| Other Health | 19.5 | 6.1 | 4.5 | 1.7 |
| Chemicals | 4.2 | 2.1 | 1.6 | 0.5 |
| Electronics | 7.1 | 2.8 | 2.1 | 0.5 |
| Mechanical | 6.5 | 2.8 | 2.2 | 0.7 |
| Computers | 14.8 | 4.5 | 3.4 | 0.6 |
| Biotechnology | 12.9 | 6.3 | 5.3 | 1.3 |
| Miscellaneous | 8.3 | 4.6 | 3.7 | 1.9 |
| Ownership Type | | | | |
| Domestic Individual | 9.4% | 4.4% | 3.5% | 1.9% |
| Domestic Unlisted | 13.7 | 5.9 | 4.2 | 1.9 |
| Small Domestic Listed | 6.3 | 5.4 | 4.1 | 1.8 |
| Large Domestic Listed | 4.8 | 2.0 | 1.5 | 0.5 |
| Foreign Listed | 2.5 | 1.4 | 1.0 | 0.3 |
| Foreign Individual | 4.2 | 1.4 | 1.1 | 0.6 |
| Foreign Unlisted | 1.4 | 0.8 | 0.7 | 0.3 |

NOTE: The distribution of population probabilities for patents with different characteristics is calculated by first computing the sample probabilities with the parameter estimates for infringement suits in Table 9. These are then adjusted to reflect population probabilities with Appendix equation (A.3.1).

probabilities are given in the first row of the table. The probability of litigation for the median patent is just under 1 percent. However, among the top 1 percent of patents (99th percentile), the probability of involvement in a suit is over 8 percent. The table shows that the rates can be far higher when the patents are segregated into different technology and ownership groups. The top percentile of patents in areas that are most at risk have probabilities of litigation over 15 percent (see Other Health, Computers, and Biotechnology). Similarly, the top 1 percent of all patents held by domestic unlisted firms or individuals have a litigation risk over 10 percent. Because most evidence, from patent renewal data and firm surveys, indicates that private value of innovations is highly skewed—with most value attributable to the top patents—it is precisely the litigation risk in these top percentiles that is relevant for determining R&D incentives.

We now turn to the econometric analysis of post-suit outcomes. In estimating these regressions, we do not control for selection, i.e., we do not use a (filing) selection equation together with the outcomes equation. Selection bias arises if there is significant covariance between the disturbances in the filing and outcome equations. We ask: Given the selection that occurs at filing, is there any remaining association between patent and patentee characteristics and the outcomes? For purposes of assessing ex ante litigation risk (e.g., for patenting decisions or insurance pricing), this is the relevant question. Controlling for selection in the analysis of outcomes (see, e.g., Somaya, 2003) is appropriate if one wanted to infer the effects of characteristics in *a random sample at the outcomes stage*. In any event, the evidence that there is any sample selection bias is mixed (Somaya, 2003).

The evidence presented in the previous section indicated that the main characteristics of patents and their owners do not affect the probability of settlement after a suit is filed or the plaintiff win rates for cases that reach trial. The probit regressions for settlement and win rates confirm this conclusion. For brevity, we summarize the findings but do not present the parameter estimates. The settlement regression has a meager pseudo-$R^2$ of 0.01. The null hypothesis that the regression as a whole is insignificant is not rejected [$\chi^2(29) = 39.7$; $P$-value = 0.089]. The only positive finding is that the coefficients on three technology field dummies are significant and indicate that the settlement probability is about eight percentage points higher for patents in Electronics, Mechanical, and Miscellaneous.[23] The probit regression for win rates has a pseudo-$R^2$ of 0.02. The whole regression is statistically insignificant [$\chi^2(28) = 19.7$; $P$-value = 0.90], as is each individual coefficient. On the basis of our discussions with staff at the FJC, there is no reason to believe that the data on settlements and plaintiff win rates are systematically bad (these outcome data are recorded at different times and in many different courts). We are confident that the "insignificance" of these regressions is meaningful, i.e., settlement and win rate outcomes are almost completely independent of observed characteristics of patents and their owners.

The probability that the settlement of infringement suits occurs early (before the pretrial hearing) is also unrelated to most characteristics of the patent and its owner, with three noteworthy exceptions [the probit regression is significant: $\chi^2(29) = 50.5$; $P$-value = 0.008]. First, early settlement is more likely if the patent in dispute is part of a larger portfolio (Portsize). A one standard deviation increase in portfolio size (1,300 patents) raises the probability of early settlement by about 12.9 percent. This is consistent with our earlier result that portfolio size makes filing a suit less likely in the first place, because of a greater ability to "trade" intellectual property. Second, a higher technology concentration index (*C4*) makes early settlement somewhat less likely. A one standard deviation in-

---

[23]It is also interesting to note that, if we restrict attention to suits in which the original patentee is identified as the plaintiff, those suits involving smaller patentees (unlisted firms and domestic individuals) are significantly less likely to settle. These are patentees who do *not* have an exclusive licensee or late assignee litigating in their place. As plaintiffs they are more likely to be inexperienced and more attached to their innovations than owners who have licensed or sold out. Both characteristics could impede settlement.

crease (doubling) in the concentration index lowers the probability by about 2 percent. Finally, patent owners that are large relative to a representative disputant (Relsize) are also less likely to settle early. A one standard deviation rise in relative size reduces the probability of early settlement by about 5 percent.[24] Recall that the probability that a suit is filed is lower when the relative size of the patentee is larger, which we interpret as reflecting greater threat power. But if the (implicit) threats do not succeed in preventing the need to file suit, it is important for the patentee to carry out those threats to maintain credibility (post-suit "toughness"). Similarly, if the discipline of repeated interaction has failed to keep firms in a concentrated area out of court in the first place, the dispute is probably very intractable. Both could delay any post-suit settlement, and this is what we find.

## CONCLUDING REMARKS

We studied the determinants of patent infringement and declaratory judgment suits, and their outcomes, by linking detailed information from the USPTO to data from the U.S. federal court system, the Derwent database, and industry sources. This allows us to construct a suitable controlled random sample of the population of potential disputants. The data set we construct is the most comprehensive yet available, covering all patent suits in the United States reported by the federal courts during the period 1978-1999.

A major finding of the chapter is that almost all of the effect of observable characteristics on patent disputes that we examined occurs in the decision to initiate a suit. Among others, these characteristics included the technology field, the number of patent claims, the numbers of forward and backward citations, patent portfolio size, type of patentee, and technology concentration index. Major post-suit outcomes—the probability of settlement and plaintiff win rates at trial—do not depend on these characteristics. From a policy perspective, this is good news because it means that enforcement of patent rights relies on the effective *threat* of court action (suits) more than on extensive post-suit legal proceedings that consume court resources. This feature is reinforced by high post-suit settlement rates and the fact that most settlement occurs soon after the suit is filed, often before the pretrial hearing is held. These findings mean that the enforcement of patent rights minimizes the use of judicial resources for sorting out patent disputes. The bad news is that individuals and small companies are much more likely to be involved in suits, conditional on the characteristics of their patent, but they are no more likely to resolve disputes quickly in post-suit settlements.

We also provide evidence that there are considerable advantages to scale in patent enforcement. Being able to trade a portfolio of intellectual property and having other dimensions of interaction that promote "cooperative" behavior are

---

[24]Marginal changes are given in terms of standard deviations here because the distribution of these variables is very skewed after the selection for filing.

likely sources of this advantage. Thus there are two sides to aggressive patenting strategies. On one hand, the buildup of large patent portfolios and the creation of patent thickets can make disputes over intellectual property more likely. But those same patents can also make the suits easier to resolve at lower cost.

An important direction for future research is to explore the dynamic aspects of conflict between firms over intellectual property assets. This would include studying the determinants of the filing and outcomes of multiple (sequential) suits on the same patent with different parties and multiple suits on different patents involving the same parties. Initial work along these lines for a sample of cases has been done by Somaya (2003). Proceeding further requires matching the names of litigants across all cases, a project that is under way. When completed, these data will provide information about the role of reputation building in the area of patent enforcement and allow a more detailed assessment of litigation risk and its associated costs.

## REFERENCES

American Intellectual Property Law Association. (2001). *Report of the Economic Survey*. Arlington, VA: American Intellectual Property Lawyers Association.

Bebchuk, L. (1984). "Litigation and Settlement Under Imperfect Information." *RAND Journal of Economics* 15: 404-415.

Cohen, W., R. Nelson, and J. Walsh. (2000). "Protecting Their Intellectual Assets: Appropriability Conditions and Why U.S. Manufacturing Firms Patent (or Not)," NBER Working Paper, No. 7552.

Cooter, R., and D. Rubinfeld. (1989). "Economic Analysis of Legal Disputes and Their Resolution." *Journal of Economic Literature* 27: 1067-1097.

Danish Ministry of Trade and Industry. (2001). "Economic Consequences of Legal Expense Insurance for Patents," report prepared for the Danish Patent Office by the Economic Analysis Group. Copenhagen.

Eisenberg, R. (1999). "Patents and the Progress of Science: Exclusive Rights and Experimental Use." *University of Chicago Law Review* 56: 1017-1055.

Federal Judicial Center, Federal Court Cases: Integrated Data Base, 1970-89. Ann Arbor, MI: Inter-university Consortium for Political and Social Research. Tapes updated to 1999.

Graham, S., B. Hall, D. Harhoff, and D. Mowery. (2003). "Patent Quality Control: A Comparison of U.S. Patent Re-examinations and European Patent Oppositions." In W. Cohen and S. Merrill, eds., *Patents in the Knowledge-Based Economy*. Washington, D.C.: National Academy Press.

Grindley, P., and D. Teece. (1997). "Managing Intellectual Capital: Licensing and Cross-Licensing in Semiconductors and Electronics." *California Management Review* 39(2): 8-41.

Hall, B., and R. Ziedonis. (2001). "The Patent Paradox Revisited: An Empirical Study of Patenting in the Semiconductor Industry, 1979-1999." *RAND Journal of Economics* 32(1): 101-128.

Harhoff, D., and M. Reitzig. (2000). "Determinants of Opposition against EPO Patent Grants—The Case of Biotechnology and Pharmaceuticals." CEPR Discussion Paper No. 3645, Centre for Economic Policy Research, London.

Jaffe, A., and M. Trajtenberg. (1999). "International Knowledge Flows: Evidence from Patent Citations." *Economics of Innovation and New Technology* 8: 105-136.

Lanjouw, J. O., A. Pakes, and J. Putnam. (1998). "How to Count Patents and Value Intellectual Property: Uses of Patent Renewal and Application Data." *Journal of Industrial Economics* 46(4) (December): 405-432.

Lanjouw, J. O., and J. Lerner. (2001). "Tilting the Table? The Predatory Use of Preliminary Injunctions." *Journal of Law and Economics* 44(2): 573-603.

Lanjouw, J. O., and M. Schankerman. (2001). "Characteristics of Patent Litigation: A Window on Competition." *RAND Journal of Economics* 32(1): 129-151.

Lerner, J. (1994). "The Importance of Patent Scope: An Empirical Analysis." *RAND Journal of Economics* 25: 319-333.

Lerner, J. (1995). "Patenting in the Shadow of Competitors." *Journal of Law and Economics.* 38: 463-96.

P'ng, I. P. L. (1983). "Strategic Behavior in Suit, Settlement and Trial." *Bell Journal of Economics* 14: 539-550.

Priest, G., and B. Klein. (1984). "The Selection of Disputes for Litigation." *Journal of Legal Studies* 13: 1-55.

Schankerman, M. (1998). "How Valuable Is Patent Protection: Estimates by Technology Field." *RAND Journal of Economics* 29(1): 77-107.

Scotchmer, S. (1991). "Standing on the Shoulders of Giants: Cumulative Research and the Patent Law." *Journal of Economic Perspectives* 5: 29-41.

Shapiro, C. (2001). "Navigating the Patent Thicket: Cross Licenses, Patent Pools and Standard-Setting." In A. Jaffe, J. Lerner, and S. Stern, eds., *Innovation Policy and the Economy.* Cambridge: MIT Press for the NBER, vol. 1, pp. 119-150.

Siegelman, P., and J. Waldfogel. (1999). "Toward a Taxonomy of Disputes: New Evidence Through the Prism of the Priest/Klein Model." *Journal of Legal Studies* 18(1): 101-130.

Somaya, D. (2003). Strategic Decisions not to Settle Patent Litigation," *Strategic Management Journal* 24(1): 17-38.

Spier, K. (1992). "The Dynamics of Pretrial Negotiation." *Review of Economic Studies* 59(1): 93-108.

Tirole, J. (1994). *The Theory of Industrial Organisation.* Cambridge, MA: MIT Press.

Waldfogel, J. (1998). "Reconciling Asymmetric Information and Divergent Expectations Theories of Litigation" *Journal of Law and Economics* XLI (October): 451-476.

**APPENDIX 1** Reporting and Truncation Rates for Case Filings (percent)

| Cohort | Reporting | Lag | Truncation |
|---|---|---|---|
| 1978 | 15.9 | 1 | 97.6 |
| 1979 | 25.0 | 2 | 91.3 |
| 1980 | 26.6 | 3 | 82.4 |
| 1981 | 30.2 | 4 | 75.3 |
| 1982 | 29.4 | 5 | 67.8 |
| 1983 | 33.9 | 6 | 60.2 |
| 1984 | 36.8 | 7 | 52.8 |
| 1985 | 33.7 | 8 | 44.9 |
| 1986 | 38.7 | 9 | 37.7 |
| 1987 | 43.0 | 10 | 30.0 |
| 1988 | 48.5 | 11 | 23.7 |
| 1989 | 49.5 | 12 | 18.1 |
| 1990 | 61.2 | 13 | 12.5 |
| 1991 | 60.0 | 14 | 7.2 |
| 1992 | 57.6 | 15 | 3.7 |
| 1993 | 50.0 | 16 | 1.2 |
| 1994 | 54.4 | 17 | 0.2 |
| 1995 | 53.6 | 18 | 0.0 |
| 1996 | 55.2 | | |

NOTES: The reporting rate is computed as the number of cases reported in Derwent divided by the number in the Federal Judicial Center data. The truncation rate is computed from the lag structure of filings for cohorts 1982-1986. The reporting rate for 1996 is used for 1997-1999, because data are not available.

**APPENDIX 2** Computing Population Filing Probabilities and Their Variance

Let $L_{gz}$, $M_{gz}$, and $N_{gz}$ denote, respectively, the number of patents in the litigated and matched samples and in the population that are in portfolios of size $z$ and from group $g$, where the latter is defined by technology field, cohort, and ownership type. The observed filing probabilities in the sample are $L_{gz}/(L_{gz} + M_{gz})$. The filing probabilities in the population are $q_{gz} = [L_{gz}/N_{gz}]$. We cannot calculate these directly because $N_{gz}$ is unobserved. However, because the matched sample is random with respect to portfolio size, we can use the sample share of the patents in group $g$ that are in portfolios of size $z$, $\hat{s}_{gz} = [M_{gz}/M_g]$, as an unbiased estimator of the population share $[N_{gz}/N_g]$. Using this, our estimator is:

$$\hat{q}_{gz} = \frac{L_{gz}}{N_g} \frac{1}{\hat{s}_{gz}}.$$

Now, treating the population itself as a random sample from an underlying distribution, $L_{gz}/N_g$ will also be an estimate of an underlying probability, say $p$, with an associated sampling variance. Taking a Talyor expansion, we can capture both sources of error in the following approximation:

$$Var(\hat{q}_{gz}) = Var\left(\hat{p}\frac{1}{\hat{s}_{gz}}\right) \approx \left[\frac{-\hat{p}}{\hat{s}_{gz}^2}\right]^2 \frac{\hat{s}_{gz}(1-\hat{s}_{gz})}{M_g} + \left[\frac{1}{\hat{s}_{gz}}\right]^2 \frac{\hat{p}(1-\hat{p})}{N_g},$$

where the covariance terms are zero because the two sources of sampling error are independent. This simplifies to:

$$Var(\hat{q}_{gz}) = \hat{q}_{gz}^2 \left[\frac{(1-\hat{s}_{gz})}{m_{gz}}\right] + \frac{\hat{q}_{gz}}{N_g}\left[\frac{1}{\hat{s}_{gz}} - \hat{q}_{gz}\right].$$

Filing probabilities at a more aggregated level are calculated as a weighted average of these rates, with weights based on $M_g$.

**APPENDIX 3** Deriving Population Litigation Probabilities and Marginal Effects

### Population Litigation Probabilities

We define classes by using characteristics with respect to which the sampling was nonrandom: USPC groups, cohort, infringement suits, and declaratory judgment suits. Let $P(X_c)$ denote the population probability of litigation for a patent in class $c$ with a vector of other characteristics $X_c$, and let $Q(X_c)$ be the corresponding probability in the pooled (litigated and matched) sample. $P(X_c)$ and $Q(X_c)$ differ because the matched sample was constructed so that the overall litigation probability is 50 percent, controlling for technology and cohort. We want to infer $P(X_c)$ from the estimated value of $Q(X_c)$.

First we determine the extent to which we must inflate the matched sample for a given class to have it reflect the number of unlitigated patents in that class in the population. Let $Q$ and $P$ represent the aggregate sample and population litigation probabilities for a given class:

$$Q = L/(L + M)$$

Where $L$ and $M$ denote the number of litigated and matched patents in the sample. The population probability is

$$P = L/N$$

The number of litigated patents is the same in both cases because the sample contains all (reported) litigated patents, and $N$ is the number of unlitigated patents in the class in the population. Using these equations, we get

$$N = \{Q/(1 - Q)P\}M = KM$$

Within a class, the matched patents are random draws so the distribution of characteristics in the matched sample is the same as the population. Thus the expected number of matched patents with characteristics $X_c$ in the population, $N(X_c)$, is greater than in the sample by the inflation factor, $K$, and so equals $KM(X_c)$. Letting $L(X_c)$ be the number of litigated patents with characteristics $X_c$, the expected population probability of litigation for such patents is

$$P(X_c) = L(X_c)/[KM(X_c)].$$

Similarly, $Q(X_c) = L(X_c)/[L(X_c) + M(X_c)]$. Solving for $M$ and substituting, we get the result:

$$P(X_c) = Q(X_c)/[K(1 - Q(X_c))] \qquad (A.3.1)$$

## Population Marginal Effects

For each characteristic $X_k$, the population marginal effect is

$$\partial P(X_c)/\partial X_{kc} = [dP(X_c)/dQ(X_c)]\, \partial Q(X_c)/\partial X_{kc}$$

The last term is the sample marginal effect computed from the probit regression. From the expression for $P(X_c)$ we get

$$dP(X_c)/dQ(X_c) = 1/K[1 - Q(X_c)]^2$$

Measuring $Q(X_c)$ by the sample probability of litigation in the class, $Q$, we get the result:

$$dP(X_c)/dQ(X_c) \approx P/Q(1 - Q)$$

We measure $P$ for each class as follows. For the denominator, we take the total number of patents in the class during 1978-1995. In the numerator we use the number of infringement or declaratory judgment suits that can be directly identified as such and include all others as infringement suits. These are inflated for underreporting and for truncation as described in Appendix 1. We then calculate marginal adjustment factors by USPC groups, infringement and declaratory judgment suits. Separate classes defined by cohort are not needed because of the maintained hypothesis that the litigation model applies to all cohorts, making nonsystematic sampling in this dimension unimportant. Results are at the bottom of Table 9. Because $dP(X_c)/dQ(X_c)$ is the same for all $X_k$ for a given class $c$, all sample marginal effects are adjusted by the same factor to convert them to population marginals.

# Patent Litigation in the U.S. Semiconductor Industry[1]

Rosemarie Ham Ziedonis
University of Michigan Business School

## INTRODUCTION

Firms in many industries utilize and build on the innovations of others, often in the face of short product life cycles. Recognizing this, scholars and industry representatives alike have started to question whether changes in the U.S. patent system over the past two decades are, in effect, hindering rather than promoting this cumulative process of innovation. Record numbers of patents are issuing from the U.S. Patent and Trademark Office (USPTO) in areas ranging from semiconductors and computer software to business methods and human gene sequences, raising concerns about the costs and feasibility of navigating through mazes of overlapping patent rights in these areas (Shapiro, 2001; Heller and Eisenberg, 1998). At the same time, the past two decades have witnessed a noticeable rise in patent litigation in the United States (Merz and Pace, 1994; Moore, 2000) as well as an escalation in the costs associated with enforcing patent rights in court (Ellis, 1999; AIPLA, 1999). Calling for reform, some have started to question whether the direct and indirect costs associated with obtaining and enforcing U.S. patent rights have started to outweigh the benefits provided by this system (Barton, 2000; Pooley, 2000; Mazzoleni and Nelson, 1998).

This chapter aims to shed additional light on the operation of the U.S. patent system by tracing the incidence and nature of patent-related legal disputes over the past three decades in one important cumulative technological setting—semiconductors. Much like software or computer firms, semiconductor firms typically require access to a "thicket" of external intellectual property to advance technol-

---

[1]This study received financial support from the GE Fund of the Wharton School's Reginald H. Jones Center for Management Policy, Strategy, and Organization. I gratefully acknowledge exceptional research assistance provided by Leslie Schafer, Yelena Slutskaya, and Owen Smith of Wharton. I also thank Jim Bessen, Wesley Cohen, Judge T. S. Ellis, III, Bronwyn Hall, Robert Merges, Stephen Merrill, Kimberly Moore, David Mowery, Cecil Quillen, Leslie Schafer, Deepak Somaya, Jim Walsh, Arvids Ziedonis, and participants in the October 2001 STEP Board conference for helpful comments and suggestions. All remaining errors and omissions are of course my own.

ogy or to legally manufacture and sell their products. In contrast to software, business methods, or biomedical inventions, however, innovation in semiconductors was already highly cumulative and subject to patent protection *prior to* the 1980s "pro-patent" shift in the United States.[2] For example, over 20,000 U.S. patents had been issued on inventions pertaining to semiconductor devices and manufacturing processes by 1981 (USPTO, 1995). In contrast, few software or biotechnology-related patents had been awarded before 1980 in part because of the legal uncertainty over patentable subject matter in these emerging areas (see Graham and Mowery, 2003 on software; Merges, 1997 on biotechnology-related inventions). The extent to which changes in the U.S. patent landscape during the 1980s have altered patterns of cooperation and conflict over patented technologies in semiconductors remains unclear.

The semiconductor industry is also an important empirical context within which to examine the broader incentives generated by the patent system in cumulative technological settings. In surveys on appropriability conducted in 1983 and 1994, (the "Yale" and "Carnegie Mellon" surveys, respectively), R&D managers in semiconductors consistently report that patents are among the *least* effective mechanisms for appropriating returns to R&D investments (Levin et al., 1987; Cohen et al., 2000).[3] Driven by a rapid pace of technological change and short product life cycles, semiconductor manufacturers tend to rely more heavily on lead time, secrecy, and manufacturing or design capabilities than patents to recoup investments in R&D.

However, in a recent study on patenting in semiconductors, Hall and Ziedonis (2001) find that the strengthening of patent protection in the United States in the 1980s had two divergent effects on dedicated U.S. semiconductor firms. On the

---

[2]Throughout this chapter, the term "pro-patent" refers to a series of legal reforms and rulings discussed in the second section that tilted the judicial treatment of patents in the United States more in favor of the patentee (see Merges, 1997 and Jaffe, 2000 for a review of related studies and empirical evidence). It is important to point out, however, that this term does *not* imply that the patent regime was "strengthened" in the sense of awarding patents more selectively or ensuring that only the rights of "stronger" patents are upheld. In fact, Quillen and Webster (2001) find that the USPTO has screened out a remarkably low percentage of patent applications since the early 1980s (as little as 5-10 percent). Others emphasize that the Federal Circuit's interpretation of the nonobviousness standard has effectively "lowered the bar" of patentability (see, e.g., Quillen, 1993 and Hunt, 1999) and, in doing so, has generated additional uncertainty in the enforcement process (as discussed by Lunney, 2001).

[3]The 1994 Carnegie Mellon Survey on Industrial R&D in the U.S. Manufacturing Sector (Cohen et al., 2000) updated and extended the influential "Yale" survey conducted in 1983 (Levin et al., 1987). Respondents in both surveys were R&D lab managers in a variety of "focus industries." Both surveys found that R&D managers in only a handful of industries, including pharmaceuticals, chemicals, and (more recently) biotechnology and medical devices, considered patents to be an effective mechanism by which to appropriate the returns to R&D. These results echo the findings of Scherer et al. (1959), Taylor and Silberston (1973), and Mansfield (1986). As discussed below, the Carnegie Mellon survey extends upon the Yale survey by asking questions on why firms seek patent protection.

one hand, the "pro-patent" shift induced capital-intensive firms to "ramp up" their portfolios of patents more aggressively—largely to deter threats of litigation and to improve their bargaining positions in negotiations with external patent owners.[4] On the other hand, the strengthening of U.S. patent rights also appeared to facilitate entry into the industry by firms specializing in chip design. Interviews with representatives from design firms suggest that these firms (often relatively small in size) enforce their patent rights quite aggressively in court vis-à-vis direct rivals, primarily to establish proprietary rights in niche product markets.[5] If this is true, any apparent increase in patent litigation within this sector may simply reflect the emergence of these specialized firms and their reliance on U.S. courts to bar use of their intellectual assets. Combined, these findings underscore the importance of considering the multifaceted effects of the U.S. patent system even among firms within an industry.

With this in mind, this study seeks to address several basic empirical questions. How do the characteristics of semiconductor firms involved in legal patent disputes over the past three decades compare with those of nonlitigating semiconductor firms? To what extent have firms in this sector been involved in more legal disputes over intellectual property during the so-called "pro-patent" era? Is patent litigation in this industry still fairly "uncommon" as sometimes claimed? Finally, do semiconductor design firms and manufacturers differ in their propensity to enforce patents or in the characteristics of their patent-related legal disputes?

To address these questions, the study examines the characteristics of patent cases filed in U.S. District Courts and the U.S. International Trade Commission (USITC) from January 1, 1973, through June 30, 2001, that involve 136 dedicated U.S. semiconductor firms as plaintiff, defendant, or owner of a litigated patent. Firms in the sample include the universe of publicly traded U.S. firms during 1973-2000 that either (a) list semiconductors and related devices (SIC3674) as their primary line of business or (b) were identified by industry sources as dedicated U.S. semiconductor firms. In 2000, sample firms collectively generated over $88 billion in revenues, spent $12 billion in R&D, and had been awarded roughly 31,000 U.S. patents. An unfortunate weakness of this approach is the exclusion from the sample of large U.S. "systems" manufacturers (e.g., IBM,

---

[4]Cohen et al. (2000) report similar findings based on responses to the Carnegie Mellon survey. In industries characterized by "cumulative" (or "complex") innovation, respondents consistently reported that prevention of lawsuits and blocking of patenting by others were among the most important reasons for patenting.

[5]This information is based on a series of structured interviews conducted in 1998 with intellectual property managers and executives from seven U.S. semiconductor firms (three specialized design firms and four dedicated manufacturers). Although this small sample of firms is not representative of the industry as a whole, consistent views emerged in roughly 30 informal interviews with outside legal counsel and others involved in managing and evaluating intellectual property within this industry (see Ziedonis, 2000).

AT&T, or Motorola) and non-U.S. firms (e.g., Mitsubishi, Samsung, or Siemens). Although these firms are important patent owners and users of semiconductor technologies, it is not possible to isolate the share of R&D expenditures directed toward semiconductor technologies for these diversified firms.

This approach offers several methodological advantages. First, it enables the identification of a fairly large sample of firms whose R&D expenditures are primarily directed toward semiconductor-related innovation, regardless of whether the firms are involved in legal disputes over patents. By focusing on the patent acquisition and enforcement histories at the level of individual firms, it is possible to examine changes in the propensity of firms to enforce their own patents while also observing changes in the propensity of firms of different sizes and types to encounter patent lawsuits initiated by others. The sample also includes a mix of U.S. semiconductor manufacturers and design firms, including 81 "manufacturers" (i.e., firms like Intel, Texas Instruments, and Micron Technologies, which design *and* manufacture the majority of their products in-house) and 55 "design" firms (i.e., firms like Altera, Xilinx, and SonicBlue, which specialize in chip design but contract out the manufacture of products to third parties).[6] Even though most of the design firms in the sample commercialize and sell products of their own, they are typically much smaller in size (in terms of number of employees or sales revenues) than semiconductor manufacturers in the sample and they invest more heavily in R&D. The results of this study may therefore help inform the underlying factors driving patterns of litigation involving small firms—at least in this sector.

Finally, comparing patent litigation and patent issuance trends yields "litigation rates" that are somewhat difficult to interpret in the context of the semiconductor industry. Typically, patent litigation rates are calculated by denominating the number of filed patent cases with the number of patents "at risk" for litigation (Lerner, 1995; Lanjouw and Schankerman, 2001, 2003; Somaya, 2003). Yet, as mentioned above, the decision to patent for many semiconductor firms is, in fact, driven by a desire to deter litigation (Hall and Ziedonis, 2001; Cohen et al., 2000). This methodology enables me to offer a different perspective by calculating liti-

---

[6]Most of the design (or "fabless") firms in the sample commercialize products based on their designs (e.g., Altera and Xilinx in programmable logic devices). Toward the end of the sample period, so-called "chipless" firms entered the industry that specialize in chip design but license out their designs for others to embed into end products (see Arora et al., 2001; Linden and Somaya, 2000). Rambus, a company specializing in interconnection technologies that speed communications between memory chips and microprocessors, is a prominent example of such a company. Because these "chipless" firms are a recent phenomenon and represent less than five percent of the design firms in the sample, the litigation trends and practices discussed in this chapter refer primarily to those involving the former category of design firms (i.e., those that compete directly in semiconductor product markets). Distinguishing between the patent acquisition and enforcement strategies of "traditional" design firms and those of "chipless" firms is an interesting topic for future research.

gation rates in the industry (overall, and for manufacturers and design firms separately) using (a) firm-level patenting activity and (b) firm-level R&D expenditures.

Detailed information about all reported patent cases involving one or more of the 136 semiconductor firms in the sample is merged with information about the patents and other parties involved in the disputes. Several empirical patterns emerge, which are summarized as follows:

1. Of the 136 U.S. semiconductor firms in the sample, roughly 56 percent were involved in at least one reported patent case filed in U.S. District Courts and the USITC between January 1, 1973, and June 30, 2001. On average, sample firms involved in patent lawsuits spent more on R&D (in absolute terms and per employee), were larger (in terms of sales or number of employees), and owned more patents than "peer" semiconductor firms not involved in patent litigation during this period.

2. The number of annual cases filed that involve these firms (as a group) increased sharply around the mid-1980s and continued at a higher level throughout the 1990s. This trend is not remarkable when compared to the overall growth in U.S. patent litigation during this period documented elsewhere (Merz and Pace, 1994; Moore, 2000). It is consistent, however, with popular reports that legal disputes over intellectual property have become more common in semiconductors—despite the widespread use of cross-licenses in this industry.

3. Relative to annual R&D spending by these firms, the patent litigation rate in semiconductors rose considerably during 1986-2000 from that in the preceding decade (by as much as 93 percent). In contrast, the number of cases filed per 1,000 patents awarded to these firms (a more common metric used to estimate litigation rates) exhibited a slight decline between the two periods. The apparent decline in litigated patents per patents awarded during the latter period is driven, however, by the dramatic rise in patenting by semiconductor firms since the mid-1980s (as reported in Hall and Ziedonis, 2001). Indeed, updating the trends reported in Hall and Ziedonis (2001) reveals that the "patent portfolio races" of U.S. semiconductor manufacturers continued to accelerate through the end of the 1990s both in absolute terms and relative to firm-level R&D spending.

4. Regardless of how it is measured, the average litigation rate of specialized design firms in the sample is high and is more than twice that of manufacturers in the sample. Manufacturers, on average, are involved in disputes with a more disparate set of parties and tend to enforce patents that are almost 4 years older than the average patent in their portfolios. In contrast, design firms typically enforce their patents against other design firms and litigate over patents that are roughly the same age as the average patent in their portfolios.

In addition to these general trends, it was also interesting to observe what appears to be an active "market" for intellectual property that predates the filing of a patent lawsuit. In at least 30 percent of identified cases, legal title to a litigated patent had been reassigned from the original inventor (or assignee) to one of the litigating parties—typically, to the plaintiff in an infringement suit. Some of these disputes involved plaintiffs that had acquired the intellectual property as part of a broader acquisition of a firm or its physical assets (e.g., SGS Thomson successfully enforced Mostek's memory chip patents after acquiring the company in the mid-1980s; similarly, Atmel enforced patents awarded to Seeq Technologies after acquiring Seeq's intellectual and physical assets pertaining to nonvolatile memory). In other cases, the plaintiff appeared to be using externally generated patents in reciprocal suits. For example, after failing to reach agreement on the terms of a renewed cross-license agreement, Hyundai sued Texas Instruments in 1991 for infringing five patents—four of which Hyundai had purchased from outside inventors. There also was an apparent rise in infringement suits brought by specialized "patent licensing" companies.[7] On one hand, the emergence of specialized patent management and enforcement companies may help "tilt the table" more in favor of independent inventors or patentees from small businesses (see Lerner, 1995; Lanjouw and Lerner, 2001) or may represent the continued development of markets for technology (Arora et al., 2001). On the other hand, others raise concerns that an increased trade in and enforcement of so-called "paper patents" (i.e., "blocking" patents owned by inventors or companies that do not compete in the related product markets) is imposing an implicit tax on innovation (Pooley, 2000). These are interesting issues that warrant future investigation.

Before turning to the rest of the chapter, it is important to acknowledge two inherent limitations of this research. First, like any study of litigation events, this study is inherently limited by its examination of the proverbial "iceberg's tip."[8]

---

[7]Some of these firms, for example, General Patent Corporation (or its affiliated IP Holdings LLC), specialize in the management and enforcement of patents on a contingency fee basis. Arora et al. (2001) refer to such companies as "technology intermediaries" and discuss their services in more detail (p. 84-85).

[8]Because most companies treat information about licensing negotiations and agreements as highly confidential information, it is unusual to observe directly how the "iceberg's tip" (in this case, patent-related disputes that involve the filing of a lawsuit) compares with the underlying set of disputes that are settled privately and in the absence of a case being filed with the courts. To my knowledge, the most direct evidence on this point is from a 1994 survey of patent and licensing practices of British and Japanese firms that asked intellectual property managers to estimate the number of complaints of infringement made against the company and to indicate how the complaints were resolved (Pitkethly, 1996). Overall, British managers estimated that 87 percent of the disputes they encounter over alleged patent infringement are settled privately—without a lawsuit being filed. Similarly, Japanese managers estimated that they resolved around 80 percent of infringement complaints either by simple notification or by private negotiations (as reported in Pitkethly, 1996, data table 57; based on responses from 50 British and 120 Japanese managers). More recently and in the context of the United States, Lanjouw and Schankerman (2001) examine this issue indirectly by comparing the characteristics of litigated and nonlitigated patents in the United States. As discussed below, they find that most patents issued

As discussed above, semiconductor firms have long licensed and cross-licensed their intellectual property and private settlement over intellectual property rights is still the rule rather than the exception (Grindley and Teece, 1997). Although future versions of this research will attempt to control more explicitly for this selection bias, this study simply summarizes the characteristics of observable case filings. A second limitation of this study is its primary reliance on data that may underestimate the number of patent cases filed in U.S. District Courts, particularly during the early period of the study (i.e., before 1984). As discussed in the third section of this chapter, several attempts were made to address this shortcoming in the data. Nonetheless, underreporting in the early period may still exist, and the results should be interpreted with this in mind.

The remainder of this chapter is organized as follows. The second section summarizes what is often referred to as the "pro-patent" shift in the U.S. legal environment during the 1980s and discusses its effects on the use of patents in the semiconductor industry. The third section presents the data and methodology used to trace patent litigation involving U.S. semiconductor firms during 1973-2001. The descriptive findings are summarized in the fourth section, and concluding remarks follow.

## THE CHANGING PATENT LANDSCAPE

The patent system has long been recognized as an important policy instrument used to promote innovation and technological progress. Two fundamental mechanisms underpin the patent system. First, an inventor discloses to the public a "novel," "useful," and "nonobvious" invention. In return, the inventor receives the right to exclude others from using that patented invention for a fixed period of time (now 20 years from the date of patent application in the United States). The rules of the patent game may differ from country to country (e.g., whether rights are assigned to the first inventor or the first to file the patent application), but the underlying principle remains the same. By providing exclusionary rights for some period of time and a more conducive environment in which to recoup R&D investments, the patent system aims to encourage inventors to direct more of their resources toward R&D than would otherwise be the case. At the same time, detailed information about the invention is disclosed to the public when the patent application is published.

---

in the United States are never involved in litigated disputes; on average, however, litigated patents tend to be more valuable than nonlitigated patents and are more likely to form the basis for a sequence of innovations by the patentee. Lanjouw and Schankerman (2001, 2003) discuss these selection issues in greater detail. See also Siegelman and Donohue (1990) and Siegelman and Waldfogel (1999) on the selection biases inherent in studies of litigation events more generally.

## The So-Called Pro-Patent Shift in the United States

The creation of the Court of Appeals for the Federal Circuit (CAFC) in 1982 is often credited with ushering in an era that reversed the judicial treatment of patent rights in the United States from the preceding decades in ways that favored patent owners.[9] From the trust-busting era of the 1930s through much of the 1970s, patents were largely viewed as anticompetitive weapons used to stifle competition.[10] For example, in 1959, Scherer and his co-authors report:

> During the past two decades a pronounced change has taken place in the policies of governmental bodies towards patents owned by corporations.... The courts have become increasingly critical of patent validity, and cases in which the exercise of patent rights conflicted with antitrust statutes have been prosecuted by denying the exclusiveness of the patent grant. Since 1941, more than 100 judgments have been entered which required corporations to license their patents to all applicants at reasonable royalties or no royalties at all. This trend was brought sharply to the public's attention in January of 1956 when two of the nation's foremost leaders in industrial technology, the American Telephone and Telegraph Co. and International Business Machines, Inc., entered into decrees requiring them to license all of their more than 9,000 patents, in most cases without receiving royalties in return." (Scherer et al., 1959, p. 2-3)

Antipatent sentiment continued through much of the 1970s. As Merges (1997) states: "It was difficult to get a patent upheld in many federal circuit courts, and the circuits diverged widely both as to the doctrine and basic attitudes toward patents. As a consequence, industry downplayed the significance of patents" (p. 12).

By the early 1980s, the pendulum started to swing away from a restrictive treatment of patents toward a view that patent rights should be construed liberally to stimulate innovation. Driven by general concerns about increased international competition in several key industries—including semiconductors—and a growing belief that stronger intellectual property rights were needed to stimulate innovation, Congress passed a series of laws in the early 1980s aimed at improving the function of the U.S. patent system and at relaxing antitrust constraints on the

---

[9]Although governmental agencies (in the United States, the USPTO) examine applications and decide whether an invention qualifies for patent protection, the courts ultimately determine the strength of patent rights once granted. By deciding whether a patent is valid or whether another party has infringed on the patent owner's rights, courts play a pivotal role in determining the strength (and hence, the value) of patent rights.

[10]As Merges (1997) states: "Unfortunately for the patent system, the identification of patents with big business meant that when big business lost favor, so too would patents.... The exclusive nature of the patent grant, coupled with the actual market power that some patents conferred on their holders, seemed closely related to many of the monopolists' oppressive practices" (p. 11).

collaborative R&D activities of firms.[11] Unique to semiconductors, the 1984 Semiconductor Chip Protection Act (SCPA) also conferred protection against theft of the "mask works," the overall layout of the chip designs (see Samuelson and Schotchmer, 2001 for a recent review of this form of protection).

No other event signaled the shift toward stronger legal protection for patents in the United States than the 1982 creation of the CAFC, a centralized appellate court with jurisdiction over all patent infringement appeals (Jaffe, 2000).[12] Although the driving force behind the legal reform was a need to unify U.S. patent doctrine, the Federal Circuit put in place a number of procedural and substantive rules that collectively favored patent owners. For example, the new court endorsed the broad, exclusionary rights of patent owners through its interpretation of patent scope, increased evidentiary standards to make it more difficult to invalidate the rights of patent owners, was more willing to halt allegedly infringing actions early in the dispute process by granting preliminary injunctions, and was more willing to sustain large damage awards and thereby penalize infringing parties more severely.[13] The plaintiff success rates in patent infringement suits also increased substantially during this period (Lerner, 1995).

Although the CAFC was created in 1982 and issued a flurry of written opinions during 1983 (Adelman 1987; Nies 1993), the impact of the CAFC on the favorable legal treatment of patent rights in U.S. courts was not widely publicized until the mid-1980s.[14] The "surprising new power of patents" was perhaps most clearly revealed by Polaroid's success in a longstanding lawsuit against Kodak for infringing certain instant photography patents awarded to Polaroid. In a 1985

---

[11]For example, the 1984 National Cooperative Research Act reduced the antitrust penalties for collaboration among firms in "pre-commercial" research, which paved the way for the subsequent formation of SEMATECH, a large ongoing research consortium in the semiconductor industry. See also Kortum and Lerner (1998) on the legislative initiatives aimed at improving the operation of the USPTO: "More patent legislation was enacted between 1980 and 1982 than had passed in the previous two decades" (p. 7).

[12]Until 1982, patent appeals were primarily heard in the court of appeals of the district in which the case was tried, which led to "forum shopping" among firms (Kortum and Lerner, 1998). Adelman (1987), however, argues that the 1982 establishment of the new court represented an "outgrowth of the dissatisfaction with the functioning of both the Supreme Court and the federal appellate courts" and a "realization by Congress that a uniform and more reliable patent system was necessary for sustained economic growth and to rise to the challenge of Japanese and German industrial competition" (p. 983).

[13]Merges (1997) and Lanjouw and Lerner (2001) provide additional information on these and other effects of the court. The main point, however, is that many of the rules and decisions of the CAFC during this period favored the rights of patent owners.

[14]A series of articles surfaced in the popular press during 1985-1986 that proclaimed the "new" legal environment for patent owners. See, for example, "A Change in the Legal Climate," *Forbes*, Oct. 7, 1985, p. 41; "A Weapon at Last [pro-patent decisions]," *Forbes*, Mar. 10, 1986, p. 46; and "The Surprising New Power of Patents," *Fortune*, June 23, 1986, p. 57.

ruling, Kodak was required to pay Polaroid almost $1 billion in damages and interest, was barred from manufacturing and selling instant cameras, and was forced to close its instant camera production line (Warshofsky, 1994). As reported by Hall and Ziedonis (2001), representatives from the semiconductor industry emphasized the importance of this case, along with Texas Instruments' successful patent infringement cases against Japanese and Korean firms during 1985-1986, in demonstrating the "new power of patents." Not only did TI and other large patent owners such as IBM and Motorola increase the price (i.e., royalty rates) charged for "rights to use" their patents, but the increased value associated with patents may have induced entry into the patent licensing business as the licensure of patents became more profitable under the new regime (as discussed below).

### The Evolving Role of Patents in Semiconductors

Not surprisingly, the use and importance of U.S. patents in semiconductors was affected by this changing patent landscape, albeit in some unanticipated ways. By the early 1980s, a broad range of semiconductor technologies, including methods for manufacturing semiconductors and integrated circuit design, had diffused widely across the industry (Levin, 1982). The "technological giants" in semiconductors, including AT&T and IBM, were effectively curtailed from aggressively enforcing their patent rights against rival firms (either merchant manufacturers or other users of semiconductor technologies) from the 1950s to the late 1970s by the antitrust constraints discussed above. As a result of its 1956 consent decree with U.S. antitrust authorities, for example, AT&T had licensed its semiconductor inventions widely to other firms in return for access to subsequent inventions by licensees. AT&T's active role in licensing and disseminating semiconductor technologies is credited with stimulating the early growth of the U.S. merchant semiconductor industry (Tilton, 1971; Grindley and Teece, 1997; Mowery and Rosenberg, 1998). Nonetheless, Tilton (1971, p. 76) concludes:

> Certainly the great probability that other firms were going to use the new technology with or without licenses is another reason for the liberal licensing policy. Secrecy is difficult to maintain in the semiconductor field because of the great mobility of scientists and engineers and their desire to publish. Moreover, semiconductor firms, particularly the new, small ones, have demonstrated over and over again their disposition to infringe on patents. The prospect of lengthy and costly litigation in which its patents might be overturned could not have been very attractive to AT&T.

Similarly, Von Hippel (1988) emphasizes that the semiconductor field was a very fast-moving one that, even by the early 1980s, contained many unexpired patents with closely related subject matter and claims. He writes:

> Since patents challenged in court are unlikely to be held valid, the result of high likelihood of infringement accompanying use of one's own patented—or unpat-

ented—technology is not paralysis of the field. Rather, firms in most instances simply ignore the possibility that their activities might be infringing the patents of others. The result is what Taylor and Silberston's interviewees in the electronics components field termed 'a jungle' and what one of my interviewees termed a 'Mexican standoff'.... The usual result is cross-licensing, with a modest fee possibly being paid by one side or the other. (p. 52-53)[15]

Two other factors contributed to the infrequent patent litigation and widespread cross-licensing that have historically characterized the semiconductor industry. First, as reported in the introduction to this chapter, semiconductor firms tend to rely on mechanisms other than patents to recoup their R&D investments, including being first to market and safeguarding the "know-how" (often through secrecy) required to manufacture commercially viable chips (Levin et al., 1987; Cohen et al., 2000). Indeed, there is little evidence that the strengthening of U.S. patent rights boosted aggregate R&D spending by firms in this sector (Hall and Ziedonis, 2001; Bessen and Maskin, 2000). Second, to reduce the risk of disruptions in supply, large customers of chips (e.g., IBM and the U.S. government) typically required dedicated manufacturers to transfer to a competing supplier the know-how and patent rights required to manufacture a compatible product (Shepard, 1987). These second source agreements further promoted cross-licensing in the industry but declined in use over the decade of the 1980s as the industry matured and built up capacity (Grindley and Teece, 1997).

Although cross-licensing continues to be an important mechanism by which firms trade access to one another's patents, the terms of these agreements appear to have changed (not surprisingly) as the rights of patent owners have grown stronger. Firms with large patent portfolios, such as Texas Instruments, IBM, AT&T, and Motorola, adopted a more aggressive licensing and litigation strategy to profit directly from their patent portfolios—both by seeking licenses from a larger number of firms and by increasing royalty rates on use of their inventions. For example, in the early 1980s, Texas Instruments launched a more aggressive patent licensing program—initially against Japanese and Korean competitors in the market for memory chips. During 1986-1993, TI earned almost $2 billion from licensing rights to its semiconductor patents (Grindley and Teece, 1997). Similarly, IBM increased its royalty rates around 1987-1988 from 1 percent of sales revenues for products using IBM patents up to a range of 1 to 5 percent (Shinal, 1988). By 2000, IBM earned over $1.5 billion in income from licensing its intellectual property portfolio, up from $646 million in 1995.[16] According to industry representatives as well as accounts in the general business press (e.g., Warshofsky, 1994; Rivette and Kline, 2000), the increased value associated with

---

[15]Similarly, writing in 1987, Levin et al. report: "In the semiconductor industry ... the cumulative nature of technology makes it difficult to participate legally without access to the patents of numerous firms. In consequence, there is widespread cross-licensing" (p. 798, fn 29).

[16]As reported in IBM annual reports, available at www.ibm.com.

patents also induced entry into the more lucrative patent licensing business by firms and individuals that had not found it worth their while to assert legal rights under the previous patent regime. As one industry representative put it, around the mid-1980s "we started receiving more knocks on the door and letters threatening infringement suits [from patent owners seeking royalty payments]."

As the effective price of purchasing legal "rights to use" patented semiconductor technologies rose during the 1980s, so too did the capital investments required to build and operate state-of-the-art manufacturing facilities. In the early 1980s, a wafer fabrication facility (fab) cost about $100 million and had an expected life span of 10 years. By the mid-1990s, however, the cost of a new fab had risen to over $1 billion, while the useful life of these capital investments had been reduced to little more than 5 years (ICE, 1995). As a result, the costs associated with halting production or altering production processes used in high-volume facilities had risen significantly, exacerbating concerns among capital-intensive firms of being "held up" by owners of patented technologies used in the design or manufacture of their products. Indeed, Hall and Ziedonis (2001) find a sharp increase in patenting rates by capital-intensive semiconductor firms around the mid-1980s. Instead of being driven by a desire to win strong legal rights to a stand-alone technological prize, these firms appear to be engaged in "patent portfolio races" aimed at reducing concerns about being held up by external patent owners and at negotiating access to external technologies on more favorable terms. In principle, such racing behavior is not an inevitable outcome of strengthening patent rights in cumulative technological areas; if patents were strictly awarded to inventors of "nonobvious," "useful," and "novel" inventions, it should become increasingly difficult to obtain a patent when a thicket of prior art exists. In line with the more general findings of Quillen and Webster (2001), however, it does not appear that the USPTO is successfully "weeding out" marginal patent applications in this sector.

Although the "pro-patent" shift induced capital-intensive semiconductor firms to amass larger portfolios of patents for trading purposes, Hall and Ziedonis (2001) also find that the strengthening of U.S. patent rights may have facilitated entry by firms specializing in chip design. In the early 1980s, the U.S. semiconductor industry comprised two main types of firms: (1) vertically integrated "systems" manufacturers (e.g., AT&T, Motorola, or IBM) that manufactured semiconductors primarily for in-house use, and (2) less diversified "merchant" manufacturers (e.g., Analog Devices, Intel, or National Semiconductor) that specialized in designing, making, and selling semiconductor products. Since the mid-1980s, however, there has been a considerable increase in entry by specialized design firms (ICE, 1995; Macher et al., 1998).[17] As discussed above, these so-

---

[17] As Macher et al. (1998) explain: "The diffusion of MOS [metal-oxide semiconductor] production technology facilitated the division of labor between device designers in fabless [i.e., design] firms, who were able to operate within relatively stable design rules, and foundries, who were able to incre-

called fabless firms design semiconductor components (e.g., graphics, communications, or networking chips) but rely on third parties, or "foundries," to manufacture their designs. Competing primarily on the basis of innovative products or functional designs, these firms appear to rely more heavily on patents to profit from innovation than appears to be true of the semiconductor manufacturing firms represented in the Yale and Carnegie Mellon surveys mentioned above. Consistent with this view, Hall and Ziedonis (2001) find that the "bargaining chip" role of patents was less apparent in interviews with representatives from design firms. Interviewees typically emphasized the importance to their firms of securing strong, "bulletproof" patents in areas surrounding their core product lines and of signaling to potential rivals the firm's commitment to protecting its intellectual property in court. Although some design firms register mask works with the U.S. Copyright Office (under the SCPA discussed above), interviews with industry representatives suggest that the lion's share of their intellectual property-related financial and managerial attention is devoted toward protecting inventions with patents (Ziedonis, 2000).

## DATA

To examine changes, if any, in the incidence and nature of legal disputes over patents in the semiconductor industry during the "pro-patent" era, one first must establish an appropriate sample of firms and compile information about their patent-related disputes. Ideally, one would use information about threats of litigation and the terms of intellectual property-related settlements over time, but such data are not publicly available. This chapter therefore relies on information contained in patent cases filed in U.S. courts, which (as the title of Lanjouw and Schankerman's 2001 paper suggests) provide a useful window through which to view competition and conflict over intellectual property. After describing the sample of semiconductor firms selected for study, this section identifies the main sources used to compile information about (1) sample firms, (2) their involvement in patent disputes filed in U.S. courts since 1973, and (3) the characteristics of the parties and technologies involved in those disputes.

---

mentally improve their process technologies to accommodate a succession of new device designs." These contracts with suppliers of manufacturing services also may have alleviated the need for design firms to negotiate separate licenses with large owners of semiconductor patents, particularly in areas pertaining to process technologies. For example, unless prohibited from doing so in cross-licensing agreements, foundries could "resell" rights to third-party patents in purchase agreements for manufacturing services (see *Intel v. ULSI Technology, Inc.* 995 F. 2d 1566, 1567; Fed. Cir. 1993). According to conversations with industry representatives, other foundry providers, such as IBM, reportedly charge a higher price for their manufacturing services in return for indemnifying firms from claims of infringement of patents used in the manufacture of their products.

## Sample Selection

The sample of semiconductor firms is drawn from two main sources. A universe of 108 publicly traded U.S-owned firms identified their principal line of business as semiconductors and related devices (SIC3674) and reported financial data in Compustat in at least one year during 1973 and 2000. Of these, nine firms were dropped from the sample because they were partially owned subsidiaries of more diversified firms. An additional set of 37 dedicated U.S. semiconductor firms was identified with annual reports from Integrated Circuit Engineering, Inc. (ICE)—a market research firm that tracks the commercial activities of semiconductor firms (ICE, 1976-1998). Most firms added to the sample from the ICE reports were specialized design firms assigned to nearby four-digit SIC classes (e.g., pertaining to storage, telecommunications, and other electronics).

The final sample includes 136 dedicated U.S. semiconductor firms that were publicly traded in the United States for one or more years during 1973-2000. Using sources discussed in Hall and Ziedonis (2001), I updated and assembled financial information and patenting data for the 96 firms in the original Hall-Ziedonis sample and assembled corresponding data for the 41 newer firms for which sufficient data were now available.[18] The resulting database contains, for all 136 firms, the following information:

- the number of U.S. patents awarded to each firm and its subsidiaries from 1965 to 2000;[19]
- detailed characteristics of those patents (e.g., the patent number and class);
- annual balance sheet and income statement data for each firm through 2000 (e.g., sales, R&D spending, number of employees);
- the founding year of the firm; and
- annual information about whether the firm owned and operated its own manufacturing facilities (manufacturer) or whether it specialized in product design alone (design firms).[20]

Summary statistics for these variables are shown in Table 1. The median firm in the sample is 17 years old (in 2000), has 420 employees, spends $7.6 million (in constant 1996 dollars) on R&D, and receives one U.S. patent a year. The distribu-

---

[18]Sources include Micropatent and the NBER/Case Western databases (for patent data); Compustat (for financial information); ICE industry reports (for manufacturing information and founding years); and annual 10-K filings and LEXIS/NEXIS business directories (to confirm manufacturing status, founding years, and ownership structures). Detailed information about these sources is provided in the on-line version of Hall and Ziedonis (2001), available at: http://jonescenter.wharton.upenn.edu/papers/2000/wp00-16.pdf.

[19]Patent portfolios were constructed based on each firm's 1996 ownership structure.

[20]Following ICE industry status reports, a firm was designated as "design" if more than 50 percent of its products sold in a given year were manufactured by third parties.

**TABLE 1** Sample Statistics U.S. Semiconductor Firms, 1973-2000

|  | All Firms (1725 Observations; 136 Firms) | | | | |
|---|---|---|---|---|---|
| Variable | Mean | Std. Dev. | Median | Min | Max |
| Age (2000-founding year) | 22.08 | 12.20 | 17.00 | 6 | 64 |
| D(Founded before 1982=1) | 0.49 | 0.50 | 0.00 | 0 | 1 |
| D(Manu=1) | 0.59 | 0.49 | 1.00 | 0 | 1 |
| Sales (Constant 1996 $M) | 368.31 | 1,844.78 | 45.30 | 0.00 | 36,056.47 |
| Employees (1,000s) | 3.10 | 10.46 | 0.42 | 0.01 | 89.88 |
| Prop, Plant & Equip (Constant 1996 $) | 271.11 | 1,440.95 | 19.45 | 0.00 | 30,205.28 |
| R&D (Constant 1996 $M) | 51.46 | 201.11 | 7.60 | 0.00 | 3,747.09 |
| R&D Intensity (R&D/Employee) | 26.81 | 31.26 | 15.88 | 0.00 | 295.29 |
| R&D Intensity (R&D/Total Assets) | 0.15 | 0.19 | 0.11 | 0.00 | 2.95 |
| # Annual US Patents Received | 14.95 | 79.16 | 1 | 0 | 1,463 |

tion of these variables is, however, highly skewed. One firm was awarded 1,463 U.S. patents in a single year (Texas Instruments in 1998), whereas another spent over $3.7 billion in R&D in one year (Intel in 2000; again, based in 1996 dollars).

Many firms enter and exit the sample over the 29-year period.[21] As seen in Figure 1, the annual number of firms in the sample grew from 18 to 119 between 1973 and 1994, primarily because of two waves of entry by design firms in the

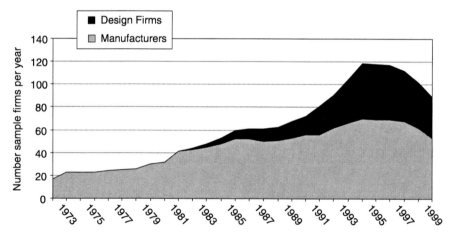

**FIGURE 1** Number of sample firms by year (U.S. semiconductor manufacturers and design firms.

---

[21] The resulting database, which includes 1,728 firm-year observations, is therefore an unbalanced panel. The list of firms included in this study is available from the author on request.

mid-1980s and early 1990s. The number of sample firms fell to 91 by 2000, however, as an economic downturn forced consolidation in the industry. As discussed above, design firms in the sample are more R&D intensive than sample manufacturers. More specifically, design firms spent on average 19-20 percent of their revenues on R&D during 1985-2000, whereas the R&D intensity of manufacturers hovered around 12 percent during the same period (see Figure 2). Deflating R&D spending by number of employees (instead of revenues) reveals similar trends.

## Linking Firms to U.S. Patent Cases

As mentioned in the introduction to this chapter, the sample includes a large number of firms whose R&D investments are primarily directed toward semiconductor-related innovation. In the next phase of this research, I hope to investigate the economic effects, if any, of patent-related litigation on the R&D and patenting activities of firms in the industry. With this longer-term objective in mind, it is important to identify all reported patent disputes that involve each firm in the sample—regardless of whether the patents involved pertained to semiconductors or to broader classes of inventions.[22] I therefore relied on sources that identify the

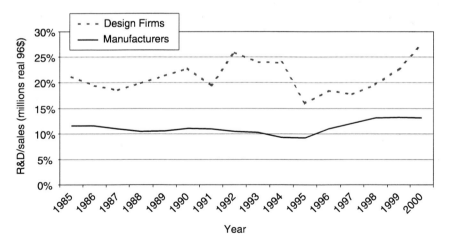

**FIGURE 2** Average annual R&D intensity of sample firms: design firms v. manufacturers, 1985-2000.

names of all litigating parties involved in patent cases filed with either the USITC or U.S. District Courts. For cases filed with the USITC, I reviewed the universe of 464 investigations filed under Section 337 between January 1973 and June 2001, as reported on the USITC website (http://info.usitc/gov/337).[23] From this search, I identified 28 cases that involved semiconductor firms as either plaintiffs or defendants. Most of these cases also identified (by U.S. patent number) the inventions involved in the dispute.

For patent cases filed in U.S. District Courts, I retrieved information about the litigating parties and patents from the *LitAlert* database produced by Derwent.[24] This private database, which is available on Westlaw, reports information about U.S. patent and trademark suits filed in U.S. District Courts that are reported to the USPTO Commissioner. The database includes cases filed (and reported to the USPTO) from January 1973 to date and is updated weekly (see http://www.derwent.com/data/specs/lita.pdf). To identify patent lawsuits that involved firms in my sample, I searched the *LitAlert* database using the following search terms for each firm: (1) its name; (2) common misspellings of that name; (3) former names attributed to the firm (e.g., as cross-listed in *Hoover's Business Directory* or reported in industry reports); and (4) names of major subsidiaries, if any. A list of 318 patent cases were identified that involved one or more of the semiconductor firms in my sample as plaintiff, defendant, or patent assignee and that had been filed in U.S. District Courts between January 1973 and June 30, 2001. Two duplicative records (where identical information was filed under different *LitAlert* case identification numbers) and 57 sequential cases (where a change in venue or an outcome of a previously filed case was announced but the

---

[22]A more common approach, followed by Lanjouw and Schankerman (2001, 2003) and Somaya (2003), is to focus on the litigated patent as the unit of analysis and to define technological areas (e.g., "semiconductors" or "electronics") according to the U.S. patent classification scheme. Although suitable for cross-industry studies, such an approach would jeopardize the ability to identify lawsuits involving firms that fell into non-semiconductor-related classes (e.g., Lemelson's lawsuits against manufacturers for use of machine vision technologies). Similarly, I sought to examine the characteristics of semiconductor firms that have *not* been involved in patent lawsuits filed in the United States over the past three decades. Searching litigation records and databases by firm name instead of technology class helps rule out the possibility that some firms were simply involved in litigation over a broader range of inventions. More similar to my approach is a study by Farn (1996), which traces firm-level patterns of patent acquisition and enforcement for eight semiconductor firms and eight computer companies during 1986-1995.

[23]Under Section 337 of the Tariff Act of 1970, a firm can challenge the importation of products that infringe its U.S. patent rights. See Mutti (1993) and Mutti and Yeung (1996) for empirical analyses of cases brought before this alternative U.S. forum for intellectual property-related disputes.

[24]Lanjouw and Schankerman (2001, 2003) use *LitAlert* to examine the factors driving patent-related conflict and settlement across a broad range of industries. Similarly, Somaya (2003) uses information from *LitAlert* to examine the probabilities of suing and settling patent-related disputes in computers and research medicines.

patents and litigated parties involved in the lawsuit were the same) were omitted from the sample.

In total, I identified 287 unique patent cases filed from January 1973 through June 2001 that involved at least one sample firm. Of these, 259 cases were filed in U.S. District Courts and 28 were filed with the USITC. For each case, I recorded or assembled information about:

- the parties named in the dispute (all plaintiffs, defendants, and patent assignees)
  - –type of entity (firm, independent inventor, university/government agency)
  - –nationality (by headquarters of firms or address listed on disputed patents for independent inventors)
  - –primary and secondary SICs for all firms and the parent company, if any, of the named litigants in the year of the dispute
- when and where the case was filed (filing date; name of U.S. District Court)
- the litigated patents (by U.S. patent number)
  - –"front page" information from the published patent document [e.g., year applied for and issued, name of inventors and original assignee (if any), patent classes, etc.]
  - –whether the invention pertained primarily to semiconductor-related products or manufacturing processes (Appendix A describes how I defined these categories and coded the inventions)

To determine the type of dispute involving the patent, I followed the convention of Lanjouw and Schankerman (2001) and classified cases as (1) an infringement suit if the plaintiff was the original assignee of one or more of the patents or (2) a suit for declaratory judgment if the defendant was the original assignee of one or more of the disputed patents.[25] Using this approach, I identified 146 infringement suits and 23 declaratory judgment suits. However, I was unable to classify a large number of cases (118 cases, or 41 percent of the total) using this information alone.[26] On closer examination, it was clear that the plaintiffs in some

---

[25]In suits for declaratory judgment, a firm typically seeks a ruling that it is not infringing another party's patent—either because the patent is invalid or because the firm is not guilty of infringement (see Moore, 2000). It is not possible using Derwent data alone to determine how many of these suits involve challenges to a patent's validity, claims of noninfringement, or both.

[26]Lanjouw and Schankerman (2001) also report a high percentage of unclassified patents using this approach. For electronics-related cases, they classified 58.1 percent as infringement suits and 10.5 percent as third-party disputes, while 31.4 percent were unclassified.

of the "unclassified" cases had acquired the patent rights of other companies (e.g., the well-known enforcement by Harris Corporation of patents acquired from RCA). By searching the patent "reassignment" data (at the U.S. Patent Depository of the Free Library of Philadelphia) I discovered—somewhat to my surprise—that 82 of the 118 unclassified cases (70 percent) involved situations in which the original assignee had transferred, or "reassigned," legal title to one or more of the litigated patents and registered the transaction with the USPTO. In almost all of these cases, title to the disputed patents was reassigned to the plaintiff in an infringement suit. In the end, the additional information about patent reassignments enabled me to classify 221 cases (77 percent of the total) as infringement suits and 32 cases (11 percent) as declaratory judgment suits; 34 cases (12 percent) remained unclassified. As discussed in Appendix A, roughly 75 percent of the patents involved in these disputes pertained to product-related inventions either for semiconductor devices or downstream products; approximately 23 percent of the litigated patents pertained solely to production processes.

Finally, recognizing the potential downward bias in the number of cases reported in Derwent during the early period of my study (discussed in the introduction),[27] I searched the trade press and "litigation" sections of 10-K reports filed during 1973-1985 for the 38 sample firms that were publicly traded in those years. Using these sources, I identified only four patent lawsuits filed in U.S. District Courts during this period that were not also reported in Derwent.[28] On one hand, the lack of reported legal disputes over patents before the mid-1980s is consistent with historical accounts of the industry discussed in the second section of this chapter.[29] On the other hand, it is possible that the numbers I report below still suffer from an underreporting bias in the early period despite my attempts to

---

[27]Although the U.S. federal courts are required to report to the USPTO cases that involve a U.S. patent, Lanjouw and Schankerman (2001) and Somaya (2003) report that they sometimes fail to do so. More troublesome, the underreporting bias appears to be more egregious in the pre-1984 period.

[28]For simplicity, these cases are included in the counts presented above.

[29]Other accounts corroborate this received wisdom. For example, Von Hippel, characterizing the semiconductor industry through the early 1980s, concludes that "...in the semiconductor field—except for a very few patent packages that have been litigated, that have been held valid, and that most firms license without protest—the patent grant is worth very little to the inventors who obtain it" (1988, p. 53; note that von Hippel goes on to acknowledge the value of defensive patenting). Moreover, Mel Sharp, the former General Counsel of Texas Instruments who led TI's aggressive enforcement strategy, has been quoted as follows: "[In the early 1980s], I clearly was able to convince the company that they needed to accelerate the budgets and the internal procedures to acquire and protect our intellectual property rights. I was not able during that period to convince management that we ought to aggressively enforce our patents. Management simply wasn't ready to change at that time" (as quoted in Warshofsky, 1994, p. 117).

mitigate the problem. In future versions of this research, I plan on investigating this issue further by examining archival records in selected U.S. District Courts.[30]

## MAIN FINDINGS

How do the characteristics of semiconductor firms involved in legal patent disputes over the past three decades compare with those of nonlitigating firms in the industry? To what extent have semiconductor firms been involved in more patent lawsuits during the "pro-patent" era? Is patent litigation in this industry still fairly "uncommon," as sometimes claimed? Finally, do semiconductor design firms and manufacturers differ in their propensity to enforce patents or in the characteristics of their patent-related legal disputes? This section presents the descriptive findings that pertain to these main questions.

**TABLE 2A** Mean Values of Firm Characteristics: Litigating Versus Nonlitigating Sample Firms, 1725 Observations (136 Firms), 1973-2001[a]

| | All Firms (mean values) | | |
|---|---|---|---|
| Variable | Litigating Firms | Nonlitigating Firms | $p$-Value, Test of Equality[b] |
| Age (2000-founding year) | 20.45 | 24.15 | 0.093 |
| D(Founded before 1982=1) | 0.47 | 0.52 | 0.622 |
| D(Manu=1) | 0.53 | 0.68 | 0.074 |
| Sales (M Constant 1996$) | 619.72 | 34.46 | 0.000 |
| Employees (1,000s) | 5.11 | 0.40 | 0.000 |
| Prop, Plant & Equip (Constant 1996 $) | 463.22 | 57.13 | 0.000 |
| R&D (M Constant 1996 $) | 82.45 | 4.59 | 0.000 |
| R&D Intensity (R&D/Employee) | 32.70 | 17.75 | 0.000 |
| R&D Intensity (R&D/Total Assets) | 0.16 | 0.14 | 0.077 |
| # Issued US Patents | 25.35 | 0.84 | 0.000 |
| $n$ | 76 | 60 | |

[a]Based on annual financial information and patent counts, 1973-2000, and litigation events, 1973-June 2001.

[b]$P$-values based on a two-tailed test of equality are presented.

---

[30]I welcome additional suggestions on how to obtain more reliable information about patent cases filed before 1984 that reveals the identities of all parties involved in the dispute. The most comprehensive archival source of U.S. patent case filings is a database available from the Federal Judicial Center (see discussion in Lanjouw and Schankerman, 2001 and Somaya, 2003). Because this database reveals only the names of the *first* plaintiff and defendant in a case, its usefulness is limited if one seeks to identify litigation events at the level of individual firms (regardless of whether firms appear as first plaintiff or defendant in a case).

## Characteristics of Litigating vs. Nonlitigating Firms

Roughly 56 percent of the semiconductor firms in the sample are involved in one or more U.S. patent lawsuits filed during 1973-2001. Table 2A compares the mean characteristics of the sample firms involved in litigation ("litigating firms") with those of sample firms that were not listed in patent lawsuits ("nonlitigating firms") during the sample period. On average, litigating firms spend more on R&D (in absolute terms and per employee), are much larger (as measured by sales, number of employees, or capital expenditures), and have larger patent portfolios than "peer" semiconductor firms not involved in legal patent disputes during this period. In a study of intellectual property-related case filings before the USITC during 1976-1990, Mutti and Yeung (1996) report similar findings.

To see whether these results differ between design firms and manufacturers, Table 2B divides the sample into these two main types of firms. Again, a similar pattern emerges. For both subsets of firms, those involved in patent litigation (on average) tend to be larger, invest more heavily in R&D, and own more patents compared to nonlitigating firms. It should also be noted from Table 2B that a higher share of design firms in the sample are involved in patent litigation over the sample period than is true for manufacturers: Whereas 65 percent of the design firms (36 of 55 firms) appear in at least one reported legal dispute over patents between 1973 and June 2001, less than half of the manufacturers in the sample (40 of 81 firms) are involved in reported patent cases during the same period. Whether these disputes represent design firms enforcing their own patent rights or defending against lawsuits initiated by others is an issue that I return to below.

## Overall Litigation Trends and "Rates"

Figure 3 plots the annual number of reported cases that list at least one sample firm as plaintiff, defendant, or patent assignee from 1973 through June 2001. For perspective, the annual number of patents issued to sample firms (collectively) is also reported for 1973-2000.

Two prominent trends emerge. First, in general, the number of cases involving these firms has increased over time, with relatively infrequent litigation activity until the mid- to late 1980s. As acknowledged above, the lack of reported cases in the early period could reflect an underreporting bias in the data (Lanjouw and Schankerman, 2001, 2003). Even so, the upward trend in litigation for firms in this sample is unremarkable when compared with trends reported by other studies using the more comprehensive (but less detailed) data from the Federal Judicial Center. For example, Merz and Pace (1994) report that annual patent case filings neither increased nor decreased between 1971 and 1982 but rose steadily by an average rate of 25 percent per year from around 1983 through 1991.

**TABLE 2B** Mean Values of Manufacturing and Design Firms Characteristics: Litigating Versus Nonlitigating Firms, 1725 Observations (136 Firms), 1973-2001[a]

Manufacturers (*n*=81)

| Variable | Litigating Firms | Nonlitigating Firms | *p*-Value, Test of Equality[b] |
|---|---|---|---|
| Age (2000-founding year) | 25.33 | 29.23 | 0.000 |
| D(Entered before 1982=1) | 0.68 | 0.68 | 1.000 |
| Sales (Constant 1996 $M) | 872.85 | 34.05 | 0.000 |
| Employees (1,000s) | 7.41 | 0.45 | 0.000 |
| Prop, Plant & Equip (Constant 1996 $) | 683.76 | 16.62 | 0.000 |
| R&D (Constant 1996 $M) | 113.35 | 4.07 | 0.000 |
| R&D Intensity (R&D/Employee) | 19.26 | 9.66 | 0.000 |
| R&D Intensity (R&D/Total Assets) | 0.13 | 0.09 | 0.000 |
| # Issued US Patents | 32.70 | 0.64 | 0.000 |
| *n* | 40 | 41 | |

Design Firms (*n*=55)

| Variable | Litigating Firms | Nonlitigating Firms | *p*-Value, Test of Equality[b] |
|---|---|---|---|
| Age (2000-founding year) | 15 | 14 | 0.120 |
| D(Entered before 1982=1) | 0.25 | 0.21 | 0.746 |
| Sales (Constant 1996 $M) | 138.09 | 36.98 | 0.000 |
| Employees (1,000s) | 0.46 | 0.15 | 0.000 |
| Prop, Plant & Equip (Constant 1996 $) | 44.25 | 15.10 | 0.000 |
| R&D (Constant 1996 $M) | 24.04 | 6.56 | 0.000 |
| R&D Intensity (R&D/Employee) | 59.65 | 50.44 | 0.009 |
| R&D Intensity (R&D/Total Assets) | 0.21 | 0.33 | 0.000 |
| # Issued US Patents | 4.89 | 1.21 | 0.000 |
| *n* | 36 | 19 | |

[a]Based on annual financial information and patent counts, 1973-2000, and litigation events, 1973-June 2001.
[b]*P*-values based on a two-tailed test of equality are presented.

Even more striking in Figure 3 is the growth in patenting by semiconductor firms, which continued to accelerate through the last half of the 1990s.[31] As reported by Hall and Ziedonis (2001), the propensity of semiconductor firms to patent (relative to their R&D spending) more than doubled between 1982 and

---

[31]The number of annual patents awarded to sample firms rose from approximately 1,750 in 1995 to 5,430 in 2000.

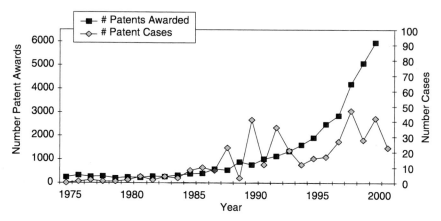

**FIGURE 3** Number of annual patent awards v. patent case filings (U.S. semiconductor firms, 1973-2001).

1992, from about 0.3 to 0.6.[32] Similar calculations for the expanded sample suggest that the upward trend reported in Hall and Ziedonis did *not* level off during the decade of the 1990s: Between 1992 and 1997, patenting per million real R&D dollars in semiconductors continued to climb from 0.6 to almost 0.8.

Has the *litigation rate* in semiconductors increased? Because I have information about R&D spending and patenting trends related to these firms, I approach this question from several angles. In brief, the answer depends on how you measure "litigation rate." The answer ranges from "no" to "slightly" (when case filings are deflated by patent counts) to "yes" (when compared with firm-level R&D spending).

The most common way to estimate patent litigation rates is to deflate the number of annual patent cases filed (or number of patents involved in disputes) with a measure of relevant patents "at risk" of litigation that year. Using a similar approach, I find that the average litigation rate for semiconductor manufacturers *fell slightly* (by 5 percent) between 1973-1985 and 1986-2000, from 9.5 to 9.0 cases filed per thousand patents—as reported in Table 3, column 3.[33] Given the dramatic growth rates in patenting by firms in this industry since the mid-1980s (discussed in the second section of this chapter and revealed clearly in Figure 3), these results are not particularly surprising. On the one hand, they may echo the econometric findings of Lanjouw and Schankerman (2003) that firms with large

---

[32] During the same period, the patent yield for manufacturing as a whole was fairly stagnant and that for pharmaceuticals actually declined (see Figure 1 in Hall and Ziedonis, 2001).

[33] The sample was divided into periods that predate and follow 1985, in line with the discussion in the second section of this chapter about key "demonstration events" around 1985-1986 that signaled the importance of the new patent regime to firms in this industry.

**TABLE 3** Patent Litigation "Rates," Before and After 1986, U.S. Semiconductor Manufacturers Versus Design Firms

| Variables[a] | Column 1[b] 1973-1985 | Column 2 1986-2000 | Column 3[c] (% change, 2 vs. 1) |
|---|---|---|---|
| **Manufacturers** | | | |
| Number of cases/1,000 patents awarded | 9.477 | 9.020 | −5% |
| Number of litigated patents/1,000 patents awarded | 7.190 | 9.262 | 29% |
| Number of cases/million real R&D dollars | 0.0028 | 0.0040 | 43% |
| Number of litigated patents/million real R&D dollars | 0.0021 | 0.0041 | 94% |
| **Design Firms** | | | |
| Number of cases/1,000 patents awarded | na | 40.75 | na |
| Number of litigated patents/1,000 patents awarded | na | 39.98 | na |
| Number of cases/million real R&D dollars | na | 0.012 | na |
| Number of litigated patents/million real R&D dollars | na | 0.012 | na |
| **Magnitude of Difference (Design v. Manufacturing, second period only)** | | | |
| Number of cases/1,000 patents awarded | | 4.52 X | |
| Number of litigated patents/1,000 patents awarded | | 4.32 X | |
| Number cases/million real R&D dollars | | 2.93 X | |
| Number litigated patents/million real R&D dollars | | 2.80 X | |

[a]Variables in the left-hand column were calculated as follows:
- "number of cases" = the number of unique U.S. patent cases filed that involved one or more sample firms in that category (averaged across firms and within period)
- "1,000 patents owned" = based on the cumulative stock of patents awarded to sample firms in that category (averaged across firms and within period, with deletion of expired patents)
- "litigated patents" = the number of patents assigned to firms in that category that were involved in patent cases during each period
- "million real R&D dollars" = average R&D spending by sample firms in that category during each period (based in 1996 dollars using NSF R&D Deflators).

[b]The average litigation rate for 1973-1985 was calculated with the subset of manufacturing firms that existed throughout both periods; few design firms were involved in patent cases filed before 1986.

[c]Adjusting the two periods by 1-2 years did not substantively alter the results.

portfolios more successfully avoid litigation; semiconductor manufacturers may have deterred litigation events more effectively or reached agreement more easily in licensing negotiations as their patent portfolios grew larger. Although this is perhaps true, the disproportionate growth in patenting by these firms (the denominator) could mask important underlying trends or variation within the sample.

If we compare case filings with an "input" measure of innovation (R&D spending), a different pattern emerges. Here, the average rate of litigation for manufacturers rises noticeably between the two periods—with a 45 percent increase in the number of patent cases filed and almost twice as many patents being litigated per R&D dollar in the post-1985 period (see again Table 3, column 3). Numerous studies report that the direct and indirect costs associated with preparing, negotiating, filing, and (for the subset of cases that proceed to trial) litigating patent cases have risen over time (e.g., see Ellis, 1999; AIPLA, 1999). If true, this suggests that—relative to their investments in R&D—semiconductor manufacturers on average have been devoting far more financial resources toward enforcing, defending, and challenging patents in court since 1985 than was true in the preceding period.

### A Closer Look at Litigation Patterns for Design Firms vs. Manufacturers

Regardless of how it is measured, the average litigation rate for design firms is consistently higher—by an order of magnitude—than that for manufacturers. This is particularly interesting because the design firms in this sample are (as a group) relatively small in size. During the sample period, the median design firm had less than 250 employees; in 1990, roughly 80 percent of the design firms in this sample had less than 500 employees.

Table 3 (column 2) reveals that during 1986-2000, design firms were involved (as either plaintiff or defendant) in approximately 4 patent cases for every 100 patents in their portfolios—a "litigation rate" that is more than 4.5 times that of manufacturers based on similar calculations. Restricting the sample to cases in which the firms enforce their *own* patents suggests that (on average) design firms in the sample enforce approximately 4 of every 100 patents they own. Although litigation rates of similar magnitude have been reported for new biotechnology firms during 1990-1994 (Lerner, 1995), they are unusual across technological sectors; for example, they are roughly three times the litigation rates reported by Lanjouw and Schankerman (2001) in electronics classes more generally and more than twice those reported in computers by Somaya (2003).[34] The high propensity of design firms to enforce their patents is similarly revealed in the R&D-based measures reported in Table 3 (column 2). Relative to manufacturers, design firms were involved (on average) in almost 3 times as many cases and enforced over 2.5 times as many patents per R&D dollar during the 1986-2000 period.

To illuminate factors that might underpin these divergent litigation propensities of manufacturers and design firms, Tables 4A and 4B compare the characteristics of cases that involve these two types of firms. For clarity, the tables distinguish among cases in which firms are enforcing their own patent rights against others (plaintiffs in infringement suits), are defending themselves against claims

---

[34]Lanjouw and Schankerman (2003) and Somaya (2003) use a slightly different approach to calculate litigation rates by basing the estimates on earlier cohorts of patents. If anything, this suggests that my approach *underestimates* the "true" litigation propensity of design firms given the relatively young age of these firms and their patent portfolios.

**TABLE 4A** Profile of Cases Involving Sample Manufacturing Firms, by Type of Case

| Panel A. Cases Involving Manufacturers | Total (includes 3rd party) | Infringement Suits | | Declaratory Judgment Suits | |
|---|---|---|---|---|---|
| | | As Plaintiff | As Defendant | As Plaintiff | As Defendant |
| **Overview** | | | | | |
| Number of cases[a] | 209 | 101 | 91 | 13 | 14 |
| Number, excluding Texas Instruments | 170 | 83 | 81 | 12 | 9 |
| Number of unique parties involved | 173 | 97 | 82 | 28 | 22 |
| Average number of parties/case (median) | 2.61 (2) | 2.47 (2) | 2.78 (2) | 2.61 (2) | 2.21 (2) |
| Number of patents involved | 372 | 182 | 156 | 51 | 62 |
| Average number of patents/case (median) | 2.48 (1) | 2.46 (1) | 2.09 (1) | 3.92 (1) | 4.43 (1.5) |
| **By characteristics of opposing party:**[b] | | | | | |
| percent cases "within sample" | | | | | |
| percent with other US semiconductor manufacturers | 16.75% | 28.71% | 31.87% | 38.46% | 35.71% |
| percent with US Design firms | 11.96% | 18.81% | 4.40% | 0.00% | 7.14% |
| percent cases with foreign firms | 24.40% | 27.72% | 14.29% | 7.69% | 14.29% |
| percent cases with non-semiconductor firms[c] | 49.28% | 32.67% | 45.05% | 53.85% | 28.57% |
| percent cases with independent inventors | 8.61% | 2.97% | 6.59% | 23.08% | 0.00% |
| percent cases with univs or govt labs | 0.96% | 0.99% | 0.00% | 0.00% | 0.00% |
| **By characteristics of litigated patents:** | | | | | |
| percent in electronics-related classes (G01-G21; H- ) | 90.59% | 91.21% | 90.38% | 88.24% | 98.39% |
| percent that pertain to new or improved semiconductor devices (Invention Type=1)[d] | 53.23% | 65.93% | 55.77% | 45.10% | 48.39% |
| percent that pertain to new or improved manufacturing processes (Invention Type=3)[d] | 23.66% | 21.98% | 21.15% | 41.18% | 17.74% |

[a]Overall, manufacturers were involved in 209 cases and design firms were involved in 90 cases during the sample period. This exceeds the number of cases in the sample (287) because of 12 cases that involve both manufacturers and design firms in the sample.

[b]Percentages exceed 100% since multiple parties (of different types) can be involved in a case.

[c]Defined as firms for which SIC3674 is not listed as a primary or secondary class among its lines of business.

[d]See Appendix A for information about invention types and how they were coded.

**TABLE 4B** Profile of Cases Involving Sample Design Firms, by Type of Case

| | | Infringement Suits | | Declaratory Judgment Suits | |
|---|---|---|---|---|---|
| Panel B. Cases Involving Design Firms | Total (Includes 3rd party) | As Plaintiff | As Defendant | As Plaintiff | As Defendant |
| **Overview** | | | | | |
| Number of cases[a] | 90 | 43 | 48 | 5 | 4 |
| Number of unique parties involved | 92 | 57 | 47 | 10 | 4 |
| Average number of parties/case (median) | 2.41 (2) | 2.30 (2) | 2.27 (2) | 3 (2) | 2.75 (3) |
| Number of patents involved | 134 | 60 | 80 | 19 | 6 |
| Average number of patents/case (median) | 2.19 (1) | 1.91 (1) | 2.12 (1) | 4.8 (4) | 3.25 (3.5) |
| **By characteristics of opposing party:[b]** | | | | | |
| Percent cases "within sample" | | | | | |
| Percent with other US semiconductor manufacturers | 27.78% | 9.30% | 39.58% | 20.00% | 0.00% |
| Percent with US Design firms | 18.89% | 37.21% | 33.33% | 0.00% | 0.00% |
| Percent cases with foreign firms | 11.1% | 13.95% | 6.25% | 0.00% | 0.00% |
| Percent cases with non-semiconductor firms[c] | 43.3% | 34.88% | 25.00% | 60.00% | 100.00% |
| Percent cases with independent inventors | 2.2% | 0.00% | 0.00% | 20.00% | 0.00% |
| Percent cases with univs or govt labs | 0.0% | 0.00% | 0.00% | 0.00% | 0.00% |
| **By characteristics of litigated patents:** | | | | | |
| Percent in electronics-related classes (G01-G21; H- ) | 98.51% | 100.00% | 98.75% | 100.00% | 100.00% |
| Percent that pertain to new or improved semiconductor devices (Invention Type=1)[d] | 79.9% | 93.33% | 82.50% | 57.89% | 100.00% |
| Percent that pertain to new or improved manufacturing processes (Invention Type=3)[d] | 6.77% | 0.00% | 8.75% | 5.26% | 0.00% |

[a]Overall, manufacturers were involved in 209 cases and design firms were involved in 90 cases during the sample period. This exceeds the number of cases in the sample (287) because of 12 cases that involve both manufacturers and design firms in the sample.

[b]Percentages exceed 100% since multiple parties (of different types) can be involved in a case.

[c]Defined as firms for which SIC3674 is not listed as a primary or secondary class among its lines of business.

[d]See Appendix A for information about invention types and how they were coded.

of infringement (defendants in infringement suits), or are engaged in declaratory judgment suits. Because of the limited number of declaratory judgment suits in the sample, the discussion below focuses on the trends reported in the overall and patent infringement columns.

In general, we see from Tables 4A and 4B that cases involving manufacturers tend to include a more disparate set of parties and inventions than is true of disputes involving design firms. Almost 25 percent of cases involving manufacturers are against foreign firms, and almost 10 percent of the cases include an independent inventor as an opposing party. In contrast, less than 12 percent of cases involving design firms are against non-U.S. firms, and only 2 percent are in opposition with an independent inventor. Similarly, opposing parties in disputes involving design firms are more heavily concentrated among other design firms or U.S. semiconductor manufacturers. Design firm disputes also involve a more focused set of technologies more targeted toward product-related inventions (as revealed in the lower rows of Tables 4A and 4B). Many of these trends are, of course, not surprising given the broader range of technological and commercial activities in which manufacturers are involved. Table 4B also reveals, however, that design firms (on average) tend to enforce their patent rights most frequently against other design firm rivals. Interestingly, they most commonly defend themselves in litigation initiated by domestic semiconductor manufacturers, followed by disputes initiated by other design firms.

To the extent that design firms are using the courts to protect market share (as suggested in interviews discussed in the second section of this chapter), we should expect them to enforce their patents relatively early in the patent's lifetime to block competition in related markets. Consistent with this view, I find that design firms enforce patents that are roughly the same age as the average patent in their portfolios. In contrast, manufacturers in the sample enforce patents that are, on average, almost 4 years older than the average patent in their portfolios.[35] Somaya (2003) argues that the "strategic stakes" are higher for patents enforced early in their lifetimes. In a study comparing case filings and settlements in computers and research medicines, Somaya finds that the patentee's strategic stakes renders settlement less likely in both sectors. My descriptive results are consistent with this finding, albeit in a different setting.

## CONCLUDING REMARKS

This chapter examines the enforcement of U.S. patents in semiconductors—an industry characterized by a rapid, cumulative process of innovation. Starting with a sample of 136 dedicated U.S. semiconductor firms, the study compares the

---

[35]More specifically, patents enforced by design firms were (on average) issued within 1 year of the average patent in the litigating design firm's portfolio. These results do not appear to be driven by outliers.

characteristics of litigating and nonlitigating firms and explores the incidence and nature of litigation events involving these firms from 1973 through June 30, 2001. Despite active cross-licensing in this industry, the results suggest that litigation events over patented technologies have become more frequent during the period associated with stronger U.S. patent rights. Previous research suggests that the aggressive patenting by manufacturing firms in this sector is driven by a desire to deter such litigation and to negotiate more favorable access to external technologies (Hall and Ziedonis, 2001; Cohen et al., 2000). Indeed, this study finds that the "patent portfolio races" of U.S. semiconductor manufacturers identified by Hall and Ziedonis continued apace throughout the decade of the 1990s, dwarfing overall patent litigation trends in this sector. Although the number of patent cases involving these firms has declined slightly relative to their patenting activity since the mid-1980s, I nonetheless find that it has increased relative to the R&D investments of these firms during the same period. Assuming that the direct and indirect costs associated with litigation have also increased over time (Barton, 2000; AIPLA, 1999), these trends suggest that semiconductor firms have been directing a larger share of their innovation-related resources toward defending, enforcing, and challenging patents in court since the mid-1980s than was true in the preceding period.

The descriptive findings yield somewhat mixed results regarding the litigation behavior of small firms—a matter explored in detail in recent studies by Lanjouw and Lerner (2001) and Lanjouw and Schankerman (2003). On one hand, I find that semiconductor firms involved in litigation over the sample period are larger, invest more in R&D, and own more patents than nonlitigating firms in the industry. This result holds both across the sample and within each group of design and manufacturing firms. Yet the high average litigation propensity of design firms in the sample is quite striking. These firms, which typically employ less than 500 employees, enforce an average of 4 out of every 100 patents they own—a litigation rate that is not only high relative to semiconductor manufacturers but closely resembles that of dedicated biotechnology firms in the early 1990s (as reported by Lerner, 1995). As "technology specialists" lacking complementary manufacturing assets of their own, semiconductor design firms appear to rely quite heavily on U.S. courts to protect their intellectual assets—primarily against other design firm rivals. At least within this sector, these results call into question whether the propensity of small firms to enforce patents stems from a desire to aggressively defend technological niches ("high stakes") or from a lack of large portfolios with which to trade (as Lanjouw and Schankerman's 2003 study suggests). The next step in this research is to test between these competing explanations econometrically, which will also enable us to investigate a broader range of factors shaping cooperation and conflict over intellectual property rights in this sector.

Although suggestive, these results highlight several important (but unresolved) questions about the role of patents, and patent portfolios, in the complex process of innovation. What is the effect of litigation—and threats of litigation—on the R&D and patenting behavior of firms? Do firms in cumulative technological settings "avoid the shadows" of better-capitalized rivals (or those with larger patent portfolios), as Lerner (1995) finds in the biotechnology sector? Alternatively, if "mutual blocking" conveys value to these firms in terms of more favorable cross-licensing deals or implicit design freedom, firms may seek to "race into" the patent thickets by amassing larger portfolios of their own with which to trade. Addressing these questions would enrich our understanding of the underlying incentives generated by the patent system.

Finally, to what extent is the emergence of patent thickets deterring entry or "tilting the tables" more in favor of large firms (or, somewhat separately, firms with large patent portfolios)? This question extends well beyond the scope of this study but is important from a policy perspective and interesting to consider within the context of semiconductors.[36] As mentioned in the second section of this chapter, the early success of the U.S. semiconductor industry is often attributed to the liberal licensing terms offered by firms such as AT&T and IBM for rights to use their portfolios of patents in the 1950s through 1970s—in part because of constraints imposed by antitrust authorities (Tilton, 1971; Levin, 1982; Grindley and Teece, 1997). Yet, as demonstrated by the emergence of design firms within this sample, widespread entry continued to characterize this industry during the period associated with stronger U.S. patent rights and a less restrictive antitrust regime. Several recent studies provide partial insights into this apparent paradox. For example, by modeling the decision of patent owners to invest in monitoring potential infringers, Crampes and Langinier (2002) show that strengthening the rights of patent owners may deter entry if an incumbent already had incentives to negotiate licenses with the entrant (to settle rather than sue) but now imposed a higher license fee. Alongside this conventional finding, however, they also find that strengthening the rights of patent owners may (1) expand the "settlement area" under which it is becomes profitable for the patent owner to license rather than sue (and thereby induce entry) and (2) induce entry by firms with highly differentiated products (where the bargaining surplus is greatest). Recent empirical findings by Gans et al. (2002) similarly suggest that, in the shadows of stronger patent protection, the relationships between incumbents and start-up firms may become more cooperative than competitive in nature, whereas others emphasize the importance of patent rights in sustaining entry by specialized firms (e.g., Arora et al., 2001). In summary, these studies highlight the importance of

---

[36]To empirically estimate the effects, if any, of incumbent portfolios on observed patterns of entry, one would need some way of establishing the counterfactual: Absent such portfolios (or given portfolios of smaller size), what pattern of entry would we expect to observe? Studies along these lines would also need to take into account the demand-side factors and technological opportunities that also influence the entry decision (as discussed in Cohen, 1995).

understanding the economic incentives generated by the patent system but also underscore the multifaceted effects that may arise even within one technological sector.

## REFERENCES

Adelman, M. J. (1987). "The New World of Patents Created by the Court of Appeals for the Federal Circuit." *Journal of Law Reform* 20(4): 979-1007.

American Intellectual Property Law Association (AIPLA). (1999). *Report of Economic Survey 1999.* Arlington, VA: AIPLA.

Arora, A., A. Fosfuri, and A. Gambardella. (2001). *Markets for Technology: The Economics of Innovation and Corporate Strategy.* Cambridge, MA: MIT Press.

Barton, J. H. (2000). "Reforming the Patent System." *Science* 287: 1933-1934.

Bessen, J., and E. Maskin. (2000). "Sequential Innovation, Patents, and Imitation," Working Paper No. 00-01, Department of Economics, Massachusetts Institute of Technology.

Cohen, W. M. (1995). "Empirical Studies of Innovative Activity," in P. Stoneman, ed., *Handbook of Economics of Innovation and Technological Change.* Oxford, UK: Blackwell Publishers Inc.

Cohen, W. M., R. R. Nelson, and J. Walsh. (2000). "Protecting Their Intellectual Assets: Appropriability Conditions and Why U.S. Manufacturing Firms Patent (or Not)," NBER Working Paper 7552.

Crampes, C., and C. Langinier. (2002). "Litigation and Settlement in Patent Infringement Cases." *RAND Journal of Economics* 33(2): 258-274.

Ellis, Judge T. S., III. (1999). "Distortion of Patent Economics by Litigation Costs." In K.M. Hill, T. Takenaka, and K. Takeuchi, eds., *Streamlining International Intellectual Property: Enforcement and Prosecution, University Technology Transfer, and Incentives for Inventors.* Seattle, Washington: University of Washington School of Law Center for Advanced Study and Research on Intellectual Property, publication series No. 5.

Farn, M. (1996). "A Quasi-Statistical Analysis of the Acquisition and Enforcement of Patent Rights in the Computer and Semiconductor Industries for the Period 1986-1995," Stanford Law School directed research, Fall 1996.

Gans, J. S., D. H. Hsu, and S. Stern. (2002). "When Does Start-Up Innovation Spur the Gale of Creative Destruction?" *RAND Journal of Economics* 33(4): 571-586.

Graham, S., and D. C. Mowery. (2003). "Intellectual Property Protection in the U.S. Software Industry." In W. Cohen and S. Merrill, eds., *Patents in the Knowledge-Based Economy.* Washington, D.C.: The National Academies Press.

Grindley, P. C., and D. J. Teece. (1997). "Managing Intellectual Capital: Licensing and Cross-Licensing in Semiconductors and Electronics." *California Management Review* 39(2): 1-34.

Hall, B. H., and R. H. Ziedonis. (2001). "The Patent Paradox Revisited: An Empirical Study of Patenting in the US Semiconductor Industry, 1979-95." *RAND Journal of Economics* 32(1): 101-128.

Heller, M. A., and R. S. Eisenberg. (1998). "Can Patents Deter Innovation? The Anticommons in Biomedical Research." *Science* 280: 698-701.

Hunt, R. M. (1999). "Nonobviousness and the Incentive to Innovate: An Economic Analysis of Intellectual Property Reform," working paper no. 99-3, Economic Research Division, Federal Reserve Bank of Philadelphia.

Integrated Capital Monitor. (1999). "Company Analysis: Rambus, Inc." 1(1): 6-7.

Integrated Circuit Engineering Corporation (ICE). (1995). *Cost Effective IC Manufacturing, 1995.* Scottsdale, AZ: Integrated Circuit Engineering Corporation.

Integrated Circuit Engineering Corporation (ICE). (1976-1998). *Status: A Report on the Integrated Circuit Industry.* Scottsdale, AZ: Integrated Circuit Engineering Corporation.

Jaffe, A. (2000). "The U.S. Patent System in Transition: Policy Innovation and the Innovation Process." *Research Policy* 29: 531-557.

Kortum, S., and J. Lerner. (1998). "Stronger Protection or Technological Revolution: What Is Behind the Recent Surge in Patenting?" Carnegie-Rochester Conference Series on Public Policy 48: 247-304.

Lanjouw, J. O., and M. Schankerman. (2001). "Characteristics of Patent Litigation: A Window on Competition." *RAND Journal of Economics* 32(1): 129-151.

Lanjouw, J. O., and M. Schankerman. (2003). "Enforcement of Patent Rights in the United States." In W. Cohen and S. Merrill, eds., *Patents in the Knowledge-Based Economy.* Washington, D.C.: National Academies Press.

Lanjouw, J., and J. Lerner. (2001). "Tilting the Table? The Use of Preliminary Injunctions." *Journal of Law and Economics* 44(2): 573-603.

Lerner, J. (1995). "Patenting in the Shadow of Competitors." *Journal of Law and Economics* 38: 563-595.

Levin, R. C. (1982). "The Semiconductor Industry." In Richard R. Nelson, ed., *Government and Technical Progress: A Cross-Industry Analysis.* Oxford, UK: Pergamon Press.

Levin, R. C., A. K. Klevorick, R. R. Nelson, and S. G. Winter. (1987). "Appropriating the Returns from Industrial Research and Development." *Brookings Papers on Economic Activity* 3: 783-820.

Linden, G., and D. Somaya. (2000). "System-on-a-Chip Integration in the Semiconductor Industry: Industry Structure and Firm Strategies," Walter A. Haas School of Business working paper, University of California, Berkeley.

Lunney, G. S. (2001). "E-Obviousness." *Michigan Telecommunications and Technology Law Review* 7(363): 363-422.

Macher, J., D. C. Mowery, and D. Hodges. (1998). "Reversal of Fortune? The Recovery of the U.S. Semiconductor Industry." *California Management Review* 41(1): 107-136.

Mansfield, E. (1986). "Patents and Innovation: An Empirical Study." *Management Science* 32(2): 173-181.

Mazzoleni, R., and R. R. Nelson. (1998). "Economic Theories About the Benefits and Costs of Patents." *Journal of Economic Issues* 32(4): 1031-1052.

Merges, R. P. (1997). *Patent Law and Policy: Cases and Materials,* second edition. Charlottesville, VA: The Mitchie Company.

Merz, J. F., and N. M. Pace. (1994). "Trends in Patent Litigation: The Apparent Influence of Strengthened Patents Attributable to the Court of Appeals for the Federal Circuit." *Journal of the Patent and Trademark Office Society* 76: 579-590.

Moore, K. A. (2000). "Judges, Juries, and Patent Cases—An Empirical Peek Inside the Black Box." *Michigan Law Review* 99(281): 365-409.

Mowery, D. C., and N. Rosenberg. (1998). *Paths of Innovation: Technological Change in 20th-Century America.* Cambridge, UK: Cambridge University Press.

Mutti, J. (1993). "Intellectual Property Protection in the United States under Section 337." *The World Economy* 16: 339-357.

Mutti, J., and B. Yeung. (1996). "Section 337 and the Protection of Intellectual Property in the United States: The Complaints and the Impact." *The Review of Economics and Statistics* 78: 510-520.

Nies, H. W. (1993). "Ten Years of Patent Law Development Under the U.S. Court of Appeals for the Federal Circuit." *IIC* 24(6): 797-803.

Pitkethly, R. (1996). *The Use of Intellectual Property in High Technology Japanese and Western Companies.* DPhil Thesis, Oxford University.

Pooley, J. (2000). "The Trouble with Patents." *California Lawyer,* October.

Quillen, C. D. (1993). "Proposal for the Simplification and Reform of the United States Patent System." *American Intellectual Property Law Association Quarterly Journal* 21(3): 189-212.

Quillen, C. D., and O. H. Webster. (2001). "Continuing Patent Applications and the Performance of the U.S. Patent Office." *Federal Circuit Bar Journal* 11(1): 1-21.

Rivette, K. G., and D. Kline. (2000). *Rembrandts in the Attic: Unlocking the Hidden Value of Patents.* Boston, MA: Harvard University Press.

Samuelson, P., and S. Schotchmer. (2001). "The Law and Economics of Reverse Engineering," Boalt Hall School of Law Working Paper, University of California, Berkeley.

Scherer, F. M., S. E. Herzstein, Jr., A. W. Dreyfoos, W. G. Whitney, O. J. Bachmann, C. P. Pesek, C. J. Scott, T. G. Kelly, and J. J. Galvin. (1959). *Patents and the Corporation: A Report on Industrial Technology Under Changing Public Policy*, second edition. Boston, MA: Harvard University, Graduate School of Business Administration.

Shapiro, C. (2001). "Navigating the Patent Thickets: Cross-Licenses, Patent Pools, and Standard-Setting." In A. B. Jaffe, J. Lerner, and S. Stern, eds., *Innovation Policy and the Economy, volume 1*. Cambridge, Mass: National Bureau of Economic Research.

Shepard, A. (1987). "Licensing to Enhance Demand for New Technologies." *RAND Journal of Economics* 18(3): 360-368.

Shinal, J. (1988). "IBM Again Tops Technology Rivals for Most Patents." *The San Diego Union-Tribune*, 20 January.

Siegelman, P., and J. J. Donohue III. (1990). "Studying the Iceberg from Its Tip: A Comparison of Published and Unpublished Employment Discrimination Cases." *Law and Society Review* 24 (5): 1133-1170

Siegelman, P., and J. Waldfogel. (1999). "Toward a Taxonomy of Disputes: New Evidence Through the Prism of the Priest/Klein Model." *Journal of Legal Studies* 28: 101-130.

Somaya, D. (2003). "Strategic Determinants of Decisions Not to Settle Patent Litigation." *Strategic Management Journal* 24(1): 17-38.

Taylor, C. T., and Z. A. Silberston. (1973). *The Economic Impact of the Patent System: A Study of the British Experience.* Cambridge, UK: Cambridge University Press.

Tilton, J. E. (1971). *International Diffusion of Technology: The Case of Semiconductors,* Washington, D.C.: Brookings Institution.

U.S. Patent and Trademark Office (USPTO). (1995). *Technology Profile Report: Semiconductor Devices and Manufacture, 1/1969-12/1994.* Washington, DC: U.S. Department of Commerce.

Varchaver, N. (2001). "The Patent King." *Fortune*, May 14, pp. 203-216.

Von Hippel, E. (1988). *The Sources of Innovation.* Oxford, UK: Oxford University Press.

Warshofsky, F. (1994). *The Patent Wars: The Battle to Own the World's Technology.* New York, NY: John Wiley & Sons, Inc.

Ziedonis, R. H. (2000). *Firm Strategy and Patent Protection in the Semiconductor Industry*, unpublished doctoral dissertation, Walter A. Haas School of Business, University of California, Berkeley.

# APPENDIX A

## CLASSIFYING LITIGATED PATENTS AS PROCESS OR PRODUCT-RELATED TECHNOLOGIES

In surveys on appropriability (Levin et al., 1987; and Cohen et al., 2001), R&D managers consistently report that patents for new or improved products are generally more effective at preventing duplication or securing royalty income than those for new or improved innovations used in the manufacturing process. The general argument is that process innovations are less subject to public scrutiny and, therefore, are more easily kept secret than new or improved features of a final product. Similarly, infringement of process innovations may be more difficult to detect if they are embedded in difficult-to-observe manufacturing operations.

At the same time, some of the most publicized patent infringement suits in the semiconductor industry over the past two decades involved manufacturing-related inventions. For example, Texas Instruments earned millions in royalties on its patents covering methods to transport wafers after successfully suing Korean and Japanese firms in the mid-1980s. More notorious are the successful lawsuits by Jerome Lemelson, an independent inventor who owned patents covering scanning technologies used to monitor and control production systems. By the end of 2000, the Lemelson Foundation had collected over $1.5 billion in licensing fees from users of more modern bar code scanners, including manufacturers in the automobile and semiconductor industries (Varchaver, 2001).

Has the composition of patent cases filed in this industry tilted more toward these "royalty-seeking" disputes in which the value of the exclusionary right increases as the technology becomes more widely adopted throughout the industry? I hope to investigate this question econometrically in future versions of this research by examining the subset of cases that involve manufacturers and comparing the age and type of technologies involved in those disputes before and after the mid-1980s. Although we have made progress in this direction, we need to resolve two issues before understanding how much credence we should lend to our results. First, as discussed in the text, I need to confirm whether I have a representative sample of cases in the early (1973-1984) period. Secondly, I need to verify that our coding of inventions as product- or process related is reasonable and accurate. I would also need to normalize any findings based on the litigated patents with patents that are not involved in filed disputes. Otherwise, I could simply be picking up the fact that the stock of semiconductor-related patents "at risk" for litigation may have grown older or more toward process innovation as the technology has matured. With these caveats and longer-term objectives in mind, this appendix summarizes the approach I used to classify "product"- and "process"-related inventions within the litigation sample.

**TABLE A.1** Litigated Patents by Invention Type

| Code | Innovation Type | Definition and Notes | Examples | # Litigated Patents by Type (% total) |
|---|---|---|---|---|
| 1 | Product: Semiconductor Devices | *Definition:* inventions that pertain to new or improved semiconductor devices *Note:* Must include claims that cover the device but may also cover methods claims | US3138742: Miniaturized Electronic Circuits (TI Kilby patent, issued 1964); US4609986: Programmable Logic Device Using EPROM Technology (Altera Corp, issued 1986) | 277 (55%) |
| 2 | Process: Manufacturing | *Definition:* inventions that pertain to new or improved manufacturing processes *Note:* includes inventions related to materials and equipment used in semiconductor manufacture or more general manufacture. Processes | US4256534: Device Fabrication by Plasma Etching (Bell Labs, issued 1981); US3735350: Code Scanning System (Jerome Lemelson, issued 1973) | 118 (23%) |
| 3 | Product: Broader Electronics or Computer-Related | *Definition:* inventions that pertain to new or improved products that use or embed semiconductor devices but that are not stand-alone semiconductor products *Note:* dominated by inventions related to data processing and computing | US4074351: Variable Function Programmed Calculator (Boone calculator patent, TI, issued 1978) | 103 (20%) |
| 4 | Other/Unclear | *Definition:* inventions that pertain to new or improved products or processes that appear unrelated to semiconductors | US4553515: Cylinder Head for Spark Ignition Internal Combustion Engines (BL Technology Limited; issued 1985) | 6 (1%) |

Although conceptually straightforward, determining whether the claims of a patent cover product- or process-related improvements is far from simple. As a first approach, I identified patents pertaining to semiconductor devices ("products") or their manufacture ("processes") with the U.S. patent classification scheme. The results, however, were unsatisfactory. Of 504 litigated patents involving semiconductor firms in the sample, less than 30 percent fell into the three-digit U.S. patents classes that cover device inventions (classes 326, 327, 365, and 257) and less than 10 percent of the litigated patents fell into the three-digit class for "semiconductor device manufacturing: process" (438). These results were not entirely surprising for the process-related inventions given the wide range of materials, methods, and equipment used in semiconductor manufacturing.

To overcome this problem, I manually classified each litigated patent into one of four mutually exclusive categories. The main objective was to classify a patent as "product" if it pertained primarily to new or improved semiconductors or related devices (e.g., memory chips, logic devices, or analog-digital converters) and "process" if it pertained to new or improved manufacturing processes but did not claim rights to specific devices per se. Categories 1 and 2 include semiconductor-related "product" and "process" inventions, respectively. In reviewing the patents, however, it was clear that many of the patents pertained to products that *embed* semiconductors (e.g., computer systems, hand-held calculators) instead of stand-alone semiconductor devices. Category 3 includes these "downstream" product-related inventions. Finally, Category 4 contains patents that were difficult to classify into one of the above categories or seemed to pertain to innovations unrelated to semiconductors.

Table A.1 summarizes each "invention type" category and the number of litigated patents assigned to each category. Of the 504 litigated patents, I identified 277 that made specific claims to new or improved semiconductor devices (Category 1), 188 that pertained solely to methods, equipment, or materials used in the manufacturing process (Category 2), and 103 that pertained to innovations in downstream products (Category 3). A small number of patents (6) seemed unrelated to semiconductors or did not fall clearly within one of these categories. This suggests that 75 percent of the litigated patents in the sample pertain, broadly, to product innovation and 25 percent focus solely on processes used in manufacturing. In follow-on studies, I hope to use these results to inform whether concerns about "hold up" by manufacturers are illuminated by a change in the age or types of patents enforced by or against these firms in the period associated with stronger patent rights.

# Patents in Software
and Biotechnology

# Intellectual Property Protection in the U.S. Software Industry[1]

Stuart J. H. Graham and David C. Mowery
Haas School of Business
University of California, Berkeley

## INTRODUCTION

The software industry is a knowledge-intensive industry whose output is information, the coded instructions that guide the operations of a computer or a network of computers. Both the inputs and much of the output of this industry consist of intangibles, the prices of which contain considerable Schumpeterian rents. The rewards to innovators in the software industry of the 1980s and 1990s were extraordinary, as illustrated by the meteoric rise of William Gates III to control of the largest personal fortune in the world. The modern computer software industry thus is an extreme example of an industry in which the returns to innovators' investments, and in many cases market structure, are influenced by the ownership of intellectual property. As such, it is hardly surprising that the legal framework establishing and regulating ownership of such property has attracted considerable attention and debate.

The "modern" computer software industry of the twenty-first century differs from the software industry of the 1950s or 1960s, most notably in the growth of mass markets for so-called packaged software. These differences are reflected in the central importance of formal protection of intellectual property. The increased importance of formal intellectual property rights protection, as well as the changing economic and legal importance of different instruments for such protection, create significant challenges for U.S. intellectual property rights policy.

---

[1]We are grateful to participants in the STEP Board conference on "The Operation of the Patent System," participants in the U.C. Berkeley Innovation Seminar, and to Professors Rosemarie Ziedonis, Wesley Cohen, and Brian Silverman for comments on the paper. We also appreciate assistance with our analysis of patenting data from Arvids Ziedonis. This chapter draws on research supported by the Andrew Mellon Foundation and the National Research Council.

Although the computer software industry is a global industry, significant differences remain among the software industries and the associated intellectual property regimes of the industrial economies. Domestic lobbying for the creation or modification of legal regimes covering this relatively new form of intellectual property has contributed to differences in the level and characteristics of intellectual property rights for computer software among major industrial economies. The recent controversies over business methods patents and the response by both Congress and the U.S. Patent and Trademark Office (USPTO) to these controversies (see below) are only the latest examples of this endogenous character of national intellectual property rights regimes.

This chapter surveys intellectual property rights policies and controversies in the U.S. computer software industry. Immediately below, we discuss the historical development of the U.S. software industry, highlighting the ways in which the role, structure, and importance of formal intellectual property rights have changed over the course of the industry's development. We then present data on the (limited) portion of the software industry for which reliable indicators of the intensity of patenting activity during the 1980s and 1990s can be computed, focusing on patenting by specialized packaged software firms. These indicators cover the "propensity to patent" (patents per R&D dollar) and provide some evidence on change over time in the "importance" of these firms' patents. We also discuss patenting by large electronics systems firms in the same patent classes and compare the patenting behavior (and the "importance" of their patents) of the electronics systems firm that for many years was also the leading vendor of software, IBM, and the largest specialized packaged software firm, Microsoft. After a brief discussion of the changing prominence of U.S. universities as patenters in software, we examine the changing importance of copyright and patent protection of software-related intellectual property during the 1980s and 1990s. Our conclusion considers some of the policy implications of this analysis.

## THE HISTORICAL DEVELOPMENT OF THE COMPUTER SOFTWARE INDUSTRY

The growth of the global computer software industry has been marked by at least four distinct eras spanning the 1945-2001 period. The first era (1945-1965) covers the development and commercialization of the computer. The gradual adoption of "standard" computer architectures in the 1950s supported the emergence of software that could operate on more than one type of computer or in more than one computer installation. In the United States, the introduction of the IBM 650 in the 1950s, followed by the even more dominant IBM 360 in the 1960s, provided a large market for standard operating systems and application programs. The emergence of a large installed base of a single mainframe architecture occurred first and to the greatest extent in the United States. Nonetheless, most of the software for mainframe computers during this period was produced by their manufacturers and users.

During the second era (1965-1978), independent software vendors (ISVs) began to appear. During the late 1960s, producers of mainframe computers "unbundled" their software product offerings from their hardware products, separating the pricing and distribution of hardware and software. This development provided opportunities for entry by independent producers of standard and custom operating systems, as well as independent suppliers of applications software for mainframes. Unbundling occurred first in the United States and has progressed further in the United States and Western Europe than in the Japanese software industry.

Although independent suppliers of software began to enter in significant numbers in the early 1970s, computer manufacturers and users remained important sources of both custom and standard software in Japan, Western Europe, and the United States during this period. Some computer "service bureaus" that had provided users with operating services and programming solutions began to unbundle their services from their software, providing yet another cohort of entrants into the independent development and sale of traded software. Sophisticated users of computer systems, especially users of mainframe computers, also created solutions for their applications and operating system needs. A number of leading suppliers of traded software in Japan, Western Europe, and the United States were founded by computer specialists formerly employed by major mainframe users.

During the third era (1978-1993), the development and diffusion of the desktop computer produced explosive growth in the traded software industry. Once again, the United States was the "first mover" in this transformation, and the U.S. domestic market became the largest single market for packaged software. Rapid adoption of the desktop computer in the United States supported the early emergence of a few "dominant designs" in desktop computer architecture, creating the first mass market for packaged software. The independent vendors that entered the desktop software industry in the United States were largely new to the industry. Few of the major suppliers of desktop software came from the ranks of the leading independent producers of mainframe and minicomputer software, and mainframe and minicomputer ISVs are still minor factors in desktop software.

Rapid diffusion of low-cost desktop computer hardware, combined with the emergence of a few "dominant designs" for this architecture, eroded vertical integration between hardware and software producers and opened up opportunities for ISVs. Declines in the costs of computing technology have continually expanded the array of potential applications for computers; many of these applications rely on software solutions for their realization. A growing installed base of ever-cheaper computers has been an important source of dynamism and entry into the traded software industry, because the expansion of market niches in applications has outrun the ability of established computer manufacturers and major producers of packaged software to supply them.[2]

Estimates of the relative size of the "packaged" and "custom" software markets are extraordinarily scarce, reflecting the failure of public statistical agencies to collect reliable data on this rapidly growing component of the "information economy." Nonetheless, the few existing estimates suggest that the market for "packaged" software exceeded that for "custom" software by the mid-1980s. Data reported in Mowery (1996), which summarize surveys compiled by the OECD and the International Data Corporation (IDC), indicate that global consumption of "packaged" software amounted to roughly $18 billion in 1985 (current dollars) versus $11.6 billion for "custom" software. U.S. consumption of "packaged" and "custom" software, both of which were overwhelmingly domestic in origin, amounted to $12.6 billion and $4.2 billion, respectively, in 1985. Global consumption of packaged software in 1996 reached $109 billion, according to IDC estimates published in the Department of Commerce's 1998 *U.S. Industry and Trade Outlook*, and the Department estimated that global consumption would amount to more than $221 billion by 2002 (U.S. Department of Commerce, 1998, p. 28-3 et seq.). More recent estimates of the size of U.S. or global consumption of "custom software" unfortunately are unavailable; but most studies of the computer software industry (e.g., OECD, 1998) suggest that consumption and shipments of packaged software have grown much more rapidly than those for custom software during the 1985-2002 period.

The packaged computer software industry now has a cost structure that resembles that of the publishing and entertainment industries much more than that of custom software—the returns to a "hit" product are enormous, and production costs are low. And like these other industries, the growth of a mass market for software has elevated the importance of formal intellectual property rights. An important contrast between the software industry and the publishing and entertainment industries, however, is the importance of product standards and consumption externalities in the software market. Users in the mass software market often resist switching among operating systems or even well-established applications because of the high costs of learning new skills as well as their demand for an abundant library of applications software to complement an operating system. These switching costs typically are higher for the less-skilled users who dominate mass markets for software and support the development of "bandwagons" that create de facto product standards. As the widespread adoption of desktop computers created a mass market for software during the 1980s, these de facto product standards in hardware and software became even more important to the commercial fortunes of software producers than was true during the 1960s and 1970s.

---

[2]Bresnahan and Greenstein (1996) point out that a similar erosion of multiproduct economies of scope appears to have occurred among computer hardware manufacturers with the introduction of the microcomputer.

The fourth era in the development of the software industry (1994 to the present) has been dominated by the growth of networking among desktop computers within enterprises through local area networks linked to a server and/or the Internet, which links millions of users. Networking has opened opportunities for the emergence of new software market segments,[3] the emergence of new "dominant designs," and, potentially, the erosion of currently dominant software firms' positions. Like previous eras in the industry's development, the growth of network users and applications has been more rapid in the United States than in other industrial economies, and U.S. firms have maintained dominant positions in these markets (see Mowery and Simcoe, 2001).

How has the growth of the Internet changed the economics of intellectual property protection in the software industry? At least three different effects are apparent thus far in the Internet's development. First, the widespread diffusion of the Internet has created new channels for low-cost distribution and marketing of packaged software, reducing the barriers to entry into the packaged software industry that are based on the dominance of established distribution channels by large packaged software firms. In this respect, the Internet expands the possibilities for rapid penetration of markets by a "hit" packaged software product—in the jargon of the software industry, a "killer app[lication]"—which enhances the economic importance of protection for these types of intellectual property. The Internet also is an important factor in the growth of patents on software-embodied "business methods," many of which concern tools or routines employed by on-line marketers of goods and services.

But the Internet has also provided new impetus to the diffusion and rapid growth of a very different type of software, "open source" software. Although so-called shareware has been important throughout the development of the software industry, the Internet's ability to support rapid, low-cost distribution of new software and, crucially, the centralized collection and incorporation into that software of improvements from users has made possible such widely used operating systems as Linux and Apache (see Kuan, 1999 and Lerner and Tirole, 2000). The Internet thus has increased the importance of formal protection of some types of software-related intellectual property while simultaneously supporting the growth of open source software, which does not rely on such formal instruments of intellectual property protection.

## THE EVOLUTION OF INTELLECTUAL PROPERTY RIGHTS POLICY AND PRACTICE IN THE U.S. SOFTWARE INDUSTRY

This study is primarily concerned with intellectual property rights in software that combine some grant of limited monopoly in exchange for an element of

---

[3]For example, the operating system software that is currently installed in desktop computers may reside on the network or the server.

disclosure or public use. As such, it is appropriate to examine copyright and patent protection, because software has been brought underneath the umbrella of each of these regimes during the last several decades. In the near future, however, the use by software innovators of legal protections in the areas of trade secret,[4] misappropriation,[5] trademark,[6] and even the Semiconductor Chip Protection Act[7] will remain important.

## Copyright

Copyright protection for software innovation was singled out by policymakers during the 1970s as the preferred means for protecting software-related intellectual property (Menell, 1989). In its 1979 report, the National Commission on New Technological Uses of Copyrighted Works (CONTU), charged with making recommendations to Congress regarding software protection, chose copyright as the most appropriate form of protection for computer software (CONTU, 1979). Because copyright protection adheres to an author-innovator with relative ease and has a long life—now upwards of 120 years for works created for hire—the Commission determined that copyright was the preferred type of intellectual property protection for software. Congress adopted the Commission's position when it wrote "computer program" into the Copyright Act in 1980.[8]

The federal judiciary's application of copyright to software in the aftermath of the CONTU initially promised strong protection for inventors. *Apple Computer, Inc. v. Franklin Computer Corp.*[9] is an early and important case of copyright litigation in packaged software. Although the federal judiciary had long held that copyright protected only "expression" in works,[10] the court in *Apple Computer* held that Apple's precise code was protected by its copyright. The

---

[4]A trade secret is formally some information used in a business that, when secret, gives one an advantage over competitors. The secret must be both novel and valuable. *Metallurgical Industries, Inc. v. Fourtek, Inc.*, 790 F.2d 1195 (1986).

[5]Collectors of valuable information can prevent competitors from using the information. *International News Service v. Associated Press*, 248 U.S. 215 (1911).

[6]Protects names, words, and symbols used to identify or distinguish goods and to identify the producer. *Zatrains, Inc. v. Oak Grove Smokehouse, Inc.*, 698 F.2d 786 (5th Cir., 1983).

[7]Protection is available for software embodied in semiconductor chips—so-called mask works. *E.F. Johnson v. Uniden Corp. of America*, 653 F. Supp. 1485 (D. Minn. 1985).

[8]17 U.S.C. sec. 101, sec. 117 (as amended 1980). For a more complete discussion, see Menell (1989).

[9]714 F.2d 1240 (3rd Cir. 1983). Consistent with its position as a leading firm in the packaged software industry Microsoft, which supported stronger formal protection for software-related intellectual property, filed an *amicus curiae* brief on behalf of Apple in this case.

[10]Historically, a major distinction in the copyright law has been that ideas are not protected, only expressions are. *Baker v. Selden*, 101 U.S. 99 (1879).

court concluded that efforts by a "follower" firm to use the copyright holder's code for purposes of achieving compatibility with the original software were inconsequential to the determination of whether infringement had occurred. This decision strengthened copyright protection considerably, making it possible for one firm's copyrighted software to block the innovative efforts of others. Subsequent decisions—the so-called "look and feel" cases—extended traditional copyright protection of "expression" to such "nonliteral" elements of software as structure, sequence, and organization.[11]

Subsequent court decisions, however, narrowed the protection provided by copyright for software-related intellectual property. The sweeping interpretation of copyright protection in *Apple Computer* was narrowed and weakened considerably in a series of copyright infringement cases brought by Lotus Development. Lotus successfully sued Paperback Software International over the latter's alleged imitation of the "look and feel" of Lotus's spreadsheet software in a case that Lotus won in 1990. Lotus then sued Borland International over the alleged infringement by Borland's "Quattro" software of the "look and feel" of Lotus's 1-2-3 spreadsheet software in a case that lasted for six years, producing four opinions in a federal District Court and appeals to both the Court of Appeals and the U.S. Supreme Court. The District Court found that Borland had infringed Lotus's 1-2-3 spreadsheet software. Borland rewrote its software to achieve partial compatibility with elements of Lotus's 1-2-3 software, but this modification also was met with infringement findings by the District Court and a permanent injunction banning its sale.[12]

The Court of Appeals ultimately reversed some of the District Court's conclusions, arguing that "second movers" in the software industry must be allowed to emulate and build on parts of the innovator's code and methods.[13] The decision of the Court of Appeals was affirmed in 1996 by the Supreme Court in a 4-4 decision.[14] The *Borland* decision weakened the strong protection for software inventions provided by *Apple Computer, Inc. v. Franklin Computer Corp*, and along with other decisions affirming the strength of software patents may have

---

[11]*Computer Associates Int'l v. Altai, Inc.*, 982 F.2d 693 (2d Cir. 1992); *Whelan Associates v. Jaslow Dental Laboratory*, 797 F.2d 1222 (3rd Cir. 1986).

[12]*Lotus Development Corp. v. Borland Int'l, Inc.*, 788 F. Supp. 78 (D. Mass. 1992)(finding Quattro a virtual copy of Lotus's menu structure); *Lotus Development Corp. v. Borland Int'l, Inc.*, 799 F. Supp 203 (D. Mass. 1992); *Lotus Development Corp. v. Borland Int'l, Inc.*, 831 F. Supp. 202 (D. Mass. 1993); *Lotus Development Corp. v. Borland Int'l, Inc.*, 831 F. Supp. 223 (D. Mass. 1993).

[13]*Lotus Development Corp. v. Borland Int'l, Inc.*, 49 F.3d 807 (1st Cir. 1995).

[14]116 S. Ct. 804 (1996).

contributed to increased reliance by some U.S. software firms on patents in the 1990s.[15]

## Patents

In contrast to copyright, federal court decisions since 1980 have broadened and strengthened the economic value of software patents. Although some early cases during the 1970s supported the initial stance of the U.S. Patent and Trademark Office (USPTO) in stating that software algorithms were not patentable,[16] judicial opinions have shifted since then to support the use of patents in software (Samuelson, 1990).[17] In the cases of *Diamond v. Diehr*[18] and *Diamond v. Bradley*,[19] both decided in 1981, the Supreme Court announced a more liberal rule that permitted the patenting of software algorithms, strengthening patent protection for software (Merges, 1996). The economic value of these patents was highlighted in several high-profile cases during the 1990s. For example, a 1994 court decision found Microsoft liable for patent infringement and awarded $120 million in damages to Stac Electronics. The damages award was hardly a crippling blow to Microsoft, but the firm's infringing product had to be withdrawn from the market temporarily, compounding the financial and commercial consequences of the decision (Merges, 1996).

As the USPTO adopted a more favorable posture toward applications for software patents, the ability of patent examiners to identify "novelty" in an area of technology in which patents historically had not been used to cover major innovations was criticized well before the surge of "business methods" software patent applications in 1998 and 1999. The celebrated "multimedia" patent issued by the USPTO to Compton Encyclopedias in 1993 is one example of the difficulties associated with a lack of patent-based prior art. On November 15, 1993, Compton's Newmedia announced that it had won a "fundamental" patent for its

---

[15]Ironically, in light of subsequent controversies over the role of software patents, Menell's influential 1989 analysis of intellectual property protection of software, written in the wake of the strong judicial interpretation of copyright embodied in *Apple Computer, Inc. v. Franklin Computer Corp.*, argued that patents had significant advantages over copyright as a means for protecting computer applications software: "The patent system's threshold requirements for protection—novelty, utility, and nonobviousness—are better tailored than the copyright standard to rewarding only those innovations that would not be forthcoming without protection" (Menell, 1989, p. 47). As we note below (see also Merges, 1999), the debate over software patents centers on precisely these issues—Is the USPTO able to apply these requirements with sufficient rigor to prevent the issue of low-quality patents?

[16]*Gottschalk v. Benson*, 409 U.S. 63 (1972).

[17]Samuelson (1990) argues that the USPTO was at odds with the Court of Customs and Patent Appeals (CCPA) throughout the 1970s over the patentability of software and concludes that the CCPA's views in favor of patentability ultimately triumphed.

[18]450 U.S. 175 (1981).

[19]450 U.S. 381 (1981).

multimedia software that rapidly fetched images and sound.[20] The patent was quite broad, covering

> a database search system that retrieves multimedia information in a flexible, user-friendly system. The search system uses a multimedia database consisting of text, picture, audio and animated data. That database is searched through multiple graphical and textual entry paths.[21]

Compton's president, Stanley Frank, suggested that the firm did not want to slow growth in the multimedia industry, but he did "want the public to recognize Compton's Newmedia as the pioneer in this industry, promote a standard that can be used by every developer, and be compensated for the investments we have made." Armed with this patent, Compton's traveled to Comdex, the computer industry trade show, to detail its licensing terms to competitors, which involved payment of a 1 percent royalty for a nonexclusive license.[22]

Compton's appearance at Comdex launched a political controversy that culminated in an unusual event—the USPTO reconsidered and invalidated Compton's patent. On December 17, 1993, the USPTO ordered an internal re-examination of Compton's patent because, in the words of Commissioner Lehman, "this patent caused a great deal of angst in the industry."[23] On March 28, 1994, the USPTO released a preliminary statement declaring that "[a]ll claims in Compton's multimedia patent issued in August 1993 have been rejected on the grounds that they lack 'novelty' or are obvious in view of prior art."[24] This declaration was confirmed by the USPTO in November of 1994.[25]

## Patents in "Business Methods"

Recent federal judicial decisions have continued to support the rights of patentholders and have expanded the definition of "software" subject to protection by patent. On August 23, 1998, the Court of Appeals for the Federal Circuit (CAFC) upheld the validity of a "business methods" software patent in *State*

---

[20]Peltz, J. "Compton's wins patent covering multimedia," *Los Angeles Times,* November 16, 1993, D:2. The Compton's patent was entitled "Multimedia Search Systems Using a Plurality of Entry Path Means Which Indicate Interrelatedness of Information." Markoff, J. "Patent Office to Review A Controversial Award," *The New York Times,* December 17, 1993, D:2.

[21]Abstract, United States Patent Number 5,241,671, August 31, 1993.

[22]Abate, T. "Smaller, faster, better; Tech firms show off their latest wonders at trade show and foretell a user-friendly future," *San Francisco Examiner,* November 21, 1993, E:1.

[23]Markoff, J. "Patent Office to Review A Controversial Award," *The New York Times,* December 17, 1993, D:2.

[24]Riordan, T. "Action Was Preliminary On a Disputed Patent," *The New York Times,* March 30, 1994, D:7.

[25]Orenstein, S. "U.S. Rejects Multimedia Patent," *The Recorder,* November 1, 1994, 4.

*Street Bank v. Signature Financial Group*.[26] In ruling that the software was patentable, the court announced that

> the transformation of data, representing discrete dollar amounts, by a machine through a series of mathematical calculations into a final share price, constitutes a practical application of a mathematical algorithm, formula, or calculation, because it produces "a useful, concrete, and tangible result."[27]

The opinion has been criticized for supporting the patentability of common methods and systems previously considered unpatentable.[28]

Since the *State Street* decision, "business methods" patenting has expanded rapidly, especially for Internet-based transactions and marketing techniques. Undersecretary of Commerce and USPTO Director Dickinson noted in March 2000 that the number of applications for such patents had expanded from 1,275 in fiscal 1998 to 2,600 in fiscal 1999, resulting in the issue of 600 business methods patents in 1999. As in the case of the Compton's patent, the proliferation of Internet-based "business methods" patents has been facilitated by a lack of patent-based prior art available for review by USPTO examiners.[29] Although the doubling in business methods patent applications during fiscal 1998-1999 is noteworthy, issued patents in this class accounted for less than 0.5 percent of all issued patents in 1999.[30]

Political reactions to the surge in business methods patents and the controversy surrounding their validity were swift and involved both Congress and the USPTO. In late 1999, Congress passed the American Inventor Protection Act (AIPA). The AIPA was originally drafted to revise the U.S. patent system to be consistent with the World Trade Organization (WTO) agreements that concluded the Uruguay Round of trade negotiations, but additional provisions were added specifically to address the business methods patent controversy. One provision of

---

[26] 149 F.3d 1368 (CAFC, 1998).

[27] 149 F.3d 1368 (CAFC, 1998).

[28] Tim Berners-Lee, developer of the HTML software code that is widely used for the creation of websites, argues that some of the Internet business-methods patents "combine well-known techniques in an apparently arbitrary way, like patenting 'going shopping in a yellow car on a Thursday.'" (Waldmeir and Kehoe, 1999). A patent attorney has suggested that the opinion is so sweeping as to allow Newton to patent the calculus (National Public Radio, 1998).

[29] "'Now we're dealing with a much broader universe of "prior art,"' says J.T. Westermeier, a Washington D.C. internet attorney with Piper and Marbury, pointing out that many allegedly novel Internet business methods may already have been in use at universities or elsewhere" (Waldmeir and Kehoe, 1999).

[30] These data count only applications and issued patents in U.S. patent class 705 ("Data Processing: Financial, Business Practice, Management or Cost/Price Determination") as "business methods" patents. Depending on one's definition of this elusive concept, the number of applications and issued patents could in fact be substantially greater.

the AIPA that brought U.S. patent policy into conformity with WTO requirements stipulated the publication of most U.S. patent applications within 18 months after their submission to the USPTO. This publication requirement should make it easier for a would-be inventor to verify that he or she is not infringing pending patents. A second provision of the AIPA that was inserted in response to the business methods patenting controversy created a "first-to-invent" defense against infringement claims. Defendants who can show that they were practicing the relevant method or art one year or more before the filing of the patent application are protected against infringement suits. This provision also should reduce the exposure of inventors to infringement suits based on their use of long-established, nonpatented prior art.

Administrative responses to the business methods controversy included the USPTO's "Business Methods Patent Initiative," unveiled in the spring of 2000. The Initiative included several provisions:

1. Hiring more than 500 new patent examiners specializing in software, computer, and business methods applications

2. Tripling the number of examiners assigned to examine applications in Class 705, the primary locus of business methods patenting activity

3. Expanding the number of nonpatent "prior art" databases to which these examiners have access

4. Requiring that nonpatent and foreign prior art be searched systematically for all applications in Class 705

5. Requiring examination of all applications in Class 705 by a second examiner in addition to the primary examiner assigned the application

This administrative initiative has raised the level of scrutiny devoted to business methods patent applications and may have reduced the rate of issue of new patents in this class. The USPTO reported in 2001 that the number of examiners assigned to business methods patents increased from 45 at the beginning of fiscal 2000 to 82 by the end of fiscal 2001. The same report predicted that roughly 10,000 applications would be filed in Class 705, which covers most business methods patents, in fiscal 2001, an increase of nearly fourfold since fiscal 1999. However, the USPTO issued approximately 433 patents in Class 705 in fiscal 2001, a decrease of more than 25 percent from the number issued in this class in fiscal 1999.[31] The lags involved in review of patent applications (18 months to 2 years) and the rapid growth in applications during fiscal 1999-2001 mean that the number of business methods patents issued by USPTO almost certainly will increase in the future. Nevertheless, the drop in the number of issued business meth-

---

[31]See http://www.uspto.gov/web/menu/pbmethod/fy2001strport.html.

ods patents during 1998-2001 in the face of swelling applications suggests that the intensified scrutiny of applications in this class may indeed have reduced the rate of issue of business patents somewhat.

The economic significance and validity of U.S. business methods patents ultimately will be determined through litigation.[32] The "re-examination" system instituted in 1980 allows for interested parties to request that an issued patent be re-examined by the USPTO, but this procedure bears little resemblance to the more elaborate "oppositions" process of the European Patent Office (EPO) and a number of European countries. In particular, re-examinations affect a smaller share of issued patents and result in the invalidation or amendment of a smaller share of challenged patents than is true of the EPO oppositions process (see Graham et al., 2003; Merges, 1999).[33]

Although litigation provides rigorous scrutiny of patent claims and validity, it is a costly system for maintaining "patent quality"—the costs of a "typical" infringement suit are estimated to run to $1 million to 3 million. Moreover, litigation is a lengthy process,[34] meaning that the validity of key "foundational" patents in software or business methods, those on which subsequent inventors may rely (and for which they are either paying royalties or risking costly infringement penalties), may take years to be established. In fields that are evolving as rapidly as software, such delays could contribute to high uncertainty, high transactions costs, and impediments to innovation.

The nonlitigation avenues to ascertain the validity of business methods patents in the United States thus are limited, and the ultimate effectiveness of the Congressional and administrative initiatives described above cannot yet be ascertained. The possibility nonetheless exists that the global nature of the markets in which business methods patents are applied, especially those that rely on the Internet for their operation, may limit the proliferation of "junk patents." Given the footloose nature of the Internet,[35] global recognition of Internet-based busi-

---

[32]The Internet vendor of books and other products, Amazon.com, filed suit in 1999 against Barnes & Noble over the latter's alleged infringement of its patent on "one-click" order methods. Although Amazon was granted an injunction against Barnes & Noble's alleged infringement of its "one-click" patent by the District Court for the Western District of Washington State in December 1999, the CAFC reversed the judge in February 2001 and remanded the case to the District Court. Given the CAFC's central role in establishing the patentability of business methods, its reversal of an injunction in this case is noteworthy.

[33]According to Merges (1999), EPO opposition proceedings result in the invalidation of roughly one-third of the opposed patents, whereas the U.S. re-examination process invalidates only 12% of the patents for which re-examinations are requested.

[34]One estimate suggests that the duration of the "average" patent suit in District Court is 31 months (Magrab, 1993).

[35]An Internet enterprise can be established virtually anywhere in the world that has a reasonably well-developed infrastructure.

ness methods patents may be necessary to establish their economic value. At present, most European patent systems do not grant validity to business methods patents that do not have a "technical effect" (Hart et al., 1999). The precise meaning of this distinction is subject to considerable debate and interpretation, suggesting that at least some but by no means all business methods patents issuing in the United States will be upheld as valid within Europe. The value of many U.S. business methods patents therefore may be limited, although much uncertainty remains concerning their validity in foreign jurisdictions.

## PATENTING TRENDS IN THE U.S. SOFTWARE INDUSTRY

In this section, we examine the limited data on trends in software patenting in the United States during the 1980s and 1990s. As with most other elements of the software industry, definitional issues loom large—What is a software patent? In addition, the rapid growth in the number of software-related USPTO patents complicates longitudinal analysis: We wish to examine change over time in the number of software patents rather than change that may reflect a reclassification of patents from "all other" to a "software-related" category. Lacking a clear a priori definition of "software-related" patent classes, we focused on the following 11 main groups in the International Patent Class (IPC) classification scheme:[36]

**G06F** **Electric Digital Data Processing:**

3/ Input arrangements for transferring data to be processed into a form capable of being handled by the computer...

5/ Methods or arrangements for data conversion without changing the order or content of the data handled...

7/ Methods or arrangements for processing data by operating upon the order or content of the data handled...

9/ Arrangements for programme control...

11/ Error detection; Error correction; Monitoring...

12/ Accessing, addressing or allocating within memory systems or architectures...

---

[36]The IPC is a hierarchical classification system consisting of sections, classes, subclasses, and groups (main groups and subgroups). The IPC divides all technological fields into sections (designated by a capital letter), each section into classes (designated by a two-digit number), and each class into subclasses (designated by a capital letter). For example, "G 06 F" represents Section G, class 06, subclass F. Each subclass is in turn broken down into subdivisions called "groups" (which are either main groups or subgroups, although the former "main group" is of immediate concern in this paper). Main group symbols consist of the subclass symbol followed by a one- to three-digit number and an oblique stroke, for example, G 06 F 3/.

| | |
|---|---|
| 13/ | Interconnection of, or transfer of information or other signals between, memories, input/output devices or central processing units... |
| 15/ | Digital computers in general... |

**G06K** **Recognition of Data; Presentation of Data; Record Carriers; Handling Record Carriers**

| | |
|---|---|
| 9/ | Methods or arrangements for reading or recognising printed or written characters or for recognising patterns |
| 15/ | Arrangements for producing a permanent visual presentation of the output data |

**H04L** **Electric Communication Technique**

| | |
|---|---|
| 9/ | Arrangements for secret or secure communication |

These main groups were identified by examining overall patenting during 1984-1995 by the six largest U.S. producers of personal computer software, based on their calendar 1995 revenues.[37] These patent classes account for 57.1 percent of the more than 600 patents assigned to the 100 largest packaged software firms identified by *Softletter*, a trade newsletter, in its 1997 tabulation.[38] The 11 groups account for a higher share of the patents of these six firms during this period (72.8 percent) when we exclude unclassified design patents and IPC groups that were created after 1984 (e.g., main group G06F 17/ came into existence in 1990) (Table 1).

Because they exist throughout the 1984-1997 period, these patent classes provide a useful basis for examining time trends in U.S. software patenting. They do not map precisely to the universe of software patenting, but they do provide imperfect but reliable longitudinal coverage of the segment of the overall software industry identified by the OECD as "...the most dynamic segment of the core software industry (computer programming services, pre-packaged software, and integrated system design)" (OECD, 1998, p. 9). The data in Figure 1 indicate that the share of all U.S. patents accounted for by patents in these IPC groups more than doubled during 1987-1997, from 1.7 percent in 1987 to 3.8 percent in 1997. Moreover, growth in this share appears to accelerate after 1991, possibly as

---

[37]As reported in the *Softletter 100* (1996) this group includes Microsoft, Novell, Adobe Systems, Autodesk, Intuit, and Symantec. We chose to focus our analysis on the patents assigned to specialized, publicly traded software firms because the computation of a "software patent propensity" measure (software patents deflated by R&D spending) is meaningful only for firms reporting R&D spending for which one can assume that the bulk of this R&D spending is devoted to software development. As a result, our "definition" of software classes is somewhat narrower than that employed by Kortum and Lerner (1999), although they also found that the fraction of overall U.S. patenting accounted for by software patents increased during the 1985-1991 period.

[38]We are grateful to *Softletter* for permission to use these data.

**TABLE 1** Patenting by the Softletter 100 (1997), 1984-1997 (Total Patents = 627)

| (a) Int'l Patent Class | (b) Patent Count | (c) Share of All Firm Patents | (d) Cumulative Sum Firm Patents |
| --- | --- | --- | --- |
| G06F 15/ | 86 | 13.7% | 13.7% |
| G06F 9/ | 66 | 10.5% | 24.2% |
| G06F 3/ | 54 | 8.6% | 32.9% |
| G06F 13/ | 42 | 6.7% | 39.6% |
| G06F 11/ | 26 | 4.1% | 43.7% |
| G06F 12/ | 25 | 4.0% | 47.7% |
| G06K 9/ | 22 | 3.5% | 51.2% |
| H04L 9/ | 16 | 2.6% | 53.7% |
| G06K 15/ | 14 | 2.2% | 56.0% |
| G06F 7/ | 4 | 0.6% | 56.6% |
| G06F 5/ | 3 | 0.5% | 57.1% |

a reaction to the more expansive judicial treatment of the breadth and strength of patents in the 1990s.

### Software-Related Patenting by Packaged Software and Electronic Systems Firms, 1987-1997

This section examines the patenting behavior of large U.S. software firms during the 1980s and 1990s, focusing on large U.S. packaged software firms (based on revenues) identified in the 1997 tabulation of the 100 leading U.S.

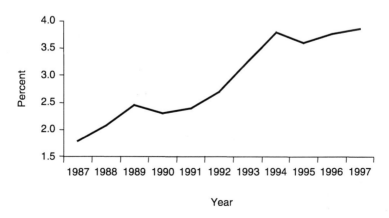

**FIGURE 1** Packaged software patents as share of all patents, 1987-1997.

packaged software firms compiled by the *Softletter*. These firms are of particular interest because packaged software is the product area in which formal intellectual property protection has become more important since 1980. These firms also are among the few U.S. firms whose publicly reported R&D spending can be treated for analytic purposes as devoted largely to software R&D, in contrast to diversified producers of electronic systems.

The 100 largest U.S. packaged software firms increased their share of software patenting during the 1987-1997 period from less than 0.06 percent in of all software patents in 1988 to nearly 3.25 percent in 1997 (Figure 2). Moreover, this trend is unchanged when Microsoft is eliminated from the ranks of the top 100 U.S. software firms (Figure 3), although the absolute magnitude of the increase in share is much smaller (from less than 0.1 percent in 1987 to slightly more than 0.7 percent in 1997). In both cases, the increase in large packaged software firms' patenting activity is most pronounced for the 1990s. However, despite the fact that the largest U.S. packaged software firms have increased their patenting activity relative to other software firms, their share of patenting within our software-related patent classes remains far smaller than that accounted for by a sample of 12 large electronic systems and component firms assembled for purposes of comparison (IBM, Intel, Hewlett-Packard, Motorola, National Semiconductor, NEC, Digital Equipment Corporation, Compaq, Hitachi, Fujitsu, Texas Instruments, and Toshiba). As Figure 4 shows, the share of overall software patenting accounted for by this group of firms grew from slightly more than 35 percent in 1987 to more than 45 percent by 1997.

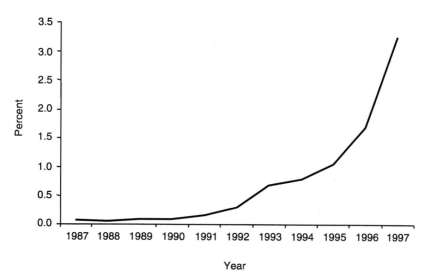

**FIGURE 2** Large packaged-software firms' software patents, share of all software patents, 1987-1997.

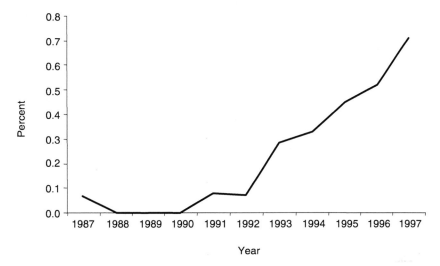

**FIGURE 3** Large packaged-software firms' software patents, share of all software patents (excluding Microsoft), 1987-1997.

These electronic systems firms thus account for a larger share of overall software patenting throughout the "pro-patent" period of 1987-1997, and their share of overall software patenting increases by nearly 10 percent, a substantially larger growth in share than that of our sample of large packaged software firms. Moreover, our relatively restrictive definition of "software patents," as well as our reliance on data from specialized producers of packaged software to develop this definition, mean that our data on patenting activity by these systems firms could understate their software-related patenting. For example, our definition of software excludes patenting activity in the "embedded software" (software that is

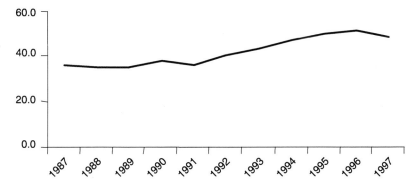

**FIGURE 4** Electronic systems firms' share of all software patenting, 1987-1997.

incorporated directly into a product and whose operation typically is not controlled by the user) that is included in such products as microprocessor chips or measurement instruments, although this class of software is not likely to be the locus of intensive patenting. Because we do not have software-related R&D spending for these systems firms, however, we cannot determine whether the increase in their share of overall software patenting reflects a reallocation in their R&D activities to focus more intensively on software-related innovations or instead is a result of an increase in their software-related patent propensity. Below, we pursue this issue in a comparison of "patent propensity" for IBM, a systems firm that reports software-related R&D spending, and Microsoft for the 1992-1997 period.

### Change in the "Patent Propensity" of Packaged Software Firms, 1987-1997

Our data for the sample of large packaged software firms enable us to analyze the "propensity to patent" of these firms, measured as the ratio of patents to constant-dollar R&D spending, during the 1987-1997 period (Figures 5 and 6).

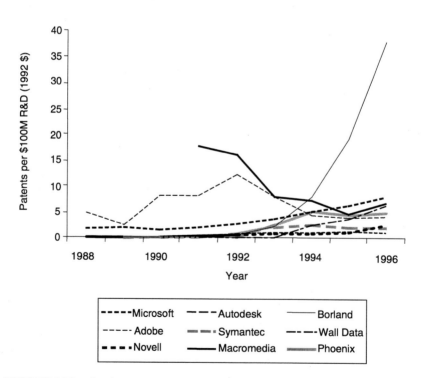

**FIGURE 5** Firm-level patent propensity, 3-year moving average, 1987-1997.

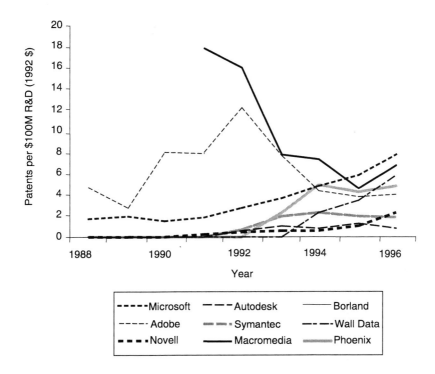

**FIGURE 6** Firm-level patent propensity, 3-year moving average, 1987-1997 (excluding Borland).

Firm-level patenting trends for a larger number of firms become almost unintelligible when presented in a single figure, and we therefore present data on trends in the average patenting propensity for the 15 largest U.S.-based packaged software firms (Figures 7 and 8). We also present data (Figure 9) on differences in the patenting propensities of "incumbents" (packaged software firms founded before 1985) and "entrants" (packaged software firms founded after that date).

Figures 5 and 6 display trends in firm-specific patenting propensities (a 3-year moving average) during 1987-1997 for the nine and eight largest U.S. personal computer software firms, respectively,[39] with significant patenting activity in 1997. All of these firms are publicly traded and therefore report annual R&D spending. Microsoft, by far the largest of these firms and the loser in the 1994 infringement suit filed by Stac Electronics, displays an upward trend (increasing

---

[39] As identified in the 1997 *Softletter* rankings of the top 100 packed software firms. Figures 5 and 6 include and exclude, respectively, Borland/Inprise. Borland/Inprise is excluded from Figure 6 to "decompress" the scaling of the figure and to facilitate the clearer depiction of trends in the patenting propensities for the other seven large packaged software firms.

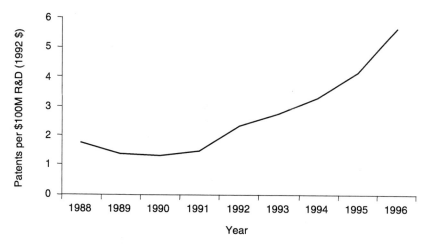

**FIGURE 7** Patent propensity, top 15 U.S. packaged software firms (1997), 3-year moving average, 1987-1997.

by roughly fourfold) in its post-1991 patenting propensity. Novell, Symantec, Wall Data, and Borland also exhibit increases in patenting propensity during the 1990s. Interestingly, the 1997 ratio of patents to R&D spending is highest for Borland, a packaged software firm with extensive experience in intellectual property litigation.

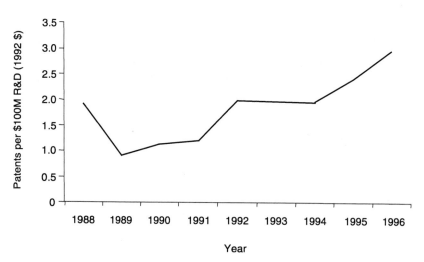

**FIGURE 8** Patent propensity, top 15 U.S. packaged software firms (1997), 3-year moving average, 1987-1997 (excluding Microsoft).

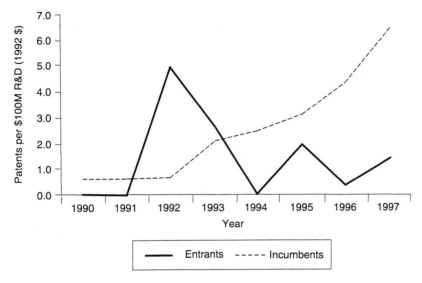

**FIGURE 9** Patent propensities of "incumbents" and "entrants," 1990-1997.

Patent propensities for the largest U.S. software firms as a group also grew during the 1987-1997 period. Figures 7 and 8 show trends in the aggregate patenting-R&D spending ratio during 1987-1997 for the 15 U.S. personal computer software firms listed by *Softletter* among the top 100 for which data are available throughout this period (once again, we use a 3-year moving average).[40] Both figures are weighted averages (weighted by R&D spending, which weights Microsoft heavily) of the patents-to-R&D spending ratios of these 15 firms. The weighted average exhibits a significant upward trend, reflecting the behavior of Microsoft. Nonetheless, excluding Microsoft from the data (Figure 8) does not change the basic conclusion; a modest increase in patent propensities is still apparent. Thus there is some evidence of increases in the aggregate patenting propensities of leading U.S. packaged software firms (as of 1997) during 1987-1997, although the size of this increase is affected by the behavior of the largest such firm.

Is increased patenting by large U.S. packaged software firms a result of entry by firms that are especially active patenters? We lack a clear basis for separating our group of large U.S. packaged software firms into "incumbents" and "en-

---

[40]The firms in the *Softletter* rankings for which 1986-1997 data are available from the Compustat Database and SEC reports include Microsoft, Adobe Systems, Novell, Autodesk, Symantec, The Learning Company, Activision, Borland, Phoenix Technologies, Quaterdeck, Micrografx, Caere, IMSI, Timberline Software, and Software Publishing.

trants," but on the basis of a visual examination of the data on founding dates for these firms, we chose 1985 to separate incumbents from entrants within the top 100 firms in 1997 (48 of these firms were founded before 1985). Figure 9 displays trends during 1990-1997 in the weighted average patenting propensities of the 15 largest incumbents and the 15 largest entrants (based on the 1997 *Softletter* ranking), defined as above.[41] There is almost no time trend in the patenting propensities of entrants (indeed, their patent propensity declines during 1992-1994), but incumbents exhibit a steady increase in their patenting propensity. Moreover, this difference between incumbents and entrants remains when Microsoft is excluded.

We pointed out above that electronic systems firms' share of overall software patenting substantially exceeded that of packaged software firms and that the increase in the systems firms' share of overall software patenting during 1987-1997 exceeded that of large packaged software firms. However, the lack of software-related R&D investment data for these systems firms means that we are unable to determine whether changes in systems firms' overall software patenting reflects a shift in their propensity to patent rather than growth in software-related R&D. IBM, however, began reporting software-related R&D investment data in its annual reports in 1992, enabling us to examine its software-related patent propensity for the 1992-1997 period. Figures 10 and 11 compare the 1992-1997 patent propensities of IBM, which for most of this period was the largest single producer of marketed software, and Microsoft, which overtook IBM in software-related revenues in 1997 (Table 2 reports software-related patents

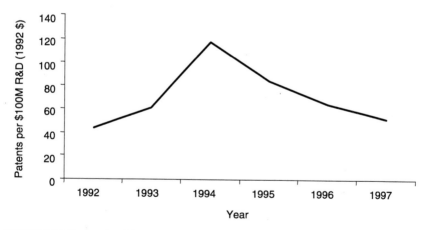

**FIGURE 10** International Business Machine patent propensity, 1992-1997.

---

[41]Our sample size and the length of the time series are limited by the need to sample only publicly traded firms, to enable us to compute the patent propensity measure.

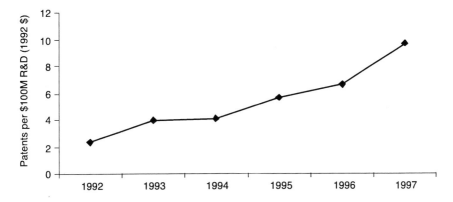

**FIGURE 11** Microsoft patent propensity, 1992-1997.

and R&D spending for the two firms). The R&D data reported by these two firms may not be strictly comparable, because a portion of Microsoft's total reported R&D investment may cover some fixed costs of maintaining an R&D facility that are not included in IBM's reported software-related R&D investment.[42] In addi-

**TABLE 2** IBM and Microsoft Software-Related R&D and Patenting, 1986-1997

| Year | IBM | | Microsoft | |
| --- | --- | --- | --- | --- |
| | R&D (MM 1992$) | Patents | R&D (MM 1992$) | Patents |
| 1986 | NA | NA | 24.9 | 1 |
| 1987 | NA | NA | 45.0 | 0 |
| 1988 | NA | NA | 79.8 | 1 |
| 1989 | NA | NA | 121.4 | 2 |
| 1990 | NA | NA | 191.6 | 3 |
| 1991 | NA | NA | 240.8 | 2 |
| 1992 | 1161.0 | 508 | 352.2 | 8 |
| 1993 | 1094.0 | 690 | 458.4 | 19 |
| 1994 | 779.0 | 965 | 581.6 | 26 |
| 1995 | 1114.0 | 1038 | 802.9 | 52 |
| 1996 | 1619.0 | 1200 | 1311.7 | 103 |
| 1997 | 1885 | 1166 | 1733.8 | 206 |

---

[42]The sharp swing in IBM's reported software-related R&D investment during 1993-1995 raises further issues of accuracy and/or reclassification of certain R&D expenses as more or less "software related."

tion, an unknown portion of Microsoft's reported R&D spending includes development programs for hardware such as the firm's recently announced "Xbox" and previous products. These data therefore may understate the Microsoft software-related patent propensity and overstate that for IBM.

Nonetheless, Figures 10 and 11 (which are not computed as moving averages) suggest that IBM is patenting in the software realm (as defined here) far more intensively than is Microsoft, which increased its patent propensity during the 1992-1997 period. IBM's patent propensity increased from slightly more than 40 patents per $100 million in R&D spending to roughly 55 per $100 million in software-related R&D investment during 1992-1997. Microsoft, on the other hand, increased its patenting per $100 million in R&D from slightly more than 2 to almost 10, a nearly fivefold increase. The reported patent propensity of IBM in 1992 is 20 times that of Microsoft, although this gap narrows by 1997, when IBM is receiving "only" 5 times as many patents per R&D dollar as Microsoft. Both IBM and Microsoft increased their patent propensity during this period, but the proportionate increase in Microsoft's patent propensity exceeds that for IBM.

## The "Importance" of Packaged Software and Electronic Systems Firms' Software Patents, 1987-1997

Increased patenting by large packaged software and electronic systems firms appears to track the trends in federal court decisions, as decisions such as *Stac Electronics* have been followed by increases in large firms' patent propensities. A closely related issue concerns the "quality" of the software patents issued to these firms, relative to all patents in our software classes, during this period of growth in software-related patenting. As we noted above, the growing use of patents for the protection of intellectual property in the software industry raises unusual challenges. The examination of patents within the USPTO for novelty, utility, and nonobviousness relies heavily on the study of patent-based prior art. Has the lack of patent-based prior art resulted in USPTO examiners approving the issue of trivial, "junk" software patents to leading software firms, as critics (Aharonian, 1993) have argued?

To examine trends in the "quality" of recent industrial software patents, we analyzed the frequency of citations to the software patents obtained by our sample of large packaged software firms, relative to citations to all software patents (defined as above). We conducted an identical analysis for the patents issued to the 12 electronic systems firms discussed above. Because of the requirement for inventors to cite prior art and the need for examiners to supplement these citations to prior art, the number of citations received by a patent serves as a crude measure of its technological importance. Moreover, recent empirical work (Trajtenberg, 1990) has found that heavily cited patents also are of greater economic value.

Our measure of "relative importance" compares the citation rates of patents issued to the *Softletter* 1997 top 100 firms over the two years after issue of the patent with the citation rates of all software patents issued in that year for three years after the year of issue. Relatively "important" patents will have citation ratios greater than one in value (i.e., they are cited more heavily than the average for all patents in the relevant classes), and relatively "unimportant" patents will have citation ratios of less than one. We compute the ratio of citation rates for firm patents to those for all software patents for two years after the date of issue. This patent citation measure is not sensitive to the truncation of the time period during which more recently issued patents can be cited, because it compares the citation rates of patents within the same cohort. Our citation measure also omits self-citations by the firms assigned the patent.

We computed this measure of patent "importance" for the patents issued during 1987-1997 to the 100 largest U.S. packaged software firms in 1997 (Figure 12) and our sample of electronic systems firms (Figure 13). This measure of patent importance for the packaged software firms (Figure 12) is greater than one in value through much of the 1987-1997 period, suggesting that the patents issued to the *Softletter* 100 software firms were cited more heavily during this period than were all software patents. Moreover, the modest upward trend in the measure through 1996 suggests that these firms' patents were being cited with growing intensity, relative to all software patents, during 1987-1996. There is little evidence of a strong trend of improvement or decline in this measure of patent importance for electronic systems firms (Figure 13), but the relatively flat time trend in Figure 13 contrasts with the upward trend in Figure 12. There is no evidence in Figure 13 of an increase in the relative intensity of citations to these electronic systems firms' software patents during the 1987-1997 period.

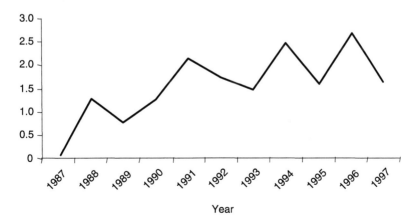

**FIGURE 12** Citations to top 100 packaged software firms patents/all software patent citations, 1987-1997.

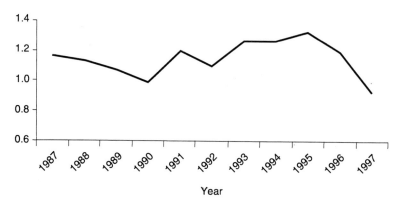

**FIGURE 13** Citations to electronic firms' software patents/all software patent citations, 1987-1997.

The data in both Figures 12 and 13 must be interpreted with caution, because it is possible that the "importance" of all software patents dropped precipitously during this period—we are able to compare only the importance of the software patents issued to the *Softletter* 100 or those issued to this sample of electronic systems firms with the importance of the software patents issued to all inventors. We also cannot compare the importance of these software patents with that of non-software patents—instead, these indicators shed light only on the "relative importance" of the software patents (as defined above) assigned to large packaged software or electronic systems firms. Nor can we exclude the possibility that packaged-software firms' patents are being cited more intensively because of the increased risk of infringement litigation involving questions of validity (Hall and Ziedonis, 2001).[43] Nonetheless, these trends indicate that the relative importance of the patents issued to large specialized producers of PC software, firms that have intensified their patenting activity during the 1990s, has not deteriorated during this recent period of significant growth in their software patenting. They also suggest that increased patenting by large electronic systems firms has not resulted in significant declines in the rate of citation to these firms' software-related patents, although these firms' patents are not being cited more intensively either during the period.

### University Software Patents

U.S. universities have long played a prominent role in the innovative activities of the U.S. software industry (Steinmueller, 1996; Mowery, 1999). Have

---

[43]Hall and Ziedonis (2001) suggest this possibility in discussing patent-citation trends in the semiconductor industry.

universities assumed a similarly prominent role in software patenting since 1980, a period that has witnessed a significant increase in overall patenting activity by U.S. universities? The number of patents issued to U.S. universities and colleges more than doubled between 1979 and 1984, more than doubled again between 1984 and 1989, and doubled yet again between 1989 and 1997 (Table 3). Figure 14, taken from the 2000 survey of member universities published by the Association of University Technology Managers (AUTM, 2000), shows a considerable increase in university-assigned patents per academic R&D dollar during the 1990s for all respondents to the AUTM survey. In other words, the overall patent propensity of U.S. universities grew steadily during the 1990s. This increased academic patenting activity is attributable to the Bayh-Dole Act of 1980, as well as the rapid growth in academic research in biomedical technologies (Mowery, et al., 2001).

Surprisingly, however, in view of the significant increases in university patenting in other fields (e.g., biomedical technologies), U.S. universities account for a small share of overall software patenting (as we have defined it) throughout the 1984-1997 period. As Figure 15 shows, university patents have never accounted for even 2 percent of the annual flow of issued software patents in the United States, less than the 3.6 percent share of overall patents accounted for by U.S. universities in the late 1990s (Mowery and Sampat, 2001). Indeed, the 1990s witnessed a slight decline in the share of software patents accounted for by universities.

We analyzed trends in the "importance" of university software patents by using the same measure that we employed for our examination of the patents issued to U.S. software firms (Figure 16). In some contrast to the patents issued to the *Softletter* 100 firms, which increase in importance relative to all software patents, the importance of university software patents displays little or no trend during the 1987-1997 period. The value of the "importance ratio" drops from a

**TABLE 3** Utility Patents Issued to U.S. Universities and Colleges, 1969-1997 (year of issue)

| Year | Number of U.S. Patents |
|------|------------------------|
| 1969 | 188 |
| 1974 | 249 |
| 1979 | 264 |
| 1984 | 551 |
| 1989 | 1228 |
| 1994 | 1780 |
| 1997 | 2436 |

SOURCE: U.S. Patent and Trademark Office (1998).

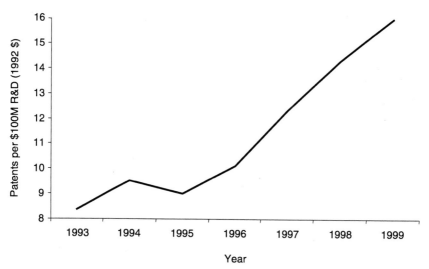

**FIGURE 14** Patents/R&D expenditures, all AUTM respondents, FY 1993-1999.

peak of nearly 3 in 1987 to a level slightly above 1, where it remains through 1996, increasing to nearly 2 by 1997.

This brief descriptive analysis of university patenting in software presents an interesting contrast to the discussion of industrial software patenting above. The

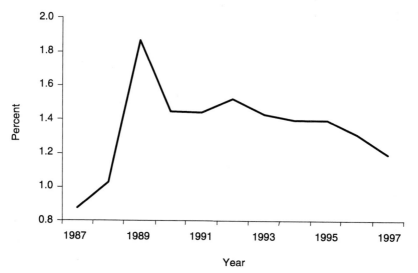

**FIGURE 15** University software patenting as a share of all software patenting, 1987-1997.

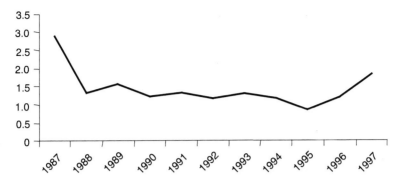

**FIGURE 16** University software patent citations/software patent citations, 1987-1997

Bayh-Dole Act appears to have increased U.S. universities' overall patent propensity during the 1990s, but we lack the necessary data to determine whether universities' software-specific patent propensity has increased.[44] Moreover, increased university patenting since 1980 is associated with a decline in U.S. universities' share of software patents, perhaps because of intensified patenting activity by software firms.

## THE RELATIONSHIP BETWEEN PATENTING AND COPYRIGHT IN SOFTWARE-RELATED INTELLECTUAL PROPERTY

As we noted above, both copyright and patent protection have been extensively employed in software-related intellectual property, and some of the current controversies over patents in software have precedents in debates over the advisability of copyrights for software. Indeed, one of the first scholarly analyses of methods for protecting software-embodied intellectual property (Menell, 1989) argued that patent protection of software was preferable because of the higher standards and more stringent reviews of prior art required for the issue of patents. Menell's analysis implicitly assumed that patents and copyright are substitutes, rather than complements, for the protection of software-related intellectual property. Lemley and O'Brien (1997) also asserted that the "primary means of legal protection for computer software has shifted from copyright to patent."

Nevertheless, little direct evidence has been adduced to support the contention that software inventors have shifted from copyright to patent. Indeed, a case can be made that copyright and patent protection are complements, rather than

---

[44]Indeed, the "user-active" character of much innovation in software, especially academic software innovation, means that innovation in software occurs in many academic departments all over university campuses, ranging from computer science to mechanical engineering to economics. As a result, obtaining the necessary data on academic "software-related" R&D funding to compute this patent propensity is nearly impossible.

substitutes, in the protection of software-related intellectual property. Copyright protection of the software code (the expression) could complement patent protection of the underlying technical advance. Although neither Menell (1989) nor Lemley and O'Brien (1997) give serious consideration to a complementary relationship between patent and copyright protection in software, it is possible that commercial software developers are indeed using both, rather than substituting patents for copyright.

In this section, we examine new data on software copyright registrations in a preliminary analysis of the changing relationship between copyright and patent protection in software. Just as we did in the examination of patent data for large packaged software firms above, we seek to develop measures of the "copyright propensity" of large packaged software firms during the 1987-1997 period. A finding that this propensity remained constant or increased would constitute evidence of complementarity between the use of copyright and the use of patents to protect software-embodied intellectual property, because these firms have increased their patent propensities during this period. A finding that the copyright propensity has declined, however, would provide preliminary support for the hypothesis that copyright and patent protection are substitutes, consistent with the Lemley-O'Brien argument cited earlier, and that commercial software firms now are relying more heavily on patents than copyrights to protect their intellectual property.

### Copyright Data

Our data on copyrighting of computer programs by packaged software firms are drawn from the U.S. Library of Congress (LOC) collection of registered U.S. copyrights. The LOC has data on all materials[45] that have been registered for copyright with the U.S. Copyright Office since 1978. Each record includes the identity of the entity requesting registration of copyright, a unique registration number, and the media type. Three dates are recorded for each registration: the date of creation of the work; its date of publication; and its date of copyright registration. As of January 2001, the LOC copyright database included over 13 million records.

Using the list of the largest packaged software firms in 1997 provided by *Softletter*, we searched these LOC records for uniquely numbered copyrights registered on "computer programs." Computer software can be designated as such by the author on the copyright registration form, and the Copyright Office assigns an internal "computer program" code to the relevant pieces of intellectual prop-

---

[45]Including books, maps, sound recordings, computer files, dramatic works, toys, games, jewelry, technical drawings, photographs, multimedia kits, sculptural works, textiles, motion pictures, and choreography, among others.

erty. We rely upon this latter internal code when defining a registered computer program copyright.

Although copyright provides some protection for a piece of written software regardless of whether it is registered with the Copyright Office,[46] there are additional incentives for pursuing registration of a copyright. The registration procedure is quick and inexpensive, and the legal strength of the resulting protection is greater for a registered copyright.[47] Registration within 5 years of original publication gives the copyright a presumption of validity under law.[48] Infringement actions cannot be brought in the courts until a copyright is registered.[49] The holder of a registered copyright is entitled to the recovery of attorney fees and statutorily defined damages, including those for willful infringement, only for the period after registration. Ordinarily, the owner cannot collect these damages for the period between the time of publication and the time of registration of the copyright, but the law offers an incentive for registering *early*: Damages are available from the date of publication *only* if the owner registers the copyright within 3 months of publication of the work.[50]

Faced with these incentives, it is plausible that rational actors in a crowded commercial space that rely on copyright to protect their software-related intellectual property will register the copyright on their software soon after creation. We therefore use data on registered copyrights to analyze trends in the use of copyright to protect software-related intellectual property. Our use of registered copyrights means that we are examining trends in the use by firms of copyrights for which some positive action and (modest) expenditure on the part of the "inventor" are required, rather than simply counting the copyrights that are created more or less automatically with the development of a new piece of software. Although all software is copyrighted at the moment of its creation, all software does not receive registered copyrights, and only registered copyrights provide a basis for the filing of a suit against an alleged infringer.

### Copyright Propensities Among the Leading U.S. Packaged Software Firms, 1987-1997

As in our analysis of software patenting among the largest U.S. packaged software firms, we restrict the sample of firms to include only firms for which R&D spending data are available, enabling us to compute "copyright propensities" for these firms. Our working definition of "software" in this analysis is

---

[46]The 1976 Copyright Act, in accord with the international Berne Convention, gives copyright protection to authors regardless of registration status.
[47]As of March 2001, registration required a two-page filing and fees totaling $30 US.
[48]17 U.S.C.A. §410 (2000).
[49]17 U.S.C.A. §411 (2000).
[50]17 U.S.C.A. §§412, 504, 505 (2000).

much broader than that employed in the examination of patent propensities, because the LOC does not provide any disaggregated copyright class information for its registered software copyrights. As before, however, we limit our sample of firms to those for which we can obtain R&D spending data, to compute a "copyright propensity."

The data in Figure 17, a weighted 3-year moving average of the "copyright propensity" for the same 15 large packaged software firms for which patent propensity data were plotted in Figures 7 and 8, tends to support the Lemley-O'Brien assertion that copyright protection has been supplanted by the use of patents in software, at least among these leading producers of packaged software. As Figure 18 shows, excluding Microsoft from this sample does not substantially alter the conclusion that the copyright propensity of these firms has declined. In data not displayed in these figures because of space limitations, the copyright propensity data for Novell, Microsoft, and Adobe all display declines in the number of copyrights registered per $100 million of (constant-dollar) R&D spending during 1987-1997. Novell and Microsoft in particular exhibit sharply contrasting trends in patents/R&D$ and copyrights/R&D$; both firms display increases during this time period in patenting propensity and a downward trend in the propensity to copyright their intellectual property. Adobe, which exhibited little or no consistent time trend in its patent propensity, also displays a downward trend in its copyright propensity. A comparison of the copyright behavior of "incumbent" and "entrant" firms among the *Softletter* 100 (defined as above) also yields little indication of contrasting behavior among these two groups in their copyright propensities. Both incumbents and entrants decreased their use of copyright, relative to R&D spending, to protect their intellectual property during the 1980s and 1990s.

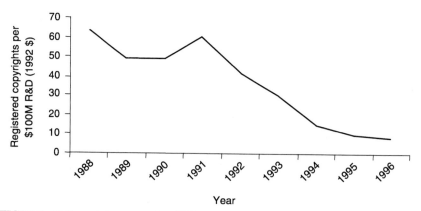

**FIGURE 17** Copyright propensity, 15 largest packaged software firms (1997), 3-year moving average, 1987-1997.

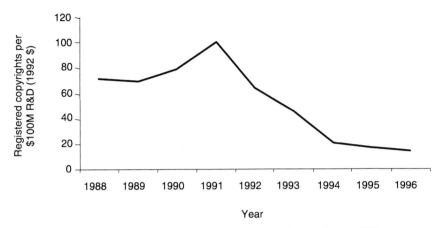

**FIGURE 18** Copyright propensity, 15 largest packaged software firms (1997), excluding Microsoft, 3-year moving average, 1987-1997.

Because we lack software-related R&D investment data for our sample of electronic systems firms, we are not able to examine changes during 1987-1997 in these firms' copyright propensities. However, we do have software-related R&D investment data for IBM for the 1992-1997 period, and Figure 19 compares trends during 1992-1997 in the copyright propensities of Microsoft and IBM. As in the case of these firms' patent propensities, IBM obtains substantially more registered copyrights per $100 million in R&D than does Microsoft throughout this period, and in contrast to their patent propensities, the gap has widened by 1997 (IBM's copyright propensity is roughly twice as large as that of Microsoft in 1992 and more than four times as great in 1997). But both firms are reducing their copyright propensity through this time period, consistent with the "substitute" relationship posited above.

Because our coverage of software copyrights differs somewhat from that of our software patents data, direct comparison of these trends in patent and copyright propensity must be interpreted cautiously. Nevertheless, our preliminary analysis of packaged software firms' use of copyright to protect software-related intellectual property suggests that patents have increased in importance relative to copyright as a means for the protection of software-related intellectual property during 1987-1997. Moreover, a decline in copyright propensity during the 1990s is apparent as well in our limited comparison of a leading packaged software firm and a leading electronic systems firm. As we noted above, a shift from copyright to patent protection was once seen as an important step to raise the threshold for protection of software-related intellectual property, and it is ironic that increased patenting by software firms has been accompanied by a chorus of concern over "junk patents." Junk patents may indeed be a problem (although our limited evidence on citations does not support this claim for the patents of large

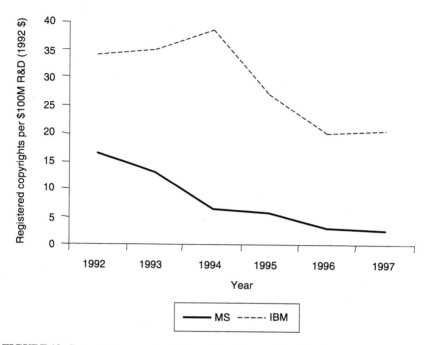

**FIGURE 19** Copyright propensity, Microsoft and International Business Machine, 1992-1997.

packaged software firms), but any such problem might have been more severe had firms continued to rely heavily on copyright in preference to patents.

Why might firms have shifted from copyright to patent protection? As we noted above in our discussion of the evolution of the software intellectual property "regime," the treatment of copyright by the U.S. federal judiciary has changed to limit the sweeping rights originally claimed by copyright holders.[51] This shift in judicial opinion may reflect the lack of a specialized appeals court that would support copyright holder rights as vigorously as the CAFC has done for patent-holders. Certainly, software patents have enjoyed a more supportive judicial climate during the past decade than copyright. In addition, patents may better support the types of "defensive" intellectual property strategies that Hall and Ziedonis (2001) describe in the semiconductor industry—cross-licensing of portfolios of

---

[51]Lemley and O'Brien note in their discussion that "...the courts have cut back the scope of protection rather dramatically in the past five years" (1997, p. 280).

patents may be less difficult than similar transactions in copyrighted material.[52] The use of software-related patents to support markets in intellectual property (suggested by Lemley and O'Brien, 1997) and/or as a complement to defensive intellectual property strategies remains an important issue for future research. Nonetheless, to the extent that transactions in intellectual property are facilitated by reliance on patent rather than on copyright, and to the extent that the (admittedly limited) quality controls imposed by the USPTO on the issue of patents enforce a higher average "quality level" among software patents than is true of copyrighted material, a shift from copyright to patent protection may well be a desirable development.

## CONCLUSION

The U.S. and global computer software industries have been transformed during the past 20 years as a result of the explosive diffusion of the microcomputer and the development of the Internet. No longer are the business activities and revenues of leading firms dominated by sales of products that incorporate high levels of user-specific customization. Instead, the dominant firms in the U.S. software industry, enterprises that account for a leading global market share as well, rely on sales of packaged software to mass markets. Accordingly, formal instruments for intellectual property protection have assumed much greater importance, despite the hazy and evolving legal status of these instruments. In the United States, which can be broadly categorized as an economy characterized in recent years by relatively strong protection for intellectual property rights, copyright protection for software-related intellectual property has been supplemented, and appears to have been supplanted, by patent protection.

The U.S. judicial and legislative arenas have strengthened the rights of owners of intellectual property in a number of industries since 1980, including computer software. The strong protection for intellectual property provided in the United States is followed by that in Western Europe, where the European Commission has provided somewhat more lenient treatment for "reverse engineering" of software for purposes of complementary invention, and Japan, where protection for software-related intellectual property historically has been relatively weak (see Merges, 1996). These contrasting regional or national systems of intellectual property policy have evolved in parallel with the software industries in each area. Indeed, the furor over the Compton's multimedia patent, as well as the more

---

[52]It is possible that software firms are choosing not to register copyrights because such early registration no longer is necessary to support litigation against alleged infringers, a possibility that would indicate greater judicial deference to copyright. This possibility seems unlikely, however, in view of the more circumscribed role accorded to copyright by the federal bench since the late 1980s that we described in earlier sections of this paper.

recent controversy over business methods patents, provides additional evidence of the influence of industry-led political action on U.S. patent policy.[53] Although U.S. intellectual property rights policy has influenced the development of its software industry, the reverse also is true. In other words, the relationship between the development of the domestic software industries and the intellectual property rights regimes of the United States, Western Europe, and Japan is best characterized as one of "coevolution," involving mutual causation and influence (Nelson, 1994; Merges, 1996; Khan and Sokoloff, 2001).[54]

Relatively large firms in the U.S. packaged software industry are shifting toward a more "patent-intensive" approach to the protection of their intellectual property, as the largest firms increase their patent propensities. Moreover, the evidence of increased patenting is strongest for older (and, in most cases, larger) firms within the U.S. industry. We observe no tendency for entrants to seek patent protection more intensively than incumbent firms. No evidence suggests that entry by specialized packaged software firms has been curtailed by these policies, however, and much more information is needed on entry, profitability, and the long-term evolution of industry structure before such a conclusion is warranted. The analysis also highlights the fact that despite increased use of patent protection by packaged software firms, large electronic systems firms are more important in overall software patenting. A comparison of the patent propensities of IBM and Microsoft suggests that the "patent propensity gap" between these two firms narrowed during the 1990s, but IBM continues to patent more intensively, relative to its R&D spending, than does the world's largest packaged software specialist. The limited evidence on the "importance" of the patents obtained by the largest U.S. packaged software and electronic systems firms does not support a characterization of these patents as "junk patents," by comparison with software patents generally. Moreover, large packaged software firms appear to be

---

[53]The filing by Microsoft of an amicus brief in *Apple Computer v. Franklin Computer* was noted above. The Business Software Alliance, a group enlisting Microsoft and other large packaged software firms (its members are Adobe, Apple, Autodesk, Bentley Systems, Borland, CNC Software/Mastercam, Macromedia, Microsoft, Symantec, and Unigraphic Solutions; additional members of its Policy Council include Compaq, Dell, Entrust, IBM, Intel, Intuit, Network Associates, Novell, and Sybase), also was active during the 1990s in appearing before congressional committees and filing amicus briefs, all in favor of stronger formal protection for software-related intellectual property.

[54]The endogenous nature of software-specific intellectual property rights policy, as well as intellectual property rights policies more generally, has been widely noted. In his 1996 discussion of software-related intellectual property policy in the United States, Western Europe, and Japan, Merges notes: "...[H]ow has the intellectual property regime affected the development of the software industry in these major markets? As we shall see, the answer must be incomplete unless one considers the converse question—how has the industry affected the legal regime?" (p. 275). In their discussion of the evolution of nineteenth century patent policies in the United States, Britain, France, and other nations, Khan and Sokoloff (2001) conclude that "...scholars who try to relate patterns of invention to patent system characteristics should be cognizant of, or take care in dealing with, the likelihood that those patent system characteristics are not exogenous with respect to the invention" (p. 28).

substituting patent for copyright, based on a comparison of trends in patent and copyright propensities.

Computer software as a product of inventive effort is nearly 50 years old, but the application of intellectual property rights to these products is relatively recent. Although patents were originally viewed by some experts as preferable to the extensive reliance on copyright for protection of software-related intellectual property (Menell, 1989) because of the higher threshold for patent protection, the expanded use of patents to protect software-related intellectual property has also sparked controversy. The 1998 *State Street Bank* decision extended patent protection into the previously unexplored area of "business methods," and growth since *State Street* in this class of patenting may trigger additional litigation over validity and infringement. Software patents, especially business methods patents, raise unusual challenges to the U.S. patent system, which relies on inventors and patent examiners for searches of "prior art" rather than allowing for interested parties to challenge patents before their issue in a formal pre-grant opposition process. Because of the historical lack of software patents, a primary source of software-related "prior art" scarcely exists, and this contributes to the issue of patents (such as the "multimedia" patent discussed above) of potentially sweeping breadth and limited validity. As the multimedia patent example suggests, there are few cases thus far of such broad patents being issued and upheld by either the USPTO or the courts. But the general problem is a serious one—how can searches of prior inventions be undertaken in a technology where patents have only recently become common?

Innovation in software generally is a cumulative activity, and individual software products frequently build on components from other products. As a result, some industry experts argue that software developers may become aware of a related patent only after they have completed development of a new product.[55] But this type of problem (which is not unique to software) is associated with the transition to a new, patent-based regime of intellectual property protection in software and may decline in severity as expanded software patenting expands the body of prior art that can be searched by patent examiners. Increased publication of patent applications after 18 months also should reduce the severity of this problem somewhat, and the liberalized "prior use" defense embodied in the AIPA also could reduce the incidence of litigation over infringement. The costs of the transition to patent-based protection of software-related intellectual property nevertheless could be high, because of the reliance on litigation to establish the validity of this growing body of prior art. The leading alternative mechanism in the

---

[55]Dan Bricklin, a pioneer in the packaged software industry and developer of the first spreadsheet program, argues that a typical software product may involve literally thousands of patentable processes, which creates enormous hazards for independent or small firm inventors who may belatedly discover that important components of their newly developed product are in fact patented by others (Merges, 1997, pp. 119-120).

U.S. system for challenging the validity of patents, re-examination, is utilized less extensively than the opposition process in the EPO. Nonetheless, current evidence on the EPO opposition system does not suggest that this process operates quickly or cheaply to resolve questions of patent validity (see Graham et al., 2003).

The computer software industry provides a fascinating "laboratory" for observing the transition from a relatively open intellectual property regime to one in which formal protection, especially patents, figures prominently. The cross-national differences in domestic patent systems, combined with cross-national differences in the structure of domestic software industries and domestic software markets, provide additional rich material for comparative studies of the interaction of intellectual property systems, innovation, and industrial development. Current research, including this chapter, has scarcely scratched the surface of this rich subject.

## REFERENCES

Abate, T. (1993). "Smaller, Faster, Better; Tech Firms Show Off Their Latest Wonders at Trade Show and Foretell a User-Friendly Future." *San Francisco Examiner*, November 21, E:1.

Aharonian, G. (1993). "Setting the Record Straight on Patents." *Communications of the ACM* 36: 17-18.

Association of University Technology Managers. (2000). *The AUTM Licensing Survey: Executive Summary and Selected Data, Fiscal Year 1999*. Norwalk, CT: Association of University Technology Managers: http://www.autm.net/surveys/99/survey99A.pdf.

Bresnahan, T., and S. Greenstein. (1996). "The Competitive Crash in Large Scale Commercial Computing." In R. Landau, T. Taylor, and G. Wright, eds., *The Mosaic of Economic Growth*. Stanford, CA: Stanford University Press.

CONTU (National Commission on New Technological Uses of Copyrighted Works). (1979). Final Report. Washington: USGPO.

Graham, S. J. H., B. H. Hall, D. Harhoff, and D. C. Mowery. (2003). "Patent Quality Control: A Comparison of U.S. Patent Re-examinations and European Patent Oppositions." In W. Cohen and S. Merrill, eds., *Patents in the Knowledge-Based Economy*. Washington, D.C.: The National Academies Press.

Hall, B. H., and R. M. Ziedonis. (2001). "The Patent Paradox Revisited: An Empirical Study of Patenting in the U.S. Semiconductor Industry, 1979-1995." *Rand Journal of Economics*, 32(1): 101-128.

Hart, R., P. Holmes, and J. Reid. (1999). "The Economic Impact of Patentability of Computer Programs: A Report to the European Commission." London: Intellectual Property Institute.

Khan, B. Z., and K. L. Sokoloff. (2001). "The Innovation of Patent Systems in the Nineteenth Century: A Comparative Perspective." Presented at the Franco-American conference on the Economics, Law, and History of Intellectual Property Rights, UC Berkeley, October.

Kortum, S., and J. Lerner. (1999). "What Is Behind the Recent Surge in Patenting?" *Research Policy* 28: 1-28.

Kuan, J. (1999). "Understanding Open Source Software: A Nonprofit Competitive Threat," unpublished manuscript, Haas School of Business, UC Berkeley.

Lemley, M., and D. O'Brien. (1997). "Encouraging Software Reuse." *Stanford Law Review* 49(2): 255-305.

Lerner, J., and J. Tirole. (2000). "The Simple Economics of Open Source," unpublished manuscript, Harvard Business School.

Magrab, E. B. (1993). "Patent Validity Determinations of the ITC: Should U.S. District Courts Grant Them Preclusive Effect?" *Journal of the Patent and Trademark Office Society* 75(125): 127-135.

Markoff, J. (1993). "Patent Office to Review A Controversial Award." *The New York Times*, December 17, D:2.

Menell, P. (1989). "An Analysis of the Scope of Copyright Protection for Application Programs." *Stanford Law Review* 41: 1045-1096.

Merges, R. P. (1996). "A Comparative Look at Intellectual Property Rights and the Software Industry." In D. C. Mowery, ed., *The International Computer Software Industry: A Comparative Study of Industry Evolution and Structure*. New York: Oxford University Press.

Merges, R. P. (1997). *Patent Law and Policy*. Charlottesville, VA: Michie.

Merges, R. P. (1999). "As Many as Six Impossible Patents Before Breakfast: Property Rights for Business Concepts and Patent System Reform." *Berkeley Technology Law Journal* 14: 578-615.

Mowery, D. C. (1996). "Introduction." In D.C. Mowery, ed., *The International Computer Software Industry: A Comparative Study of Industry Evolution and Structure*. New York: Oxford University Press.

Mowery, D. C. (1999). "The Computer Software Industry." In D. C. Mowery and R.R. Nelson, eds., *The Sources of Industrial Leadership*. New York: Cambridge University Press.

Mowery, D. C., R. R. Nelson, B. Sampat, and A. A. Ziedonis. (2001). "The Growth of Patenting and Licensing by U.S. Universities: An Assessment of the Effects of the Bayh-Dole Act of 1980." *Research Policy* 30: 99-119.

Mowery, D. C., and B. N. Sampat. (2001). "University Patents and Patent Policy Debates: 1925-1980." *Industrial and Corporate Change* 10: 781-814.

Mowery, D. C., and T. Simcoe. (2001). "The Origins and Evolution of the Internet." In B. Steil, R. Nelson and D. Victor, eds., *Technological Innovation and National Economic Performance*. Princeton, NJ: Princeton University Press.

National Public Radio. (1998). "Patenting Math Formulas." *All Things Considered* August 11, transcript 98081115-212.

Nelson, R. R. (1994). "The Co-Evolution of Industries and Institutions." *Industrial and Corporate Change* 3: 47-64.

OECD. (1998). "Measuring Electronic Commerce: International Trade in Software." Directorate for Science, Technology, and Industry.

Orenstein, S. (1994). "U.S. Rejects Multimedia Patent." *The Recorder*, November 1, 4.

Peltz, J. (1993). "Compton's Wins Patent Covering Multimedia." *Los Angeles Times*, November 16, D:2.

Riordan, T. (1994). "Action Was Preliminary On a Disputed Patent." *The New York Times*, March 30, D:7.

Samuelson, P. (1990). "Benson Revisited: The Case Against Patent Protection for Algorithms and Other Computer Program-Related Inventions." *Emory Law Journal* 39: 1025-1122.

Softletter. (1986). "The 1986 Softletter 100." *Softletter Trends & Strategies in Software Publishing* 3: 1-8.

Softletter. (1991). "The 1991 Softletter 100." *Softletter Trends & Strategies in Software Publishing* 8: 1-24.

Softletter. (1996). "The 1996 Softletter 100." *Softletter Trends & Strategies in Software Publishing* 12: 1-24.

Softletter. (1998). "The 1998 Softletter 100." *Softletter Trends & Strategies in Software Publishing* 14: 1-24.

Steinmueller, W. E. (1996). "The U.S. Software Industry: An Analysis and Interpretive History." In D.C. Mowery, ed., *The International Computer Software Industry: A Comparative Study of Industry Evolution and Structure*. New York: Oxford University Press.

Trajtenberg, M. (1990). "A Penny for Your Quotes: Patent Citations and the Value of Innovations." *The Rand Journal of Economics* 21: 172-187.

U.S. Bureau of the Census. (1995). *Service Annual Survey: 1995*. Washington, D.C.: U.S. Government Printing Office.

U.S. Department of Commerce. (1998). *U.S. Industry and Trade Outlook 1998*. Washington, D.C.: U.S. Government Printing Office.

U.S. International Trade Commission. (1995). *A Competitive Assessment of the U.S. Computer Software Industry*. Washington, D.C.: U.S. International Trade Commission.

Waldmeir, P., and L. Kehoe. (1999). "E-commerce Companies Sue to Protect Patents: Intellectual Rights Given Legal Test." *Financial Times*, October 25, 16.

# Internet Business Method Patents[1]

John R. Allison
Emerson H. Tiller
McCombs School of Business
University of Texas at Austin

## INTRODUCTION

The large number of Internet business method patents applied for and received since the mid-1990s has raised considerable concern among policymakers, academics, business, and other interested observers. That business methods are patentable subject matter seems to be beyond question after the decisions in *State Street Bank & Trust Co. v. Signature Financial Group, Inc.*[2] and *AT&T v. Excel Communications.*[3] Nonetheless, criticisms of these patents have been numerous. Some commentators attack the practice of patenting business methods rather than technology,[4] with Internet business methods taking the brunt of the criticism given that they make up the bulk of newly granted business method patents. At another level, many critics argue that granting patents on Internet-related software and business methods "closes" the Internet environment, making it more difficult for the diffusion of ideas, innovation, and entrepreneurial activity that are often associated with the Internet.[5] This criticism is especially relevant for those who argue that larger business organizations are patent mills, able to squeeze out small entrepreneurs with new property rights over Internet business activities. Others see Internet-related patents as an expansion of software patents more generally, some-

---

[1]The authors thank Thomas Bohman and Xinlei Wang of the Information Technology Services at the University of Texas for statistical consulting.

[2]149 F.3d 1368 (Fed. Cir. 1998).

[3]172 F.3d 1352 (Fed. Cir. 1999).

[4]John R. Thomas observes that recognizing the patentability of business processes opens the door for patenting new developments in all of human experience and that patents should remain grounded in science and engineering, areas that traditionally have been viewed as "technology." He fears that we may have paved the way for patenting developments in the liberal arts, social sciences, the law, and other indeterminate areas of human activity (Thomas 1999). *See also* Durham (1999) using similar reasoning to argue that software-embodied business method patents should not be patentable subject matter.

[5]Lessig (2001).

thing critics have attacked as duplicative of copyright protection and harmful to innovation. There are also concerns from the international community that U.S. firms may be gaining an unfair advantage in patenting in this area, especially over Japan and Europe, who have been slower to adopt a pro-patent stance to business methods.

Critics from all sides argue that Internet business method patents are too easily granted and are "weaker" than other patents because of inadequate reference to prior art in the patent applications. The main target of this criticism has been the U.S. Patent and Trademark Office (USPTO), the institution in charge of granting patents and ensuring the quality of the patents that eventually issue. There is special concern about whether the USPTO has adequately reviewed Internet business method patent applications and whether the prior art references in those patent applications are sufficient to warrant patent issuance. In the areas of software patents generally, and business method patents particularly, there has been much concern that the corps of patent examiners has been insufficiently populated with those qualified to seek out nontraditional sources of prior art and to knowledgeably examine these patents. Some observers argue that examiner inexperience has been and continues to be a major problem in these areas.[6] Only recently has the USPTO begun to hire examiners in software and related fields[7] and, even more recently, to institute programs for training and providing more access to literature on the business disciplines.[8]

Because much of the criticism of Internet business method patents focuses on their perceived differences from other patents granted by the USPTO, it is important to know whether these patents do in fact differ from the more general patents that issue from the USPTO and, if they do differ, in what ways. Our study compares characteristics of Internet-related patents with a random set of more general patents issued by the USPTO during a contemporaneous time period to see whether there are observable differences that would justify the criticisms. The main motivation of this study is to inform the debate over Internet business method patents with facts, rather than speculation, about the differences between these patents and more general patents granted by the USPTO. We conclude that criticisms of Internet-related patents that focus on prior art in particular should be taken with some caution, as we find the statistical differences between these patents and more general patents to be small and, if anything, to suggest that Internet-related patents are well supported by prior art references.

We note that this study looks primarily at quantitative data from patents rather than the quality of the information provided in the patents. Among other data, we collected information on the total number of patent and nonpatent prior art references, the amount of time a patent spent in the USPTO before issuance, and the

---

[6]See, e.g., Ross (2000).
[7]See, e.g., Cohen and Lemley (2001).
[8]See United States Patent & Trademark Office (2000).

country of origin of the invention. Our data on number of references, for example, do not tell us anything about the quality and relevance of the references or how well differentiated the claims are from the relevant prior art. The information we gathered would best be used in conjunction with other indices of quality. With some caution, we do provide additional measures that may further the quality of the inquiry—such as the type of nonpatent prior art reference cited (e.g., academic vs. popular press) or the type of Internet patent being examined (i.e., business models vs. business techniques vs. software techniques). The data, and the motivation behind including each type of data, are discussed below.

## THE DATA

We compared two data sets in this analysis. The first—data on a random set of general patents issued by the USPTO—was generated in a previous study by John R. Allison and Mark A. Lemley.[9] That data set (General Patent Data Set) consisted of a random sample of 1,000 patents issued by the USPTO between mid-1996 and mid-1998. For each patent in the sample, Allison and Lemley obtained a wide variety of information including, among other items (1) the number and type of prior art references cited on the face of the patent; (2) the invention's state or country of origin; (3) the time spent in prosecution; (4) small or large entity status and type of entity owning each patent; and (5) the number of inventors.

The second data set (Internet Patent Data Set) was developed especially for this study. It generally mirrors the data categories from the General Patent Data Set, with a few additions.[10] We list below the data elements collected from each Internet patent and our motivation for including each element.

---

[9] Allison and Lemley (2000).

[10] To create the Internet Patent Data Set, we used the Lexis-Nexis database of full-text patents. The word search request that we ultimately used was "Internet or World Wide Web" within three USPTO classifications, 705 ("Data Processing: financial, business practice, management, or cost/price determination"), 707 ("Data Processing: database and file management, data structures"), and 709 ("Electrical Computers and Digital Processing Systems: Multiple Computer or Process Coordinating"), with a date parameter of 1/1/90 to 12/31/99. This search produced over 2,800 patents, most of which were actually issued during 1998-1999, thus making our Internet Data Patent Set essentially contemporaneous with our General Patent Data Set. Although we found that some software patents clearly targeted at use with the Internet could be found in older computer technology USPTO classifications, such as 345, 365, 370, and 375, by far the greatest concentration of these patents was found in the more recently created "Data Processing" classifications in the 700 series. We concluded this after running the search terms across all USPTO classifications. More particularly, most patents were concentrated in classifications 705, 707, and 709. If the written description of the invention in these three classifications clearly demonstrated that the invention was targeted at the Internet, it was included; otherwise, it was discarded. We approached the study of these patents with the attitude that we would take the inventors at their word as to whether the invention was Internet-related. After discarding approximately 50% of the patents found with our search strategy, we ended with a data set of 1,423 patents to be studied further for data extraction and analysis.

## Number of Prior Art References (Backward Citations)[11]

In patent applications, the referencing of prior patents and other published resources ("nonpatent references") describing related technological advances are considered key in establishing that the invention is novel[12] and nonobvious.[13] Much of the criticism surrounding Internet business method patents relates to the inadequacy of prior art cited in these patents. Evidence in various patent litigation studies suggests that uncited prior art—prior art that was not before the patent examiner—is the most common basis for court decisions invalidating U.S. patents.[14] It would seem to follow that fewer prior art references in patents would tend to decrease the probability that they would be held valid if challenged in court. Stated differently, a larger number of prior art references may point to a more serious effort by the applicant to differentiate its invention from the prior art and perhaps to a more thorough examination in the USPTO, resulting in a stronger patent more likely to withstand challenge.[15] Some research also suggests that,

---

[11]There is stronger empirical evidence that the number of *forward* citations is a predictor of patent value (Hall et al., 1998; Harhoff et al., 1999b; Trajtenberg, 1990). The term "forward citations" refers to later patents that cite the patent in question as a reference. We did not measure forward citations because the patents in our data set were so recent, the great bulk of them having been issued during 1998-1999. Our study design and data collection began early in 2000 and ended before the middle of 2001. Collection of data on forward citations should be done when sufficient time has passed for these data to be meaningful for the entire data set, especially for those patents issued toward the end of 1999.

[12]Other types of prior art consist of evidence that an invention had been in public use or had been placed on sale either before the conception of this particular invention, 35 U.S.C. § 102(a) (2000), or more than 1 year before the patent application for this invention was filed, 35 U.S.C. § 102(b).

[13]The requirement of novelty in section 102 of the Patent Act is that the invention be different from anything previously revealed in a single piece of prior art. The requirement of nonobviousness, 35 U.S.C. § 103, is that the invention be different *enough* from what is taught by the cumulative prior art to represent a significant, "nonobvious" advance over that art.

[14]See, for example, Allison and Lemley (1998), examining litigated patents leading to final written decisions on validity or invalidity during 1989-1996.

[15]Patent references are listed in the patent by both the applicant and the examiner, but it is not feasible to determine for a large number of patents which references were cited by the applicant and which by the examiner because one must study the prosecution history in the USPTO to make this determination. However, there are reasons to believe that the great majority of prior art referenced in patents has been cited by applicants rather than by the USPTO examiner. See Allison and Lemley (2000). Allison and Lemley found that U.S. patents on foreign-origin inventions cite much more foreign-origin prior art and much less U.S.-origin prior art than do U.S. patents on U.S.-origin inventions. There is reason to believe that these foreign applicants for U.S. patents have more access to foreign-origin prior art in their language. However, if very much more prior art were cited by the U.S. patent examiner, one would expect it to be English-language prior art. The fact that this did not appear to occur supports the inference that most of the prior art, at least in this subset of U.S. patents, is cited by the applicant and not the examiner.

Another observation clearly provides strong support for this conclusion. One finds wide variations in the number of patent and nonpatent prior art references among U.S. patents in the same area of technology. We certainly found this to be true in the case of Internet patents. Unless all or most of this

on average, there is likely to be a correlation between the number of references and patent value. The number of prior art references should relate positively to the resources devoted by the applicant, and possibly by the patent examiner, to the patenting process, thus supporting an inference of greater patent value. One study found empirical support for the notion that the number of prior art references is positively correlated to patent value,[16] although others found no statistically significant relationship.[17] In addition, most observers would expect Internet business method patents to cite fewer patent references than patents in general (given the short time for which business methods have been recognized as patentable subject matter). They would also expect software-related inventions (most Internet-related patents fit into this category) to rely more on citations to other software and industry publications—nonpatent prior art references—than would more general patents given the shorter cycles of innovation involved with software. In our study, a finding that Internet business method patents contain fewer total references, and especially fewer nonpatent prior art references, would add strength to the criticisms that Internet business method patents are being granted without sufficient review by the USPTO. We consider together and separately the number of both patent and nonpatent prior art references.

Additionally, data were collected on the number of nonpatent prior art references in eight categories of nonpatent prior art for 285 of the 1,423 Internet patents (20 percent random sample). Before taking this 20 percent sample, we performed a trial study of approximately 100 patents to ascertain the different types of nonpatent prior art and ultimately group them into these eight categories. The purpose of this data collection effort is to give us a better understanding of what types of nonpatent prior art are being cited in Internet patents. Some may argue that certain types of nonpatent prior art references are "better" than others, or at least that the types of nonpatent prior art cited in Internet patents are different from those cited in general patents. We created the following eight categories of nonpatent prior art.

• Academic and Trade: This category includes academic and trade books, book chapters, articles, and proceedings papers. We did not differentiate between academic and trade publications because in this field there is much overlap and collaboration between academic and industry researchers. This category represents publications characterized by the *existence of an intermediating influence*

---

prior art is cited by applicants, patent examiners in the same technology area do not have access to the same resources, they do not communicate with each other, there is little supervision by primary examiners, or all of the above. There is absolutely no reason to believe any of these possibilities, and, therefore, most prior art is probably cited by applicants.

[16]Harhoff et al., (1999a).
[17]Lanjouw and Schankerman (2001); Lanjouw and Schankerman (1997).

such as an independent reviewer or editor to increase the probability of objectivity.

- Company and Industry: This category includes company- and industry-sponsored publications, press releases, web sites, and advertisements. These are so categorized because they have *no independent intermediating influence* to increase the probability of objectivity. This category does not include software and software documentation.
- University Publications: This category includes publications from universities or consortia of universities, such as those from university research labs, departments (such as computer science), and individual faculty, as well as theses and dissertations.
- Government Documents: This category includes government documents, publications, and web sites, except for published patent applications and searches. It includes U.S. and foreign government publications, as well as those of international government organizations such as the World Intellectual Property Organization within the United Nations (WIPO).
- Software: This category includes software programs and software documentation.
- Popular Press: This category includes not only newspapers, magazines, and other publications of general interest, but also news publications aimed at general business and legal audiences.
- Published Patent Applications and Search Reports: This category includes published patent applications from any patent office that publishes them and published patent office search reports, which are most commonly those done pursuant to the PCT (Patent Cooperation Treaty) and which often are issued by the European Patent Office (EPO).
- Other: This category includes sundry items such as individual web pages, but most references placed in this category are those in which insufficient information was provided to determine what the item really is, even after we conducted a very thorough Web search of key names and terms in the incomplete reference. One example is a reference to a partial title of an item, followed by "found on the web on x date."

The General Patent Set did not contain comparable data on nonpatent prior art references.

## Entity Status of Patent Assignee

The entity status of the owner of the patent including Individual, NonProfit (such as a university or a foundation), Small Business (500 or fewer employees), or Large Entity was collected. If large businesses receive more Internet patents than patents in general, then criticisms that the Internet is being dominated by big business and that entrepreneurs are being shut out gains some credence.

### Geographic Origin of Patent

The geographic origin of the invention (by country) was determined from the residences of a majority of the inventors (or plurality if no majority). We measured this variable partly to see whether the ratio of U.S.-origin to foreign-origin inventions receiving U.S. patents is greater for Internet-related patents than for general patents. If so, those observers suggesting that the United States is dominating Internet patents may be correct.

### Days in USPTO

We also measured the amount of time that the patent application spent in the USPTO from the original U.S. priority filing date to the time of issuance to see whether Internet patents were receiving the same amount of attention, in terms of time, as general patents. Greater pendency times may relate both to the seriousness of the applicant and the resources it is willing to devote to obtaining a patent and to the thoroughness of the examination process.

### Internet Patent Subtype

We evaluated each Internet patent for inclusion into one of three Internet patent categories that we created. Through discussions with companies that have Internet-related patents, a review of the popular press and literature on Internet patents, and a review of a subset of Internet-related patents, we came up with a typology of Internet patent subtypes. We broke the patents into three subtypes. The first two subtypes we call Internet Business Model (I-Business Model) Patents and Internet Business Technique (I-Business Technique) Patents. These two groups together constitute what most people believe to be "Internet business method patents." Well-known examples of each include Priceline.com's patent on the "Name Your Own Price" method of doing business (we identify this an I-Business Model because it can be a stand-alone business or a distinct line of business) and Amazon.com's patent on "1-Click" checkout (which we identify as an I-Business Technique because it is unlikely ever to be a stand-alone business). The final subtype is Internet Software Technique (I-Software Technique) Patents, which are clearly aimed at the Internet but which purport only to be technical software advancements. Our categorizations of patent subtypes were based on the written description of the invention contained in each patent. The description typically reveals the inventor's (and, perhaps, her supervisor's) vision of what the patent is actually projected to do.[18] The Appendix to this chapter gives a more

---

[18]Claims in such patents often, but not always, read like technical software patents. Had we been focused on validity or infringement, we obviously would have looked to the claims. In placing Internet-related patents within these categories, however, the best information source was the description.

complete description of how these Internet patent subtypes were determined, along with some examples.[19] The reason we created these subtypes was to (1) identify the Internet patents that were more business concept than technology driven, because the former may be more controversial than the latter, and (2) see whether entity size was related to the level of business concept or technology of the patent, in other words, does large business dominate the business concept patents at the expense of small business or individual inventors?

## DATA ANALYSIS

Our goal for these comparisons was to examine continuous and categorical patent attributes by patent type (Internet-related compared with General, or Internet Business Methods compared with General) or Internet patent subtype (I-Business Model, I-Business Technique, or I-Software Technique) compared with General. For continuous variables, we used models that assume a normally distributed outcome and employ nonparametric tests (e.g., the Wilcoxon, Savage, and median tests) that make less stringent assumptions about the distribution of the dependent measures. By including both, we "triangulated" the analyses, which provided more evidence of the correct statistical conclusion. The statistical methods are described more fully as the results are presented in the chapter.

### Prior Art References

We first look at our main variable of interest—prior art references. As stated above, much of the criticism surrounding Internet-related patents has been the perceived absence, or inadequacy, of prior art. Critics of Internet-related patents would expect any likely deficiency to show up in the number of prior art references, especially nonpatent prior art references. In comparing the General Patent Data Set with the Internet Patent Data Set, we look at total number of references, number of patent prior art references, and number of nonpatent prior art references. A finding that, compared to general patents, Internet-related patents have fewer total references, and/or that Internet-related patents have fewer nonpatent prior art references, would strengthen the critics' position.

#### Comparison of All Internet-Related Patents and General Patent Data Set

We first look at the full Internet Patent Data Set and compare it against the General Patent Data Set. The reason to look at the full set (which includes not

---

[19]We note here that our categorization of Internet patent subtypes should be taken with great caution. Although we believe the categories make intuitive sense, we coded the patents ourselves (because of the resource constraints of this study). Distinguishing between I-Business Models and I-Business Techniques was especially challenging. A more valid study of these patent subtypes would require multiple coders from Internet business backgrounds.

only business methods but also Internet software techniques) was to (1) address the broader debate about patenting ("closing") the Internet generally and (2) allow for an alternative, and more inclusive, set of Internet patents should our definitions and coding of each Internet patent subtype be flawed in our subsequent analyses. Table 1 shows the untransformed means and standard deviations of the three measures of patent references by patent type.

Two initial analyses were performed: an Independent Groups $t$-test and a Wilcoxon nonparametric test.[20] On the basis of the descriptive analysis of the distributions of the three measures of Patent References, log transformations were useful for normalizing Number of Patent References (PatRefs) and Total Number of References (TotRef), but not for Number of Nonpatent References (NonPatRefs). The log transformation was used to normalize the distributions in order to satisfy the normality assumption for the Independent Groups $t$-test. The homogeneity of variance assumption can be tested precisely by using the $F$-test of the difference in two or more variances. We report the results of this test and use the $t$-test with the Satterthwaite correction to the degrees of freedom when this assumption is violated.

Number of Total Prior Art References: The Independent Groups $t$-test using the log-transformed Total References showed that there was a statistically significant difference between General (mean = 2.47) and Internet (mean = 2.68) patents $[t(2255) = -6.59, p < 0.0001]$.[21]

**TABLE 1** Patent References (General Patent Data Set Compared with Internet Patent Data Set)

| Internet | $N$ | Variable | Mean | Std Dev | Min | Max |
|---|---|---|---|---|---|---|
| General Patents | 1,000 | TotRefs | 15.16 | 16.29 | 0 | 163 |
|  |  | PatRefs | 12.79 | 14.13 | 0 | 154 |
|  |  | NonPatRefs | 2.37 | 6.56 | 0 | 68 |
| Internet Patents | 1,423 | TotRefs | 23.03 | 48.53 | 0 | 457 |
|  |  | PatRefs | 14.23 | 23.30 | 0 | 353 |
|  |  | NonPatRefs | 8.80 | 34.43 | 0 | 391 |

[20]The basic assumptions of a $t$-test are (1) randomness (the units of analysis in the study must be sampled at random); (2) independence of errors (units of analysis sampled must be randomly assigned to the different groups); (3) normality of errors; and (4) homogeneity of variances (variation around mean is equivalent in the two groups). The normality assumption can be examined by visually examining the distributions of the outcome variables.

[21]The $t$-test used the Satterthwaite adjustment to the degrees of freedom to account for the unequal variances between groups $[F(1422, 999) = 1.18, p < 0.0044]$. In addition, a nonparametric Wilcoxon test was performed with the unadjusted values, which also showed that there was a statistically significant difference between the number of total prior art references cited in general patents and Internet patents [Wilcoxon test statistic = 1103312, $Z = -6.42, p < 0.0001$].

Number of Patent Prior Art References: The Independent Groups $t$-test using the log-transformed Patent References did *not* show a statistically significant difference between General (mean = 2.30) and Internet (mean = 2.35) patents $[t(2421) = -1.51, p < 0.1301]$.[22]

Number of Nonpatent Prior Art References: The Independent Groups $t$-test using the untransformed nonpatent references showed a statistically significant difference between General (mean = 2.37) and Internet (mean = 8.81) patents $[t(1567) = -6.87, p < 0.0001]$.[23]

In sum, we find that the full set of Internet-related patents are supported by more total references and more nonpatent references than General patents. However, there is no statistical evidence to show that Internet-related patents are supported by more or fewer patent references. These findings suggest that criticisms of Internet-related patents that are based on the amount of prior art cited (especially nonpatent prior art) are not supported by the data.[24]

## Comparison of Internet Business Method Patents and General Patents

Excluding Internet software patents from the data set and looking only at Internet business method patents (both I-Business Model and I-Business Technique patents), gave us similar results. Looking at only the Internet business method patents eliminates any confounding effects from the software patents and focuses the empirics on the most controversial types of Internet-related patents.

Table 2 shows the means and standard deviations of the three measures of patent references by patent type.

Based on the descriptive analysis of the distributions of the three measures of Patent References, log-transformations were useful for normalizing Number of Patent References (PatRefs) and Total Number of References (TotRef) but not for Number of Nonpatent References (NonPatRefs). We used Independent Groups $t$-tests to compare means between General and Internet business method patents for all three reference measures.

---

[22]The $t$-test used the pooled degrees of freedom because there appeared to be equal variances between groups $[F(1422, 999) = 1.01, p < 0.8083]$. In addition, a nonparametric Wilcoxon test was performed with the unadjusted values, which also showed that there was *no* statistically significant difference between the number of patent references cited in general patents and Internet patents [Wilcoxon test statistic = 1191252, $Z = -1.23, p < 0.2206$].

[23]The $t$-test used the Satterthwaite adjustment to the degrees of freedom to account for the unequal variances between groups $[F(1422, 999) = 27.54, p < 0.0001]$. In addition, a nonparametric Wilcoxon test was performed with the unadjusted values, which also showed that there was a statistically significant difference between the number of nonpatent references cited in General patents and Internet patents [Wilcoxon test statistic = 971925, $Z = -14.89, p < 0.0001$].

[24]Using multiple regression techniques, we examined other variables that may affect the total number of references beyond the mere classification of the patent as General or Internet.

**TABLE 2** Prior Art References (General Patent Set Compared with Internet Business Method Patents)

| Internet | N | Variable | Mean | Std Dev | Min | Max |
|---|---|---|---|---|---|---|
| General Patents | 1,000 | TotRef | 15.16 | 16.29 | 0 | 163 |
| | | PatRefs | 12.79 | 14.13 | 0 | 154 |
| | | NonPatRefs | 2.37 | 6.56 | 0 | 68 |
| Internet Business Method Patents | 1,093 | TotRef | 24.90 | 53.15 | 0 | 457 |
| | | PatRefs | 14.90 | 23.76 | 0 | 314 |
| | | NonPatRefs | 10.00 | 38.56 | 0 | 391 |

Number of Total References: The Independent Groups $t$-test using the log-transformed Total References showed that there was a statistically significant difference between General (mean = 2.47) and Internet business method (mean = 2.72) patents $[t(2089) = 7.22, p < 0.0001]$.[25]

Number of Patent References: The Independent Groups $t$-test using the log-transformed Patent References did show a statistically significant difference between General (mean = 2.30) and Internet business method (mean = 2.38) patents $[t(2091) = 2.27, p < 0.0235]$.[26]

Number of Nonpatent References: The Independent Groups $t$-test using the untransformed Nonpatent References showed a statistically significant difference between General (mean = 2.37) and Internet business method (mean = 10.01) patents $[t(1161) = 6.44, p < 0.0001]$.[27]

We note that there are many Internet business method patents with no nonpatent prior art and a few Internet business method patents with many nonpatent prior art references that could bias our results (that Internet business

---

[25] The $t$-test used the Satterthwaite adjustment to the degrees of freedom to account for the unequal variances between groups $[F(1092, 999) = 1.27, p < 0.0001]$. In addition, a nonparametric Wilcoxon test was performed with the unadjusted values, which also showed that there was a statistically significant difference between the number of total references cited in general patents and Internet business method patents [Wilcoxon test statistic = 948361, $Z = -7.15, p < 0.0001$].

[26] The $t$-test used the pooled degrees of freedom because there appeared to be equal variances between groups $[F(1092, 999) = 1.07, p < 0.2447]$. In addition, a nonparametric Wilcoxon test was performed with the unadjusted values, which also showed that there was a statistically significant difference between the number of patent references cited in General patents and Internet business method patents [Wilcoxon test statistic = 1016904, $Z = -2.18, p < 0.0291$].

[27] The $t$-test used the Satterthwaite adjustment to the degrees of freedom to account for the unequal variances between groups $[F(1092, 999) = 34.53, p < 0.0001]$. In addition, a nonparametric Wilcoxon test was performed with the unadjusted values, which also showed that there was a statistically significant difference between the number of nonpatent references cited in General patents and Internet business method nonpatents [Wilcoxon test statistic = 856133, $Z = -14.62, p < 0.0001$].

method patents have more prior art references than General patents) in favor of Internet business method patents. In addition to the statistical methods used,[28] we also looked at the percentage of patents in the General Patent Data Set and the Internet business method patents group that contained zero nonpatent prior art to help determine whether such bias might exist. We concluded that it did not, finding that 62.1 percent (621 of 1,000) of the General Patent Data Set had no nonpatent prior art, whereas only 32.2 percent (352 of 1,093) of the Internet business method patents had no nonpatent prior art; 1.4 percent (14 of 1,000) of the General Patent Data Set had no patent prior art, whereas 2.1 percent (23 of 1,093) of the Internet business method patents had no patent prior art; and 0.20 percent (2 of 1,000) of the General Patent Data Set had no prior art references from patent or nonpatent sources, whereas 0.27 percent (3 of 1,093) of Internet business method patents had no such prior art. In short, and most significantly, we found that Internet business method patents are less likely than general patents to have zero nonpatent prior art references in the patent.

In sum, we found that Internet business method patents are supported by more total references, patent references, and nonpatent references than general patents. The parametric and nonparametric tests were consistent in their findings. These findings suggest that those criticisms of Internet business method patents that are based on the amount of prior art cited (especially nonpatent prior art) are not supported by the data.

Our analysis does not answer all questions, however. It has been suggested that Internet business methods patents would be more likely to fail a novelty and nonobviousness hurdle in a commonsense (as opposed to legal) fashion because the patents could possibly cover practices and products (nonpatent prior art) that already exist, but not in any archived form. Our data could not address this contention. There is, however, no reason to believe that this is more likely to be the

---

[28]The data analysis strategy used first visually examined the distributions of each outcome including references (total, patent, and nonpatent) with histograms and quantile-quantile (QQ) plots. On the basis of visual examination of the reference distributions, this issue was identified and two data analysis strategies were employed to ensure that the small number of patents with large references did not overinfluence the analysis. First, the number of references was transformed with natural logs. On the basis of visual inspection of the log-transformed distribution, this appeared very effective in reducing the influence of the small subset of patents with large numbers of references. The Independent Groups $t$-test using the log-transformed values provides a valid test of the central tendencies (e.g., mean) of each distribution after minimizing the effect of the patents with large references. Second, we also used nonparametric tests that test whether the medians are different between groups. Median tests are resistant to extreme values because the median value represents the value for which 50% of the observations are above that value and 50% are below that value. Therefore, the analyses examined both means of log-transformed values and medians and typically arrived at similar conclusions, which strengthened our confidence in the results. Results also included the unadjusted means to help with interpretation, although they were never directly analyzed.

case for Internet business method patents than for patents generally. Indeed, previous research has shown that trade secret protection is preferred over patents in almost all technologies as a means for appropriating returns on R&D investment, thus indicating that there is much "secret prior art" in all areas. Although our analysis does not completely put to rest the possibility that there is a good deal of nonpatent prior art that is being missed, the same possibility exists for patents of all types, and attempting to pursue the idea of "commonsense novelty" will place one on a slippery slope that is contrary to the fundamental norms of patent law. This is simply not the way to determine novelty or nonobviousness.

### Comparison of Internet Patent Subtypes and General Patent Data Set

Taking the analysis one step deeper, we looked individually at all Internet patent subtypes: I-Business Model Patents, I-Business Technique Patents, and I-Software Technique Patents. It may be that certain types of Internet business method patents are especially controversial or problematic. For example, I-Business Models may be more controversial than I-Business Techniques because they may allow more easily for monopolies on complete lines of business or an industry. In some cases, their only innovation may be that they happen to be practiced via the Internet. They may also be more likely to involve prior art given their breadth and relationship to physical world business practices. I-Business Techniques, by contrast, may be more acceptable because they are more likely tied to particular Internet technologies.

Table 3 table shows the untransformed means and standard deviations of the three measures of patent references by patent type.

Based on the descriptive analysis of the distributions of the three measures of Patent References, log transformations were useful for normalizing Number of Patent References (PatRefs) and Total Number of References (TotRefs), but not for Number of Nonpatent References (NonPatRefs). The Independent Groups $t$-test using the log-transformed Total References showed that there was a statistically significant difference between the log TotRefs means of General patents (mean = 2.47) and I-Business Model patents (mean = 2.85) $[t(524) = -6.96, p < 0.0001]$, and I-Business Technique patents (Mean = 2.67) $[t(1529) = -5.15, p < 0.0001]$, but not I-Software Technique patents (Mean = 2.55) $[t(1328) = -1.68, p < 0.0934]$. The Independent Groups $t$-test using the log-transformed number of Patent References (PatRefs) showed a statistically significant difference between the log PatRefs means of General patents (mean = 2.30) and I-Business Model patents (mean = 2.53) $[t(1343) = -4.54, p < 0.0001]$, but not between General patents and I-Business Technique patents (mean = 2.31) $[t(1746) = -0.31, p < 0.7602]$ or I-Software Technique patents (mean = 2.25) $[t(623) = -1.10, p < 0.2721]$. Finally, the Independent Groups $t$-test using the untransformed number of Nonpatent References (NonPatRefs) showed a statistically significant difference between the log NonPatRefs of General patents (mean = 2.37) and I-Busi-

**TABLE 3** Mean of Patent, Nonpatent, and Total References (General Patents Compared with Internet Patent Subtypes)

| Patent Type | N | Variable Name | Mean | Std Dev | Minimum | Maximum |
|---|---|---|---|---|---|---|
| I-Business Model | 345 | PatRefs | 17.15 | 22.50 | 0 | 313 |
| | | TotRef | 27.25 | 48.78 | 0 | 454 |
| | | NonPatRefs | 10.10 | 35.26 | 0 | 389 |
| I-Business Technique | 748 | PatRefs | 13.86 | 24.27 | 0 | 314 |
| | | TotRef | 23.82 | 55.04 | 0 | 457 |
| | | NonPatRefs | 9.96 | 40.01 | 0 | 391 |
| I-Software Technique | 330 | PatRefs | 12.03 | 21.59 | 0 | 353 |
| | | TotRef | 16.86 | 27.48 | 1 | 376 |
| | | NonPatRef | 4.83 | 13.03 | 0 | 169 |
| General Patents | 1,000 | PatRefs | 12.79 | 14.13 | 0 | 154 |
| | | TotRef | 15.16 | 16.29 | 0 | 163 |
| | | NonPatRefs | 2.37 | 6.56 | 0 | 68 |

ness Model patents (mean = 10.10) [$t(352) = -4.05, p < 0.0001$], I-Business Technique patents (mean = 9.96) [$t(777) = -5.14, p < 0.0001$], and I-Software Technique patents (mean = 4.83) [$t(386) = -3.29, p < 0.0011$].

Restated, I-Business Model patents are supported by more total references, patent references, and nonpatent references than General patents. I-Business Technique patents are supported by more total references and nonpatent references than General patents. I-Software Technique patents are supported by more nonpatent references than General patents. The *t*-test and Wilcoxon tests were consistent for all of the above findings.[29] Those prior art-related criticisms aimed at the broadest type of Internet business method patents—the Internet business model—would seem to find no support in the data. Moreover, in terms of prior art generally and nonpatent prior art specifically, there is no evidence to support the

---

[29]The *t*-tests used the Satterthwaite adjustment to the degrees of freedom to account for the unequal variances between groups.

contention that any type of Internet-related patent is weaker than patents in general when prior art is used as the measure.[30]

## Comparison of the Types of Nonpatent Prior Art

We compared the nonpatent prior art for Internet patents based on subtypes and entity status (Tables 4 and 5). We did not have comparable data for general patents. The analysis was conducted separately for Internet Patent type and Owner Status (Table 6) and used both descriptive statistics and the Kruskal-Wallis nonparametric tests because none of the variables was normally distributed or easily transformed into a normal distribution. Among the Internet Patent subtypes, we found that there were no significant differences except for popular press (PP), where there were statistical differences between I-Business Model and I-Business Technique and between I-Business Model and I-Software Technique ($p < 0.05$). These results were statistically significant ($p < 0.05$). For Owner Status (i.e., large business, small business, and individual), we found no significant differences in the use of various types of nonpatent prior art. We did find that the largest percentage of nonpatent prior art references in the Internet patent data set was in the academic/trade publication category.

**TABLE 4** Nonpatent Prior Art References in 20 Sample

| Nonpatent Reference Category | Mean | Median |
|---|---|---|
| Acad/Trade | 4.44 | 1.0 |
| Comp/Indus | 1.9 | 0 |
| Univ. Pubs | 0.17 | 0 |
| Gov Doc | 0.12 | 0 |
| SW | 0.46 | 0 |
| PP | 0.73 | 0 |
| Pat Apps/Searches | 0.08 | 0 |
| Other | 0.13 | 0 |

---

[30]In a multivariate analyses, we considered the effects of several other variables not fully discussed here and found no significant differences between General patents and Internet patents. For example, the presence of two statistically significant interactive effects were included (Patent Type by Number of Figures and Patent Type by 4-Digit IPC). This resulted in the main Internet effect on Total References being statistically nonsignificant. Caution is warranted in reading the multivariate results because (1) we did not disaggregate the three Internet patent subtypes for comparison with the General patent set and (2) these interactions may share overlapping variation with the difference between Internet and General patents. In any case, there was no evidence to suggest that Internet-related patents were ever weaker in terms of total prior art references or nonpatent prior art references.

**TABLE 5** Nonpatent Prior Art References by Internet Patent Subtype

| Internet Patent Subtype | N | Nonpatent Category | Median | Mean | Std Dev | Minimum | Maximum |
|---|---|---|---|---|---|---|---|
| I-BusMod | 70 | AcadTrade | 1.5 | 5.07 | 13.09 | 0 | 85 |
| I-BusTech | 149 | AcadTrade | 1 | 4.89 | 21.1 | 0 | 245 |
| I-SWTech | 66 | AcadTrade | 1 | 2.74 | 7.17 | 0 | 53 |
| I-BusMod | 70 | CompIndus | 0 | 1.5 | 4.96 | 0 | 37 |
| I-BusTech | 149 | CompIndus | 0 | 1.36 | 3.83 | 0 | 34 |
| I-SWTech | 66 | CompIndus | 0 | 0.48 | 0.96 | 0 | 5 |
| I-BusMod | 70 | GovDoc | 0 | 0.13 | 0.51 | 0 | 3 |
| I-BusTech | 149 | GovDoc | 0 | 0.17 | 0.81 | 0 | 6 |
| I-SWTech | 66 | GovDoc | 0 | 0 | 0 | 0 | 0 |
| I-BusMod | 70 | Oth | 0 | 0.19 | 0.73 | 0 | 4 |
| I-BusTech | 149 | Oth | 0 | 0.11 | 0.46 | 0 | 4 |
| I-SWTech | 66 | Oth | 0 | 0.09 | 0.42 | 0 | 3 |
| I-BusMod | 70 | PatAppsSearches | 0 | 0.06 | 0.29 | 0 | 2 |
| I-BusTech | 149 | PatAppsSearches | 0 | 0.12 | 0.57 | 0 | 5 |
| I-SWTech | 66 | PatAppsSearches | 0 | 0.02 | 0.12 | 0 | 1 |
| I-BusMod | 70 | PP | 0 | 0.89 | 2.46 | 0 | 13 |
| I-BusTech | 149 | PP | 0 | 0.97 | 10.25 | 0 | 125 |
| I-SWTech | 66 | PP | 0 | 0.05 | 0.21 | 0 | 1 |
| I-BusMod | 70 | SW | 0 | 0.4 | 2.27 | 0 | 18 |
| I-BusTech | 149 | SW | 0 | 0.52 | 2.01 | 0 | 20 |
| I-SWTech | 66 | SW | 0 | 0.38 | 1.13 | 0 | 7 |
| I-BusMod | 70 | UnivPubs | 0 | 0.07 | 0.26 | 0 | 1 |
| I-BusTech | 149 | UnivPubs | 0 | 0.16 | 0.53 | 0 | 4 |
| I-SWTech | 66 | UnivPubs | 0 | 0.29 | 0.86 | 0 | 5 |

Note that a larger amount of the nonpatent prior art references in Internet business method patents are attributable to patents filed by small business. And for this group, the nonpatent prior art concentrates in three of the nonpatent prior art categories, namely "academic and trade," "company and industry," and software. This could mean that one set of Internet business method patentholders pays even greater attention to nonpatent prior art than other groups (although these other groups still have as much or more nonpatent prior art than patents generally). An obvious question that we cannot answer here is, why do small enterprises appear to pay much more attention to nonpatent prior art?

**TABLE 6** Nonpatent Prior Art References by Internet Patent Owner Status

| Owner Status | N | Nonpatent Category | Median | Mean | Std Dev | Minimum | Maximum |
|---|---|---|---|---|---|---|---|
| Individual | 33 | AcadTrade | 0 | 2.12 | 4.01 | 0 | 16 |
| LargeEnt | 204 | AcadTrade | 1 | 3.99 | 17.89 | 0 | 245 |
| SmallBus | 47 | AcadTrade | 2 | 8.11 | 18.02 | 0 | 85 |
| Individual | 33 | CompIndus | 0 | 0.33 | 0.74 | 0 | 3 |
| LargeEnt | 204 | CompIndus | 0 | 0.91 | 2.23 | 0 | 22 |
| SmallBus | 47 | CompIndus | 0 | 3.06 | 7.72 | 0 | 37 |
| Individual | 33 | GovDoc | 0 | 0.15 | 0.62 | 0 | 3 |
| LargeEnt | 204 | GovDoc | 0 | 0.08 | 0.56 | 0 | 6 |
| SmallBus | 47 | GovDoc | 0 | 0.28 | 0.93 | 0 | 5 |
| Individual | 33 | Oth | 0 | 0.3 | 0.92 | 0 | 4 |
| LargeEnt | 204 | Oth | 0 | 0.08 | 0.35 | 0 | 3 |
| SmallBus | 47 | Oth | 0 | 0.21 | 0.75 | 0 | 4 |
| Individual | 33 | PatAppsSearches | 0 | 0.12 | 0.42 | 0 | 2 |
| LargeEnt | 204 | PatAppsSearches | 0 | 0.07 | 0.47 | 0 | 5 |
| SmallBus | 47 | PatAppsSearches | 0 | 0.09 | 0.35 | 0 | 2 |
| Individual | 33 | PP | 0 | 0.12 | 0.33 | 0 | 1 |
| LargeEnt | 204 | PP | 0 | 0.75 | 8.76 | 0 | 125 |
| SmallBus | 47 | PP | 0 | 1.11 | 3.01 | 0 | 13 |
| Individual | 33 | SW | 0 | 0.3 | 1.57 | 0 | 9 |
| LargeEnt | 204 | SW | 0 | 0.32 | 0.99 | 0 | 7 |
| SmallBus | 47 | SW | 0 | 1.17 | 3.99 | 0 | 20 |
| Individual | 33 | UnivPubs | 0 | 0.09 | 0.29 | 0 | 1 |
| LargeEnt | 204 | UnivPubs | 0 | 0.2 | 0.65 | 0 | 5 |
| SmallBus | 47 | UnivPubs | 0 | 0.09 | 0.35 | 0 | 2 |

## Other Variables of Interest

### Entity Status and Size

Small businesses own a larger share of Internet business method patents (Table 7) (19.4%) than general patents (10.7%).[31] Large entities own a smaller

---

[31] Nonprofits were also an identified entity, but they received so few patents in the set that we do not include them in the table.

**TABLE 7** Patent Entity Status (General Patents Compared with Internet Business Method Patents)

| Owner Type | Number of Internet Business Patents (%) | Number of General Patents (%) | Pr > ChiSq |
|---|---|---|---|
| Individual | 179 (16.38) | 175 (17.5) | 0.494 |
| Large Entity | 690 (63.13) | 707 (70.7) | 0.0002 |
| Small Business | 212 (19.4) | 107 (10.7) | <0.0001 |

share of Internet business method patents (63.13%) than general patents (70.7%). These results were statistically significant ($p < 0.05$).[32]

When looking at Internet patent subtypes we found that, with respect to I-Business Model Patents (Table 8), individuals owned a greater share of I-Business Model patents (29.9%) compared to individual ownership of general patents (17.5%). Likewise, small business owned a greater share of I-Business Model patents (27.0%) compared to small business ownership of general patents (10.7%). Large entities owned a much smaller share of I-Business Model patents (41.2%) than large entity ownership of general patents (70.7%). We also found that individuals owned a smaller share of I-Business Technique Patents (10.2%) compared to individual ownership of general patents (17.5%). Small business owned a larger share of I-Business Technique patents (15.9%) compared to small business ownership of general patents (10.7%). Finally, with respect to I-Software Technique patents, individuals owned a smaller share of I-Software Technique

**TABLE 8** Entity Status for Internet Patent Subtypes

|  | I-Business Model | I-Business Technique | I-Software Technique |
|---|---|---|---|
| Individual | 103 | 76 | 10 |
|  | 29.86% | 10.16% | 3.03% |
| Large Entity | 142 | 548 | 287 |
|  | 41.16% | 73.26% | 86.97% |
| Small Business | 93 | 119 | 32 |
|  | 26.96% | 15.91% | 9.7% |

[32]When all Internet-related patents are combined, a smaller share of Internet patents are owned by individuals (13.3%) compared to individual ownership of general patents (17.5%). A larger share of Internet patents are owned by small business (17.1%) when compared to small business ownership of general patents (10.7%). These results were statistically significant ($p < 0.05$).

patents (3.0%) compared to individual ownership of general patents (17.5%). Large entities owned a larger share of I-Software Technique Patents (87%) compared to large entity ownership of general patents (70.7%). These results were statistically significant ($p < 0.05$).

What one could conclude from these data is that large entities are not dominating the patenting of Internet business methods. Individuals and small businesses own a larger share than they own in the General Patent Data Set. Individuals are strongly represented among the I-Business Model patents and Small Business among the I-Business Technique patents. Where large businesses do dominate is in the patenting of I-Software Techniques. In sum, the data do not support a conclusion that small business, entrepreneurs, and individuals are being squeezed out by the patenting power of large business organizations.[33]

### Geographic Origin

With respect to international competitiveness, some observers have suggested that U.S. companies are being awarded a disproportionate share of Internet-related patents. There is ample evidence to support that suggestion. Table 9 shows that inventors in Europe (Internet 2.3%; General 17.3%), Japan (Internet 5.0%; General 21.4%), and Other Foreign countries (Internet 0.5%; General 5.9%) obtain significantly fewer Internet business method patents than patents in general, whereas U.S. companies obtain more. These differences are all statistically significant at $p < 0.05$. Only Canada has no significant difference.

A similar conclusion may be drawn from an analysis of the broader set of all Internet-related patents: Inventors in Europe (Internet 2.9%; General 17.3%),

**TABLE 9** Patents Compared by Region (General Patent Set Versus Internet Business Method Patents)

| Region | Number of Internet Bus Method Patents (%) | Number of General Patents (%) |
| --- | --- | --- |
| Canada | 17 (1.6) | 17 (1.7) |
| Europe | 25 (2.3) | 173 (17.3) |
| Japan | 55 (5.0) | 214 (21.4) |
| Other Foreign | 6 (0.5) | 59 (5.9) |
| United States | 990 (90.6) | 537 (53.7) |

[33]This, of course, does not take into account the greater ability of large firms to actually litigate and enforce their patents, which, in the end, may give them more power than is evident from mere ownership numbers. However, the same would hold true for general patents, where large business organizations hold an even larger share of issued patents.

Japan (Internet 5.1%; General 21.4%), and Other Foreign countries (Internet 0.5%; General 5.9%) all obtain significantly fewer Internet business method patents in general. These differences are all statistically significant at $p < 0.05$. The same conclusion holds true for the individual patent subtypes—Japan and Europe were issued each type in significantly smaller proportions than their overall ownership of general patents. All of these results are quite expected in view of the fact that software and business methods are recognized as patentable subject matter to a much greater extent in the U.S. than elsewhere.

### Days in the USPTO

The data generally do not support the hypothesis that Internet business method patents spend less time in the USPTO than general patents, although the data are more equivocal.[34] This result also holds for the Internet patent subtypes and the full set of all Internet-related patents. The Independent Groups $t$-test using log-transformed Days in USPTO (Table 10) showed there was no statistically

**TABLE 10** Days In the USPTO

|  | N | Mean # of Days | Std Dev | Min | Max |
|---|---|---|---|---|---|
| General Patents | 1,000 | 1,011.9 | 662.5 | 243 | 6,626 |
| Internet Patents | 1,423 | 889.7 | 245.7 | 154 | 2,428 |
| Internet business method patents | 1,093 | 885.56 | 244.35 | 154 | 2,198 |
| I-Business Model | 345 | 884.94 | 245.92 | 154 | 1,692 |
| I-Business Technique | 748 | 885.84 | 243.80 | 238 | 2,198 |
| I-Software Technique | 330 | 903.40 | 249.89 | 361 | 2,428 |

[34]For the various Internet patent categories, the $t$-test for log Days In USPTO produces a different result than the nonparametric Wilcoxon or median test: the $t$-test shows that there is no statistically significant difference in the time in USPTO between Internet and General patents, whereas the two nonparametric tests show that there is a significant difference. The Wilcoxon nonparametric test assumes that the two groups' distributions are similar with a shift in the location parameter (median). Unlike the previous tests, the two distributions here do not look similar. The median test makes a less restrictive assumption about the distributions and does show a statistically significant result. However, the median test also does not take into account the fact that General patents have been in existence longer and therefore would likely show a greater length in USPTO (e.g., the extreme values). Overall, the $t$-test result is the most conservative decision basis to use, and the overall conclusion is that the data do not support that the Internet patents spend shorter time in the USPTO than General patents.

significant difference between General (log-transformed mean = 6.77) and Internet (log-transformed mean = 6.75) patents [$t(1441) = -1.03, p < 0.3049$], no statistically significant difference between General (mean = 6.77) and Internet business method (mean = 6.74) patents [$t(1572) = -1.33, p < 0.1825$], and similarly for the Internet patent subtypes.

A finding that Internet patents spend less time in pendency at the USPTO might have indicated a less thorough examination by the USPTO and less willingness by applicants to devote significant time and resources to obtaining these patents. Moreover, a finding of more or less pendency time would definitely have implications for the term of patent protection. After log-transformation of means to adjust for extreme values, however, no significant difference was found in pendency times.

## CONCLUSION

Many criticisms of Internet business method patents rely on perceived differences between Internet business method patents and the more general set of patents that issue from the USPTO. Those criticisms are focused primarily on the perception that Internet business method patents have not been properly researched for relevant prior art. For the time period we studied (primarily late 1990s), we found little support for those criticisms when we compared Internet patents with a large sample taken from the general population of patents. Internet-related patents overall, Internet business method patents, and Internet patent subtypes that we identified all proved to have as much, if not more, prior art as patents in general. The major difference in Internet patents and general patents with respect to prior art was the amount of nonpatent prior art cited in Internet patents, with those patents having significantly more nonpatent prior art citations than the general population of patents. Although some observers criticize Internet business method patents for other reasons (such as allowing them to be patentable subject matter at all), criticisms focused on prior art and the USPTO's handling of these particular types of patents are not well supported by our analysis of the data.

We also found that individuals and small companies do quite well, compared to large business organizations, in getting Internet business method patents. In other words, when compared to the distribution of a set of general patents, the results of our research do not support the contention that large business organizations are dominating Internet business method patents. We did find, however, that U.S. inventors and companies overwhelmingly dominate their Japanese and European counterparts in receiving Internet business method patents. Japanese and European inventors and companies receive a far greater share of total U.S. patents than of Internet business method patents.

# REFERENCES

Allison, J., and M. Lemley. (2000). "Who's Patenting What? An Empirical Exploration of Patent Prosecution." *Vanderbilt Law Review* 53(6): 2099-2148.

Allison, J., and M. Lemley. (1998). "Empirical Evidence on the Validity of Litigated Patents." *American Intellectual Property Association Quarterly Journal* 26: 185-275.

Cohen, J., and M. Lemley. (2001). "Patent Scope and Innovation in the Software Industry." *California Law Review* 89(1): 1-58.

Durham, A. (1999). "'Useful Arts' in the Information Age." *Brigham Young University Law Review* 1999: 1419-1528.

Hall, B., A. Jaffe, and M. Trajtenberg. (1998). "Market Value and Patent Citations: A First Look." National Bureau of Economic Research, Working Paper No. 7741.

Harhoff, D., F. M. Scherer, and K. Vopel. (1999a). "Citations, Family Size, Opposition and the Value of Patent Rights." Working paper, available at http://emlab.berkeley.edu/users/bhhall/harhoffetal99.pdf.

Harhoff, D., F. Narin, F. M. Scherer, and K. Vopel. (1999b). "Citation Frequency and the Value of Patented Inventions." *The Review of Economics and Statistics* 81(3): 511-515.

Lanjouw, J., and M. Schankerman. (2001). "Characteristics of Patent Litigation: A Window on Competition." *RAND Journal of Economics* 32(1): 129-151.

Lanjouw, J., and M. Schankerman. (1997). "Stylized Facts of Patent Litigation: Value, Scope, Ownership." National Bureau of Economic Research, Working Paper No. 6297.

Lessig, L. (2001). *The Future of Ideas: The Fate of the Commons in a Connected World.* New York: Random House.

Ross, P. (2000). "Patently Absurd." *Forbes* May 29, 180-182.

Thomas, J. (1999). "The Patenting of the Liberal Professions." *Boston College Law Review* 40(5): 1139-1185.

Trajtenberg, M. (1990). "A Penny for Your Quotes: Patent Citations and the Value of Innovations." *RAND Journal of Economics* 21(1): 172-187.

United States Patent and Trademark Office (USPTO). (2000). "Automated Business Methods." White Paper, Section III, Class 705, available at http://www.uspto.gov/web/menu/busmethp/class705.htm.

## APPENDIX: INTERNET PATENT SUBTYPES

1. *Business Model:* The described method would likely stand on its own as a business on the Internet or a distinct line of business. This is the broadest subtype. Note that we do not include patents in this category if the only likely business model is licensing out what we describe below as a business technique. The business method itself as described in the patent, rather than the licensing out of the method, must be capable of being a business model or distinct line of business.

   Example (1): Walker Asset Management Limited Partnership—*Method, apparatus, and program for pricing, selling, and exercising options to purchase airline tickets.*

   An apparatus, method, and program for determining price of an option to purchase an airline ticket and for facilitating the sale and exercise of those options. By purchasing an option, a customer can lock in a specified airfare without tying up his money and without risking the loss of the ticket price if his travel plans change. Pricing of the options may be based on departure location criteria, destination location criteria, and travel criteria.

   Example (2): IMX Mortgage Exchange—*Interactive mortgage and loan information and real-time trading system.*

   The invention provides a method and a system for trading loans in real time by making loan applications, such as home mortgage loan applications, and placing them up for bid by a plurality of potential lenders. A transaction server maintains a database of pending loan applications and their statuses; each party to the loan can search and modify that database, consistent with their role in the transaction, by requests to the server from a client device identified with their role. Brokers at a broker station can add loan applications, can review the status of loan applications entered by that broker, are notified of lender's bids on their loans, and can accept bids by lenders. Lenders at a lender station can search the database for particular desired types of loans, can sort selected loans by particular desired criteria, can bid on loan applications, and are notified when their bids are accepted. Broker stations, lender stations, and the transaction server can be coupled using multiple access methods, including Internet, intranet, or dial-up or leased communication lines.

   Example (3): NCR Corporation—*Newspaper vending machine with online connection.*

A system which comprises a self-service newspaper vending machine (2) includes an electronic control means (34) with an on-line connection (36) to a news-providing organization (38) from which a newspaper containing up to the minute news can be purchased. A customer is attracted by news stories shown on a display (6). The customer is then given the opportunity of purchasing a newspaper or part of a newspaper. Communication between the customer and the vending machine (2) is by the display (6) and a keyboard (8). The newspaper can be purchased by either inserting a banking or credit card in a card reader (52) or inserting coins into a coin slot (50). The vending machine (2) would then print out the up to the minute news requested.

2. *Business Technique:* Typically would not be a stand-alone business; rather, it is a more narrow method of doing business over the Internet.

Example (1): Amazon.com, Inc.—*Method and system for placing a purchase order via a communications network.*

A method and a system for placing an order to purchase an item via the Internet. The order is placed by a purchaser at a client system and received by a server system. The server system receives purchaser information including identification of the purchaser, payment information, and shipment information from the client system. The server system then assigns a client identifier to the client system and associates the assigned client identifier with the received purchaser information. The server system sends to the client system the assigned client identifier and an HTML document identifying the item and including an order button. The client system receives and stores the assigned client identifier and receives and displays the HTML document. In response to the selection of the order button, the client system sends to the server system a request to purchase the identified item. The server system receives the request and combines the purchaser information associated with the client identifier of the client system to generate an order to purchase the item in accordance with the billing and shipment information, whereby the purchaser effects the ordering of the product by selection of the order button.

Example (2): Lucent Technologies—*System and method for scheduling and controlling delivery of advertising in a communications network.*

A system and a method for scheduling and controlling delivery of advertising in a communications network and a communications network and remote computer program employing the system or the method. The system includes (1) a time allocation controller that allocates time available in a particular advertising region in a display device of a remote computer between at least two advertisements as a function of one of a desired user frequency, a desired time frequency, or a desired geometry, for each of at least two advertisements and (2) a data communication controller, coupled to the time allocation controller, that delivers at least two advertisements to said remote computer for display in the advertising region according to the allocating of the time.

Example (3): Citibank, N.A.—*Method for electronic merchandise dispute resolution.*

A system for open electronic commerce having a customer trusted agent securely communicating with a first money module and a merchant-trusted agent securely communicating with a second money module. Both trusted agents are capable of establishing a first cryptographically secure session, and both money modules are capable of establishing a second cryptographically secure session. The merchant trusted agent transfers electronic merchandise to the customer trusted agent, and the first money module transfers electronic money to the second money module. The money modules inform their trusted agents of the successful completion of payment, and the customer may use the purchased electronic merchandise.

3. *Software Technique:* Patent focusing on more technical Internet functionality and not conditioned on a particular business application. These patents are often targeted at making the Internet more efficient and effective for conducting electronic commerce.

Example (1): Compaq Computer Corporation—*Method and apparatus for reassigning network addresses to network servers by reconfiguring a client host connected thereto.*

The present invention provides a method and an apparatus for reassigning network addresses to a plurality of network servers by reconfiguring a client host coupled to the network servers. According to the invention, when there are changes to network connections, the IP addresses (i.e., network addresses) of the individual network servers can be reassigned automatically at the client host without powering off the network servers. According to the invention, in reassigning a new network address to a port of the network server, a bootstrap protocol (BOOTP) request is first issued by the client host to the network server. The BOOTP request is received by the network server, which then sends a BOOTP response to the client host to request a new network address. After the client host receives the BOOTP response, it sends a BOOTP reply to the network server. The BOOTP reply includes a new network address for the port of the network server. The above procedure is repeated for each port of the network server. Thus each of the network server is reassigned with a new network address. In this way, reassignment of IP addresses of network servers is more efficiently performed. Furthermore, the work efforts are substantially reduced and are centralized.

Example (2): International Business Machines Corporation—*System for checking status of supported functions of communication platforms at pre-selected intervals in order to allow hosts to obtain updated list of all supported functions.*

An apparatus for dynamically providing a host information about all functions supported by a communication platform provided in a computing network environment. The computing network environment also has a gateway device besides the associated communication platform, which can be of any specific type, as well as at least having an initiating host and at least one receiving host that are electronically connected to the gateway device. The apparatus comprises a special function table for storing all possibly available functions that can be provided for all available commercial communication platforms as well as a memory location accessible by said gateway device for storing said special function table. Determining means then will obtain a list of all supported functions provided by said particularly associated communication platform and through the use of a comparison component provides information about all supported functions in the same special function table. All supported functions are then checked by a monitoring component to modify the function table in case of additions or deletions. In this manner, any host can obtain an updated list of all available and supported functions at any time and even select an option from the list if desired.

# Effects of Research Tool Patents and Licensing on Biomedical Innovation[1]

John P. Walsh
University of Illinois at Chicago and Tokyo University

Ashish Arora
Carnegie Mellon University

Wesley M. Cohen
Duke University

### ABSTRACT

*Over the last two decades changes in technology and policy have altered the landscape of drug discovery. These changes have led to concerns that the patent system may be creating difficulties for those trying to do research in biomedical fields. Using interviews and archival data, we examine the changes in patenting and licensing in recent years and how these have affected innovation in pharmaceuticals and related biotech industries.*

*We find that there has in fact been an increase in patents on the inputs to drug discovery ("research tools"). However, we find that drug discovery has not been substantially impeded by these changes. We also find little evidence that university research has been impeded by concerns about patents on research tools. Restrictions on the use of pat-*

---

[1] We would like to thank the Science, Technology, and Economic Policy Board of the National Academy of Sciences, and the National Science Foundation (Award No. SES-9976384) for financial support. We thank Jhoanna Conde, Wei Hong, JoAnn Lee, Nancy Maloney, and Mayumi Saegusa for research assistance. We would like to thank the following for their helpful comments on earlier drafts of this chapter: John Barton, Bill Bridges, Mildred Cho, Robert Cook-Deegan, Paul David, Rebecca Eisenberg, Akira Goto, Lewis Gruber, Janet Joy, Robert Kneller, Eric Larson, Richard Levin, Stephen Merrill, Ichiro Nakayama, Pamela Popielarz, Arti Rai, and participants in the STEP Board Conference on New Research on the Operation and Effects of the Patent System October 22, 2001, Washington, D.C. and the OECD Workshop on Genetic Inventions, Intellectual Property Rights and Licensing Practices, January 24-25, 2002, Berlin, Germany, as well as the School of Information Seminar at University of Michigan.

ented genetic diagnostics, where we see some evidence of patents interfering with university research, are an important exception. There is, also, some evidence of delays associated with negotiating access to patented research tools, and there are areas in which patents over targets limit access and where access to foundational discoveries can be restricted. There are also cases in which research is redirected to areas with more intellectual property (IP) freedom. Still, the vast majority of respondents say that there are no cases in which valuable research projects were stopped because of IP problems relating to research inputs.

We do not observe as much breakdown or even restricted access to research tools as one might expect because firms and universities have been able to develop "working solutions" that allow their research to proceed. These working solutions combine taking licenses, inventing around patents, infringement (often informally invoking a research exemption), developing and using public tools, and challenging patents in court. In addition, changes in the institutional environment, particularly new U.S. Patent and Trademark Office (USPTO) guidelines, active intervention by the National Institutes of Health (NIH), and some shift in the courts' views toward research tool patents, appear to have further reduced the threat of breakdown and access restrictions, although the environment remains uncertain.

We conclude with a discussion of the potential social welfare effects of these changes in the industry and the adoption of these working solutions for dealing with a complex patent landscape. There are social costs associated with these changes, but there are also important benefits. Although we cannot rule out the possibility of new problems in the future, our results highlight some of the mechanisms that exist for overcoming these difficulties.

## INTRODUCTION

There is widespread consensus that patents have long benefited biomedical innovation. A forty-year empirical legacy suggests that patents are more effec-

---

[2]See Scherer et al. (1959), Levin et al. (1987), Mansfield (1986), and Cohen et al. (2000). For pharmaceuticals, there is near universal agreement among our respondents that patent rights are critical to providing the incentive to conduct R&D. Indeed, data from the Carnegie Mellon Survey of Industrial R&D (cf. Cohen et al., 2000) show that the average imitation lag for the drug industry is nearly 5 years for patented products, whereas for the rest of the manufacturing sector, the average is just over 3.5 years ($p < 0.01$). Moreover, recent evidence shows that the profits protected by patents constitute an important incentive for drug firms to invest in R&D (Arora et al., 2003).

tive, for example, in protecting the commercialization and licensing of innovation in the drug industry than in any other.[2] Patents are also widely acknowledged as providing the basis for the surge in biotechnology start-up activity witnessed over the past two decades.[3] Heller and Eisenberg (1998) and the National Research Council (1997) have suggested, however, that recent policies and practices associated with the granting, assertion, and licensing of patents on research tools may now be undercutting the stimulative effect of patents on drugs and related biomedical discoveries. In this chapter, we report the results of 70 interviews with personnel at biotechnology and pharmaceutical firms and universities in considering the effects of research tool patents on industrial or academic biomedical research.[4] We conceive of research tools broadly to include any tangible or informational input into the process of discovering a drug or any other medical therapy or method of diagnosing disease.[5]

Heller and Eisenberg (1998) argue that biomedical innovation has become susceptible to what they call a "tragedy of the anticommons," which can emerge when there are numerous property right claims to separate building blocks for some product or line of research. When these property rights are held by numerous claimants (especially if they are from different kinds of institutions), the negotiations necessary to their combination may fail, quashing the pursuit of otherwise promising lines of research or product development. Heller and Eisenberg suggest that the essential precondition for an anticommons — the need to combine a large number of separately patentable elements to form one product—now applies to drug development because of the patenting of gene fragments or mutations [e.g., expressed sequence tags (ESTs) and single-nucleotide polymorphisms (SNPs)] and a proliferation of patents on research tools that have become essential inputs into the discovery of drugs, other therapies, and diagnostic methods. Heller and Eisenberg (1998) argue that the combining of multiple rights is susceptible to a breakdown in negotiations or, similarly, a stacking of license fees to the point of overwhelming the value of the ultimate product. Shapiro (2000) has raised similar concerns, using the image of the "patent thicket." He notes that

---

[3]For example, in one of our interviews, a licensing director for a large pharmaceutical firm said "Patents are critical for start-up firms. Without patents, we won't even talk to a start-up about licensing."

[4]The National Research Council (1997) also considers the challenges for biomedical innovation posed by the patenting of research tools and upstream discoveries more generally. In a series of case studies, the National Research Council (1997, Ch. 5) documents pervasive concern over limitations on access due to the price of intellectual property and concern over the prospect of blocking of worthwhile innovations due to IP negotiations, but no instances of worthwhile projects that were actually blocked.

[5]Examples include recombinant DNA (Cohen-Boyer), polymerase chain reaction (PCR), genomics databases, microarrays, assays, transgenic mice, embryonic stem cells, or knowledge of a target, that is, any cell receptor, enzyme, or other protein that is implicated in a disease and consequently represents a promising locus for drug intervention.

technologies that depend on the agreement of multiple parties are vulnerable to holdup by any one of them, making commercialization potentially difficult.[6]

The argument that an anticommons may emerge to undercut innovation emphasizes factors that might frustrate private incentives to realize what should otherwise be mutually beneficial trades. Merges and Nelson (1990) and Scotchmer (1991) have argued, however, that the self-interested use of even just one patent—although lacking the encumbrances of multiple claimants characterizing an "anticommons"—may also impede innovation where a technology is cumulative (i.e., where invention proceeds largely by building on prior invention). An example of such an upstream innovation in biomedicine is the discovery that a particular receptor is important for a disease, which may make that receptor a "target" for a drug development program.[7] A key concern regarding the impact of patents in such cumulative technologies is that "unless licensed easily and widely," patents—especially broad patents—on early, foundational discoveries may limit the use of these discoveries in subsequent discovery and consequently limit the pace of innovation (Merges and Nelson, 1990).[8] The revolution in molecular biology and related fields over the past two decades and coincident shifts in the policy environment have now increased the salience of this concern for biomedical research and drug innovation in particular (National Research Council, 1997). Drug discovery is now more guided by prior scientific findings than previously (Gambardella, 1995; Cockburn and Henderson, 2000; Drews, 2000), and those findings are now more likely to be patented after the 1980 passage of the Bayh-Dole Act and related legislation that simplified the patenting of federally supported research outputs that are often upstream to the development of drugs and other biomedical products.

In this chapter, we consider whether biomedical innovation has suffered be-

---

[6]The case of beta-carotene-enhanced rice (GoldenRice™) illustrates a potential anticommons/thicket problem. This innovation involves using as many as 70 pieces of IP and 15 pieces of technical property spread over 31 institutions (Kryder et al., 2000). Under such conditions, Heller, Eisenberg, and Shapiro have all suggested that acquiring the rights to practice such an innovation may be prohibitively difficult.

[7]For example, a Yale-Harvard collaborative group and researchers at Merck discovered (nearly simultaneously) that the immunophilin receptor FKBP might be important for immunosuppression, making it a target for research programs at Merck, Vertex (a biotech start-up), and Harvard Medical School that all tried to find chemicals that would bind to the receptor and thus could be used as drugs to suppress immune response (Werth, 1994). Successful development in this case would depend on combining the knowledge of the existence of the target with other innovations, particularly compounds that could modify the action of the target receptor.

[8]Scotchmer (1991) focuses on the related issue of the allocation of rents between the holder of a pioneer patent and those who wish to build on that prior discovery, suggesting that there is no reason to believe that markets left to themselves will set that allocation in such a way that the pace of innovation in cumulative technologies is maximized. Barton (2000), in fact, suggests that the current balance "is weighted too much in favor of the initial innovator." Scotchmer (1991) has suggested that ex ante deals between pioneers and follow-on innovators can, however, be structured to mitigate the problem.

cause of either an anticommons or restrictions on the use of upstream discoveries in subsequent research. Notwithstanding the possibility of such impediments to biomedical innovation, there is still ample reason—and recent scholarship (Arora et al., 2003)—to suggest that patenting benefits biomedical innovation, especially via its considerable impact on R&D incentives or via its role in supporting an active market for technology (Arora et al., 2001). Although any ultimate policy judgment requires a consideration of the benefits and costs of patent policy, an examination of the benefit side of this calculus is outside the scope of our current study.

In the second section of this chapter, we provide background to the anticommons and restricted access problems. The third section describes our data and methods. In the fourth section, we provide an overview of the results from our interviews and assess the extent to which we witness either "anticommons" problems or restricted access to intellectual property (IP) on upstream discoveries and research tools. To prefigure the key result, we find little evidence of routine breakdowns in negotiations over rights, although research tool patents are observed to impose a range of social costs and there is some restriction of access. In the fifth section of the chapter, we describe the mechanisms and strategies employed by firms and other institutions that have limited the negative effects of research tool patents on innovation. The final section discusses our findings and our conclusions.

## BACKGROUND

### Science and Policy

Changes in the science underlying biomedical innovation, and in policies affecting what can be patented and who can patent, have combined to raise concerns over the impact of the patenting and licensing of upstream discoveries and research tools on biomedical research. Over the past twenty years, fundamental changes have revolutionized the science and technology underlying product and process innovation in drugs and the development of medical therapies and diagnostics. Advances in molecular biology have increased our understanding of the genetic bases and molecular pathways of diseases. Automated sequencing techniques and bioinformatics have greatly increased our ability to transform this understanding into patentable discoveries that can be used as targets for drug development. In addition, combinatorial chemistry and high-throughput screening techniques have dramatically increased the number of potential drugs for further development. Reflecting this increase in technological opportunity, the number of drug candidates in phase I clinical trials grew from 386 in 1990 to 1,512 in 2000.[9] The consequence of these changes is that progress in biomedical research

---

[9]We thank Margaret Kyle for making these data available to us.

is now more cumulative; it depends more heavily than heretofore on prior scientific discoveries and previously developed research tools (Drews, 2000; Henderson et al. 1999).

As the underlying science and technology has advanced, policy changes and court decisions since 1980 have expanded the range of patented subject matter and the nature of patenting institutions. In addition to the 1980 *Diamond v. Chakrabarty* decision that permitted the patenting of life-forms, and the 1988 Harvard OncoMouse patent that extended this to higher life-forms (and to a research tool), in the 1980s gene fragments, markers and a range of intermediate techniques and other inputs key to drug discovery and commercialization also became patentable. Moreover, Bayh-Dole and related legislation have encouraged universities and national labs, responsible for many such upstream developments and tools, to patent their inventions. Thus coincident changes in the science underpinning biomedicine and the policy environment surrounding IP rights have increased both the generation and patenting of upstream developments in biomedicine.

## Conceptual

When is either an "anticommons" problem or restricted access to upstream discovery likely to emerge and why, and what are the welfare implications of their emergence?

Consider the anticommons. The central question here, as posed by both Heller and Eisenberg (1998) and Eisenberg (2001), is, if there is a cooperative surplus to be realized in combining property rights to commercialize some profitable biomedical innovation, why might it not be realized? They argue that biomedical research and innovation may be especially susceptible to breakdowns and delays in negotiations over rights for three reasons. First, the existence of numerous rights holders with claims on the inputs into the discovery process or on elements of a given product increases the likelihood that the licensing and transaction costs of bundling those rights may be greater than the ultimate value of the deal. Second, when there are different kinds of institutions holding those rights, heterogeneity in goals, norms, and managerial practice and experience can increase the difficulty and cost of reaching agreement. Such heterogeneity is manifest in biomedicine given the participation of large pharmaceutical firms, small biotechnology research firms, large chemical firms that have entered the industry (e.g., DuPont and Monsanto), and universities. Third, uncertainty over the value of rights, which is acute for upstream discoveries and research tools, can spawn asymmetric valuations that contribute to bargaining breakdowns and provide opportunities for other biases in judgment. This uncertainty is heightened because the courts have yet to interpret the validity and scope of particular patent claims.

Regarding the restriction of access to upstream discoveries highlighted by Merges and Nelson (1990; 1994), one can ask why that should be a policy concern. From a social welfare perspective, nothing is wrong with restricted access to IP for the purpose of subsequent discovery so long as the patentholder (or licensee) is as able as other potential downstream users to fully exploit the potential contribution of that tool or input to subsequent innovation and commercialization.[10] This, however, is unlikely for several reasons. First, firms and, especially, universities are limited in their capabilities. Second, there is often a good deal of uncertainty about how best to build on a prior discovery, and any one firm will be limited in its views about what that prior discovery might be best used for and how to go about exploiting it. Consequently, a single patentholder or licensee is unlikely to exploit fully the research and commercial potential of a given upstream discovery, and society is better off to the extent that such upstream discoveries are made broadly available.[11] For example, if there is a target receptor it is likely that there are a variety of lines of attack, and no single firm is likely capable of mounting or even conceiving of all of them. The notion that prior discoveries should be made broadly available rests, however, on an important assumption—that broad availability will not compromise the incentive to invest the effort required to come up with that discovery to begin with (cf. Scotchmer, 1991).

In this chapter, we are therefore concerned with whether access to upstream discoveries essential to subsequent innovation is restricted. Restriction is, however, a matter of degree. If a discovery is patented at all, then it is to be expected that access will be restricted—reflecting the function of a patent. Indeed, any positive price for a license implies some degree of restriction. Therefore, we are concerned with more extreme forms of restricted access that may come in the form of exclusive licensing of broadly useful research tools, high license fees that may block classes of potential users, or decisions on the part of a patentholder to itself exploit some upstream tool or research finding that it developed.

## Historical

The possibility that access to a key pioneering patent may be blocked, or that negotiations over patent rights might break down—even when a successful resolution would be in the collective interests of the parties concerned—is not a matter of conjecture. There is historical precedent. Merges and Nelson (1990) and Merges (1994), for example, consider the case of radio technology where the Marconi Company, De Forest, and De Forest's main licensee, AT&T, arrived at an impasse over rights that lasted about ten years and was only resolved in 1919

---

[10]That patents imply some type of output restriction due to monopoly is taken as given. The question here is whether there is any social harm if only one firm holds the right to exploit the innovation.

[11]The premise of this argument, well recognized in the economics of innovation (Jewkes et al., 1958; Evenson and Kislev, 1973; Nelson, 1982), is that, given a technological objective (e.g., curing a disease) and uncertainty about the best way to attain it, that objective will be most effectively achieved to the extent that a greater number of approaches to it are pursued.

when RCA was formed at the urging of the Navy. In aviation, Merges and Nelson argue that the refusal of the Wright brothers to license their patent significantly retarded progress in the industry. The problems caused by the initial pioneer patent (owned by the Wright brothers) were compounded as improvements and complementary patents, owned by different companies, came into existence. Ultimately, World War I forced the Secretary of the Navy to intervene to work out an automatic cross-licensing arrangement. "By the end of World War I there were so many patents on different aircraft features that a company had to negotiate a large number of licenses to produce a state-of-the-art plane" (Merges and Nelson, 1990, p. 891).

Although breakdowns in negotiations over rights may therefore occur, rights over essential inputs to innovation are routinely transferred and cross-licensed in industries, such as the semiconductor industry, where there are numerous patents associated with a product and multiple claimants (Levin, 1982; Hall and Ziedonis, 2001; Cohen et al., 2000). In Japan, where there are many more patents per product across the entire manufacturing sector than in the United States, licensing and cross-licensing are commonplace (Cohen et al., 2002).

Thus the historical record provides instances of both where the existence of numerous rights holders and the assertion of patents on foundational discoveries have retarded commercialization and subsequent innovation and where no such retardation emerged. The history suggests several questions. Have anticommons failures occurred in biomedicine? Are they pervasive? To what degree do we observe restricted access to foundational discoveries that are essential to the subsequent advance of biomedicine? What factors might affect biomedicine's susceptibility (or lack thereof) to either anticommons or restrictions on the use of upstream discoveries in subsequent research?

## DATA AND METHOD

To address these issues, we conducted 70 interviews with IP attorneys, business managers, and scientists from 10 pharmaceutical firms and 15 biotech firms, as well as university researchers and technology transfer officers from 6 universities, patent lawyers, and government and trade association personnel. Table 1 gives the breakdown of the interview respondents by organization and occupation. These interviews averaged over one and a half hours each. The interviews focused on changes in patenting, licensing activity and the relations between pharmaceuticals, biotechnology firms, and universities, and how patent policy has affected firm behavior.

This purposive sampling was designed to solicit information from respondents representing various aspects of biomedical research and drug development (Whyte, 1984). We used the interviews to probe whether there has been a proliferation and fragmentation of patent rights and whether this has resulted in the failure to realize mutually beneficial trades, as predicted by the theory of anti-

**TABLE 1** Distribution of Interview Respondents, by Organization and Occupation

|                   | Pharmaceutical | Biotech | University | Other  |
| ----------------- | -------------- | ------- | ---------- | ------ |
| IP lawyer         | 12             | 7       | —          | 12 (7) |
| Scientist         | 3              | 4       | 10         | 3      |
| Business manager  | 9              | 7       | 3          | —      |

NOTE: "Other" includes outside lawyers (7) and government and trade association personnel. University technology transfer office personnel are classified as "business managers," although some are also lawyers. Also, many of the lawyers and business managers were also R&D scientists before their current position.

commons. We also looked for instances in which restricted access to important upstream discoveries has impeded subsequent research. In addition, we asked our respondents how these conditions may have changed over time, including whether the character of negotiations over IP rights have changed. Finally, we asked about strategies and other factors that may have permitted firms to overcome challenges associated with IP.

## FINDINGS

### Preconditions for an Anticommons

Do conditions that might foster an "anticommons" exist in biomedicine? The essential precondition for an anticommons is the existence of multiple patents covering different components of some product, its method of manufacture, or inputs into the process through which it is discovered.

We have no direct measure of the number of patents covering a new product. There has, however, been a rapid growth in biotechnology patents over the past fifteen years, from 2,000 issued in 1985 to over 13,000 in 2000.[12] Such rapid growth is consistent with a sizable number of patents granted for research tools and other patents related to drug development. Our interview respondents also suggest that there are indeed now more patents related to a given drug development project. One biotechnology executive responsible for IP states:

> The patent landscape has gotten much more complex in the 11 years I've been here. I tell the story that when I started and we were interested in assessing the third party patent situation, back then, it consisted of looking at [4 or 5 named firms]. If none were working on it, that was the extent of due diligence. Now, it

---

[12] http://www.bio.org/er/statistics.asp

is a routine matter that when I ask for some search for third-party patents, it is not unusual to get an inch or two thick printout filled with patent applications and granted patents.... In addition to dealing with patents over the end product, there are a multitude of patents, potentially, related to intermediate research tools that you may be concerned with as well."

Almost half of our respondents (representing all three sectors of our sample: big pharmaceutical firms, small biotech firms, and universities) addressed this issue, and all of them agreed that the patent landscape has indeed become more complex.[13] How complex is, however, an important issue. Although there are often a large number of patents potentially relevant to a given project, the actual number needed to conduct a drug development project is often substantially smaller. For example, Heller and Eisenberg (1998) use the case of "adrenergic receptor" claims as an illustration of the anticommons problem and find over 100 patents that might require a license to do research in this area. Responding to the Heller and Eisenberg article, Seide and MacLeod (1998) did a search on "adrenergic receptor" and, indeed, found 135 patents using this term. They then did an (admittedly cursory) patent clearance review and found that the vast majority would not in fact be infringed by an assay to screen for ligands against this receptor and that, at most, only a small number of licenses might be required. Another case (from agricultural biotech) was that of putting hemoglobin in maize (Warcoin, 2002). Here, 500 patent applications were initially reviewed, of which 100 were potentially of interest. In the end, 13 relevant patents were identified, including research tools, specific DNA for expression, and the technology for transforming the plant.

We asked about 10 of our industry respondents to tell us how many pieces of IP had to be in-licensed for a typical project. They said that there may be a large number of patents to consider initially—sometimes in the hundreds, and that this number is surely larger than in the past. However, respondents then went on to say that in practice there may be, in a complicated case, about 6-12 that they have to seriously address, but that more typically the number was zero. An IP lawyer at a biotech firm states:

> The head of research comes to you and says he intends to develop this product and he wants you to look into the patent situation. You get back an inch or two thick pile of patents. You go through... and make judgments, what patents are relevant? Then, you go through those more in depth.... At the next step, you are

---

[13] A few respondents noted that there is some recent backing off from mass patenting strategies. For example, over the last few years, NIH went from patenting 90 percent of their inventions to patenting only 40 percent (Freire, 2002). Some firms have also begun concentrating on their most promising targets, because of the high cost of maintaining patents and the low value of many genomic patents, particularly expressed sequence tags (ESTs), that may not give rights to downstream developments in therapeutics.

left with 5-10, maybe 20, it depends. Not hundreds. You investigate these.... In the end, there are probably 3-6 that you have to negotiate.

Thus, although most R&D executives report that the number of licenses they must obtain in the course of any given project has increased over the past decade, that number is considered to be manageable.

In addition to a larger number of patents typically bearing on a given project, the numbers and types of institutions involved have also grown. Preceding the recent growth in biotechnology patenting, the number of biotechnology firms grew rapidly in the 1980s (Cockburn et al., 2000). More recently, we observe biotechnology firms acquiring significant patent positions. Hicks et al. (2001), for example, report that the number of U.S. biotechnology firms receiving more than 50 patents in the prior six years grew from zero in 1990 to 13 by 1999.

Universities have also become major players in biotechnology, as sources of both patented biomedical inventions and start-up firms that are often founded on the strength of university-origin patents. Many respondents (14 from industry and 6 from universities) noted that this new role of universities is one of the significant changes over the last two decades in the drug and related industries. Universities have increased their patenting dramatically over the last two decades, and although still small, their share of all patents is significantly higher than before 1980. Furthermore, much of the growth in university patents tends to concentrate in a few utility classes, particularly those related to life sciences. In three of the key biomedical utility classes, universities' share of total patents increased from about 8 percent in the early 1970s to over 25 percent by the mid-1990s (NSF, 1998). Also, universities' adjusted gross licensing revenue has grown from 186 million dollars going to 130 universities in 1991 to $862 million going to 190 universities in 1999 (AUTM, 2000), with the preponderance of these sums reflecting activity in the life sciences. An eightfold increase in university technology licensing offices from 1980 to 1995 is further evidence of increasing emphasis on the licensing of university discoveries (Mowery, et al., 2001).

Contributing to the rise in patenting, particularly in genomics, is the intensification of defensive patenting. An executive with a biotechnology firm compared its patenting strategy with that of Japanese firms in industries such as telecommunications or semiconductors: "We have a defensive patent program in genomics. It is the same as in the Japanese electronics industry. There they patent every nut and screw on a copier, camera, and build a huge portfolio, so Sony never sues Panasonic and Panasonic never sues Sony. There is a little of that going on in genomics. That way, if an IP issue ever arose, we have some cards in our hand." A respondent from a large pharmaceutical firm made a similar comment about their motives for patenting research tools: "I supposed because we see everyone else doing it in part. Sort of like the great Oklahoma Land Rush. If you don't do it you're not going to have any place to set up a tent, eventually." Overall, about a third of our industry respondents claimed to be increasing their patenting of gene sequences, assays, and other research tools as a response to the patenting of others to ensure freedom to operate (see also Henry et al., 2002).[14]

Thus we observe many patents (especially on research tools) owned by different parties with different agendas. In short, the patent landscape has indeed become more complex—although not as complex as suggested by some. Nonetheless, conditions may indeed be conducive to a tragedy of the anticommons.

### Preconditions for Restricted Access to Upstream Discoveries

Our second concern is that restrictive assertion or licensing of patents on research tools—especially foundational upstream discoveries upon which subsequent research must build (such as transgenic mice, embryonic stem cells, or knowledge of a potential drug target)—may undermine the advance of biomedical research. As suggested above, the key condition for this concern holds—namely, that research tools are now commonly patented. One R&D manager, for example, states that, "there has been a pronounced surge in patenting of research tools, previously more freely available in the public domain." Academic scientists we interviewed affirmed this view, observing a shift from a regime in which findings were more likely to be placed in the public domain with no IP protection.

We do not have patent data on research tools and upstream discoveries per se, but a hallmark of the advance in molecular biology and related fields over the past two decades is a proliferation in new techniques and methods that are inputs into the discovery process. In addition to recombinant DNA, prominent examples include polymerase chain reaction (PCR) and *Taq* polymerase, OncoMouse and cre-lox technology, and countless discoveries of genes and proteins that can either be used to develop therapeutics (EPO, for example) or offer promising targets for small-molecule drugs (such as the COX-2 enzyme for pain, CCR5 receptor for HIV, or telomerase for cancer).

Restricted access to upstream technology becomes a greater concern and more limiting on downstream research activity as the claims on the upstream patents are interpreted more broadly. The complaint about Human Genome Sciences asserting its patent over the HIV receptor illustrates the concern that patent holders are able to exercise control over a broad area even when their own upstream invention is narrow and there is very little disclosed about the utility of the invention (Marshall, 2000a). At the time of the patent application, Human Genome Sciences (HGS) knew only that they had found the gene for something that was a chemokine receptor. Later work published by NIH scientists detailed how

---

[14]This growth in defensive patenting echoes the patent races observed in semiconductors and other complex product industries (cf. Cohen et al., 2000; Cohen et al., 2002; Hall and Ziedonis, 2001). As noted above, these other industries have for the most part managed to overcome any possible "anticommons" problem thus far.

this receptor (CCR5) worked with HIV, making this a very important drug target. Those who discovered the utility of this receptor for AIDS research and drug development filed patents, only to find that Human Genome Sciences' "latent" discovery had priority. The concern is that knowledge of the reach of HGS's patent could have deterred subsequent research exploring the role of the gene and the associated receptor.

Another way in which the absence of a clear written description may allow upstream patents to directly affect subsequent research is via "reach-through" patent claims (as distinct from license agreements that include royalties on the product discovered using a research tool). Here, the patent claims the target and any compound that acts on the target to produce the desired effect, without describing what those compounds are. A commonly cited case is the University of Rochester's patent on the COX-2 enzyme, which includes claims on drugs that inhibit the enzyme.[15] This claim is the basis of the lawsuit against Searle for patent infringement.[16] Again, if the patentholder is given broad rights to exclude others from pursuing research in this area, we could have the problem of no one in fact possessing the innovation (in this case, a COX-2 inhibitor), and greatly reduced incentives for non-patentholders to explore possible uses of the innovation.

Thus there is a proliferation of patents on upstream discoveries and tools, and how those patents affect downstream discovery depends heavily on the breadth of claims. Although the USPTO has permitted broad claims to issue, there remains the question of how the courts will evaluate those claims.

## Evidence of an Anticommons in Biomedical Research

Given that the preconditions for an anticommons seem to exist, we turn to our findings on the incidence and nature of the different impediments to biomedical research that an anticommons may pose. These include breakdowns in negotiations over rights, royalty stacking, and "excessive" license fees.

---

[15]USP 6,048,850: "A method for selectively inhibiting PGHS-2 activity in a human host, comprising administering a non-steroid compound that selectively inhibits activity of the PGHS-2 gene product to a human host in need of such treatment." The compounds claimed include (but are not limited to): "nucleic acid encoding PGHS-2 and homologues, analogues, and deletions thereof, as well as antisense, ribozyme, triple helix, antibody, and polypeptide molecules as well as small inorganic molecules; and pharmaceutical formulations and routes of administration for such compounds."

[16]Recently, the U.S. District Court for Western New York dismissed the University of Rochester's complaint, reasoning that the description of the discovery in the school's patent lacked the clarity necessary to support an infringement claim [*University of Rochester v. G.D. Searle & Co., Inc.*, No. 00-CV-6161L., 2003 WL 759719 (W.D.N.Y. Mar. 5, 2003)]. The University of Rochester indicated an intention to appeal the case. See http://www.urmc.rochester.edu/pr/news/news.cfm?ID=198

## Breakdowns

Perhaps the most extreme expression of an anticommons tragedy is the existence of multiple rights holders spawning a breakdown in negotiations over rights that lead to an R&D project's cessation. We find almost no evidence of such breakdowns. Although idiosyncratic (because of the role played by policymakers and the absence of clear commercial value), the case of beta-carotene-enhanced rice (GoldenRice™) shows that the holding of IP by numerous parties need not defeat the development and commercialization of an innovation. Indeed, the complexity of this case is quite extreme, involving as many as 70 pieces of IP and 15 pieces of technical property spread over 31 institutions (Kryder et al., 2000).[17] Although there was strong interest in this product from international aid agencies, they required general IP clearance before the product could be developed. After about a year of negotiations, Monsanto, Zeneca, and others agreed to provide royalty-free licenses for the development and distribution of this innovation in third world countries.[18]

Beyond the case of GoldenRice™, we asked respondents and searched the literature to identify cases in which projects were stopped because of an inability to obtain access to all the necessary intellectual property rights. In brief, respondents reported that negotiations over access to necessary IP from many rights holders rarely led to a project's cessation. Of the 55 respondents who addressed this issue (representing all three sectors), 54 could not point to a specific project stopped because of difficulties in getting agreement from multiple IP owners (the anticommons problem). For example, one respondent indicated that about a quarter of his firm's projects were terminated in the past year. Of these, none were terminated because of any difficulties with the in-licensing of tools. Instead, the key factors included pessimism about technical success and the size of the prospective market. One biotechnology executive stated: "I am hard pressed to think of a piece of research that we haven't done because of blocked access to a research tool. We have dropped products because others were ahead in proprietary position, but that is different."[19]

---

[17] However, detailed study of the proprietary landscape noted that, depending on the country and the technologies that are used, the number of patents in fact could vary from 40 (in the United States or Europe) to zero (in, for example, Thailand, Bangladesh, Myanmar, Malaysia, Iran, Iraq, Saudi Arabia, or Nigeria).

[18] Another example of a potential anticommons problem was the hepatitis B vaccine, which involved 14 pieces of IP across several organizations and produced a royalty stacking that totaled $1.47 per dose, or about 13-15 percent of sales (Hackett and Totten, 1995).

[19] Numerous respondents reported that they did not initiate or had dropped projects if they learned another firm had already acquired a proprietary position on a drug they were considering developing—that is, on the output of a drug discovery and testing process. But that is quite different from other firms having IP for the research tools—the inputs into the discovery process.

A particular concern raised by Heller and Eisenberg (1998) and the National Research Council (1997, Ch. 5) was the prospect that, by potentially increasing the number of patent rights corresponding to a single gene, patents on expressed sequence tags (ESTs) would proliferate the number of claimants to prospective drugs and increase the likelihood of bargaining breakdowns. Our respondents suggested that this has not occurred. The key concern was that patents on the partial sequence might give the patentholder rights to the whole gene or the associated protein, or at least that the patent might block later patents issued on the gene or the protein [as Doll (1998) of the USPTO suggested]. Our respondents from industry and from the USPTO reflected, however, the view of Genentech's Dennis Henner who testified before Congress that EST patents do not dominate the full gene sequence patent, the protein, or the protein's use; these are separate inventions.[20] Also, although the existence of large numbers of EST patents may have had the potential to create anticommons problems, the new utility and written description guidelines implemented by the USPTO will now likely prevent many EST patents from issuing and will grant those that do issue only a narrow scope of claims. In addition, it is likely that already-issued EST patents will be narrowly construed by the courts. Thus the consensus is that the storm over ESTs has largely passed.

## Royalty Stacking

Another way in which multiple claimants on research tool IP may block drug discovery and development is the stacking of license fees and royalties to the point of overwhelming the commercial value of a prospective product. Most of our respondents reported that royalty stacking did not represent a significant or pervasive threat to ongoing R&D projects. One respondent said that, although stacking is a consideration, "I can't think of any example where someone said they did not develop a therapeutic because the royalty was not reasonable." We only heard of one instance in which a project was stopped because of royalty stacking. We were told, however, that, in this case, there were too many claimants to royalty percentages because of carelessness by a manager, who had given away royalty percentages without carefully accounting for prior agreements.[21] One of our other biotechnology respondents suggested, however, that "the royalty burden can become onerous" and that the stacking of royalties "comes up pretty regularly now" with the proliferation of IP. Even here, the respondent said that no projects had ever been stopped because of royalty stacking. Overall, about

---

[20]Testimony before House Judiciary Committee, 7/13/00, htttp://www.house.gove/judiciary/henn0713.htm.

[21]We also had one respondent, an IP lawyer, who said such cases where projects were stopped existed, but client privilege prevented the respondent from giving details.

half of our respondents complained about licensing costs for research tools, although nearly all of those concerned about licensing costs also went on to say that the research always went forward.

Royalty stacking does not represent a significant threat to ongoing R&D projects for several reasons. First, and principally, the total of fees paid, as discussed below, typically does not push projects into a loss. Second, in the minority of cases in which the stacking of fees threatens a loss, compromises tend to be struck, often in the form of royalty offsets across the various IP holders. One respondent stated, "All are sensitive and aware of the stacking phenomenon so there is a basis for negotiation, so that you don't have excessive royalties." Finally, in the few cases in which such a problem might emerge, it also tends to be anticipated.[22] One firm executive we interviewed said they had a corporate-level committee that reviewed all such requests to make sure such problems do not occur.

### Licensing Fees for Research Tools

Although obtaining systematic data on the cost of patented research tools is difficult, half of our respondents provided enough information to allow us to approximate the range of such costs. The norm for total royalty payments for the various input technologies associated with a given drug development program is in the range of 1 to 5 percent of sales, and somewhat higher for exclusive licenses. Occasionally, royalty demands were 10 percent or higher, and these were described in such terms as "high" or "ridiculous." Firms (especially the large pharmaceutical firms) also license particular technologies—such as using a gene for screening or a vector or microarrays—for a fee ranging from $10,000 to $200,000. These fees (especially for genes) were often described (by both those buying and those selling such technologies) as small amounts that large pharmaceutical firms paid as insurance both to ensure freedom to operate and to avoid the cost of litigation. The cost of patented reagents could be two to four times as much as do-it-yourself versions (or, in the case of *Taq* polymerase, buying from an unlicensed vendor), although the overall cost to the project is generally small (at most a few percent).

Large pharmaceutical firms have also been licensing access to genomic databases, and these database fees are often tens of millions of dollars and occasionally over $100 million (*Science*, 1997). In 1997, for access to its database, Incyte

---

[22]In response to a question of whether their firm ever had a case of a project being stopped for problems with royalty stacking, a biotechnology respondent stated: "No. It would be hard to find such a case, given the reality of how decisions are made. It is not a late stage decision."

was reported to be charging $10 million to Upjohn and almost $16 million to Pfizer, as well as undisclosed amounts to eight other firms. These deals also include "low single-digit" royalties for use of patented genes in drug development. Four pharmaceutical firms paid between $44 million and $90 million each to Millennium to access their data and research tools for identifying disease genes. In 1998, Bayer agreed to a deal in which they would pay up to $465 million to Millennium to have Millennium identify 225 new drug targets within 5 years (Malakoff and Service, 2001).

Overall, our respondents noted that, although these costs were higher than before the surge in research tool patents, they believed them to be within reason largely because the productivity gains conferred by the licensed research tools were thought to be worth the price. The case of Human Genome Sciences' database is illustrative. In 1993, SmithKline signed a deal for exclusive access for $125 million. By 1996, the database had already "saturated SmithKline with [drug-target] opportunities," according to Human Genome Sciences' Haseltine. Therefore, the partners extended access to the database to three other firms, who contributed a total of $140 million (Cohen, 1997). One scientist at a large pharmaceutical firm characterized the return to paying for access to Incyte's database as follows:

> The richness in Incyte's database is quite impressive. If you are just stuck with the things in the public database, the map is interesting, it is exciting, but it is a lot harder.... I was telling my family recently that I probably could have done my 4 year 8 month Ph.D. in about 6 months with today's technology.... it is that big of a technology revolution.[23]

Thus, although the development and patenting of research tools and upstream discoveries are imposing costs on downstream users, some of those users believe that their research is substantially more productive as a consequence.

Our interviews suggested, however, that although these costs were seen as manageable by large pharmaceuticals firms, and even by established biotech

---

[23]Randall Scott of Incyte offered several revealing examples of the productivity benefits of genomics that accord with the comments of our other respondents:

"An Incyte customer stated that it had reduced the time associated with target discovery and validation from 36 months to 18 months, through use of Incyte's genomic information database. Other Incyte customers have privately reported similar experiences.... One Incyte customer stated that by using Incyte's database, it quickly discovered a new histamine receptor gene which had long eluded researchers, and which is being used to develop an effective drug that is specific for brain tissue. In fact, after isolating the gene and using high-throughput screening, a candidate drug was identified in less than a month. Again, by making new targets available to the pharmaceutical industry, Incyte helped the company go from picking a target receptor to developing a potential drug in just 18 months, a process that typically takes five years or more, clearly accelerating the drug discovery process by three-fold or more" (Scott, 2000).

firms, small start-up firms and university labs noted that such costs could be prohibitive, in effect making it impossible for them to license particular research tools. This issue of restricted access due to high prices was prominently raised in the National Academy's workshop on IP and biomedical innovation (National Research Council, 1997). One of our respondents suggested that, for example, "DNA chips are a high-investment technology. Very small labs can't afford to do it. When the technology is out of reach of small labs, they have to collaborate. But this collaboration generally means giving up IP rights. The technology forces collaboration because barriers to entry are high." This sentiment was echoed by university researchers we talked with. This was one justification for the "do-it-yourself" solution of making patented laboratory technology without paying royalties (Marshall, 1999b).[24] Similarly, the manager of a small biotech start-up told us that Incyte's licensing terms for access to their gene database was several times the firm's whole annual budget. They were forced to rely on the public databases, a viable but second-best solution. One solution for universities has been the development of core facilities to share expensive resources such as chip-making facilities or high-throughput screening.

Some firms (particularly genomics firms) holding rights over research tools did, however, offer discounted terms for university and government researchers. Celera, for example, licenses their database to firms for about $5 million to $15 million per year and to university labs for about $7,500 to $15,000 (Service, 2001). In 2000, Incyte began allowing single-gene searches of its database for free, with a charge of $3,000 or more for ordering sequences or physical clones, making its database more accessible to small users (*Science*, 2000). Myriad also offers a discount rate (less than half the market rate) for academics doing NIH-funded research on breast cancer (Blanton, 2002).

In this section, we have considered the costs of licensing research tool IP— but only the out-of-pocket, monetary costs. Costs can, however, also take non-monetary forms. The most prominent of these for university researchers are publication restrictions, which we did not examine.[25]

---

[24] Affymetrix has recently adopted an easy access plan for universities to try to shift them away from a do-it-yourself approach. Several have suggested that the ability to get others to license your patented technology depends on embedding it in a form that is more convenient, reliable, or inexpensive than do-it-yourself versions would be.

[25] We could not obtain systematic data on the license terms for research tool technologies. However, Thursby and Thursby (1999) report that 44 percent of agreements to license university technologies to firms include publication delay clauses, with the average delay specified being almost 4 months. Also, Blumenthal et al. (1997) report that 20 percent of academic biomedical researchers have delayed research publication by 6 months or more, in part because of concerns about patents and commercialization. Thus a substantial fraction of university-industry agreements about the outputs of university research include delays of publication. We do not know whether these examples generalize to the case of agreements over the inputs to university research.

## Projects Not Undertaken and Broader Determinants of R&D

Although the number of ongoing R&D projects stopped because of an anticommons problem is small, it is possible that firms avoid stacking and other difficulties in accessing IP rights by simply not undertaking a project to begin with. As a practical matter, it is difficult to measure the extent to which projects were not started or redirected because of patent-related concerns. In brief, although redirecting projects to invent around research tool patents was common, it was relatively rare for firms to move to a new research area (perhaps a new disease, or even a very different way of approaching a disease) because of concerns over one or more research tool patents. Of the 11 industry respondents who did mention IP as a cause for redirecting their research, seven, however, were primarily concerned with IP on compounds, not on research tools.[26] An IP attorney with a large biotech firm suggested that patents on research tools were rarely determinative, reporting that in the "scores of projects" that his firm considered undertaking over the years, he could remember only one where such patent rights dissuaded them from undertaking the project. Another biotech firm's lawyer, while reporting that they had never stopped an ongoing project because of license stacking, stated that considerations of patents on both compounds and research tools did preempt projects:

> We start very early on... to assess the patent situation. When the patent situation looks too formidable, the project never gets off the ground.... Once you are well into development, you get patent issues, but not the show stopper that you would identify early on.

Although we have no systematic data on projects never pursued, our findings on the absence of breakdowns is consistent with the notion that there are relatively few cases where otherwise commercially promising projects are not undertaken because of IP on research tools. Consider Heller's (1998) original article on the anticommons, which paints a vivid image of empty buildings in Moscow, unrented because the various owners and claim holders that could "veto" a rental arrangement were many and had trouble coming to agreement. Our analogue to an "empty building" is, of course, an R&D project that is stopped midway. However, if the argument that the proliferation of IP is generating an "anticommons" is correct, it follows that the rational anticipation of such difficulties would prevent the construction of some (or many) buildings. Likewise, some R&D projects may not be undertaken if firms anticipate difficulty in negotiating cost-effective access to the required IP. However, absent any visible empty buildings (i.e., ob-

---

[26]And a large number (about a third of industry respondents) said that when faced with rival patents on research tools, or even compounds, they were likely to go ahead with the research, so long as they were able to develop their own IP that would protect their compounds (see below).

served stopped projects), it is unlikely that the anticipation of breakdowns in negotiations or an excessive accumulation of claims (i.e., license stacking) prevented construction (i.e., undertaking the R&D project).[27]

Our interviews suggested that the main reasons why projects were not undertaken reflected considerations of technological opportunity, demand, and internal resource constraints, with expected licensing fees or "tangles" of rights on tools playing a subordinate role, salient only for those projects which were commercially less viable. One industrial respondent affirmed that, although other considerations were key, royalty stacking could affect decisions at the margin: "I don't want to say a worthwhile therapeutic was not developed because of stacking problems. But if we have two equally viable candidates, then we choose based on royalties." One biotechnology respondent was explicit, however, about the greater importance of expected demand and technological opportunity: "At the preclinical stage, you find you have 10 candidates, and you can afford to continue work on 3. The decision is a complex prediction based on the potential for technical success, the cost of manufacturing, the size of the market, what you can charge, what you need to put in for royalties. I am not familiar with royalty stacking being the deciding factor. The probability of technical success and the size of market are key." This last remark also implies that the firm had more viable opportunities than it had the resources to pursue. Indeed, complaints about resource constraints as impediments to progress on promising research were more common than complaints about IP. As one research manager from a pharmaceutical firm put it: "What we find limiting in our process is the number of chemists we can bring to bear. That is the most limiting resource we have. We have more targets than we have chemists to work on them."[28]

---

[27]One version of this argument is that, if the anticommons problem were widespread but few stopped projects ("empty buildings") were observed, this could be because of "anticipation and redirection" where firms or university researchers redirect R&D toward projects for which they do not anticipate an anticommons problem. However, to do so with such high success (so that projects did not, in the end, get stopped), decision makers would have to be very prescient about when they would, and when they would not, face such problems. This is unlikely given the uncertainty of early-stage R&D (when researchers do not yet know what tools they might need) and given the lag in patent issuance (so that researchers do not even know which tools are patented—see Marshall, 2000a; Merz et al., 2002). Both of these factors lead to having substantially less than perfect information on potentially blocking patents.

[28]The manager goes on to say: "Isn't it great you can identify 200 targets? Yes, but how do you do chemistry on 200 targets? Staffing for a chemical program can run to 12-15 chemists for a serious program. And, despite combinatorial chemistry, at the end of the day more traditional medicinal chemistry is needed to engineer and tailor properties, to build in selectivity, remove interactions. Those generally are the stuff of 10s of compounds to 20s of compounds synthesized with discrete changes to try to target specific things. These are not amenable to high-throughput operations."

The notion that opportunities often exceed the ability of firms to pursue them suggests that, at least under some circumstances, the social cost of not pursuing projects because of IP considerations may not be as great as one might suppose.[29] Indeed, four of our industry respondents expressed the view that redirection of research effort toward areas less encumbered by patents was not terribly costly for their firm or others because the technological opportunities in molecular biology and related fields were so rich and varied. As one biotech respondent put it, "There are lots of targets, lots of diseases." Some respondents have suggested that the value of targets has actually declined substantially because companies can't exploit all of the targets they have, and so firms are more willing to license some of their targets, or abandon some of their patents and let the inventions shift to the public domain, because maintaining large portfolios of low-value patents is expensive. On the other hand, one can also argue that even in the presence of rich opportunities, shifting may be costly to the extent that diminishing the number of firms trying to achieve some technical objective makes success less likely.

## Evidence of Restricted Access to Upstream Discoveries and Tools

Although biomedical research does not appear to be especially vulnerable to breakdowns over IP negotiations, restricted access to important research tools—especially foundational upstream discoveries—can potentially impede innovation in a field. Moreover, this has occurred in other settings (Merges and Nelson, 1990). Our question is whether the restrictions on access to such upstream discoveries, through, for example, exclusive licensing, has impeded biomedical innovation. As noted above, in contrast to the prospect of an anticommons, this is not a problem of accessing multiple rights but one of accessing relatively few—perhaps even one—patent on a key tool or discovery.

In its report, "Intellectual Property Rights and Research Tools in Molecular Biology," the NRC (1997) provided a series of case studies on the uses of patents covering a small number of important research tools in molecular biology where the question of restricted access was considered. In the case of the Cohen-Boyer technology for recombinant DNA developed at Stanford University and the University of California—"arguably the defining technique of modern molecular biology" (NRC, 1997, p. 40)—the three patents were broadly licensed on a nonexclusive basis on a sliding scale, providing the basis for the creation of the biotechnology industry as we know it. The license was available for about $10,000

---

[29]Under plausible conditions, there can also be excessive correlation in research portfolios to the degree that research bandwagons emerge around the mining of what may be considered the most promising veins (Dasgupta and Maskin, 1987). Under such circumstances, a shift to less crowded areas of art would be socially beneficial.

per year plus a royalty of 0.5 to 3 percent of sales (Hamilton, 1997). Stanford and UC eventually had several hundred licensees, and the patent generated an estimated $200 million for the universities.

The second case was that of polymerase chain reaction (PCR) technology, which "allows the specific and rapid amplification of DNA or targeted RNA sequences," and *Taq* polymerase, which is the enzyme used in the amplification. The technology was also key to subsequent innovation. It "...had a profound impact on basic research not only because it makes many research tasks more efficient but also because it ...made feasible ...experimental approaches that were not possible before" (NRC, 1997). In addition to being a discovery tool, the technology also provides a commercial product in the form of diagnostic tests. Developed by Cetus Corporation, the technology was sold to Roche in 1991 for $300 million. As the NRC (1997) reports, the controversy over the sale of the technology has been primarily over the amount of the licensing fees and the fees charged for the material (*Taq* polymerase) itself. Although Roche licensed the technology widely, particularly to the research community, they did charge high royalty rates on their licenses for diagnostic service applications. Also, small firms complained about Roche's fees for applications of the technology outside of diagnostics, which ranged between $100,000 and $500,000 initially with a royalty rate of 15 percent. The high price likely restricted access for some, especially small biotech firms.[30]

The CellPro case, described in detail by Bar-Shalom and Cook-Deegan (2002),[31] also illustrates the potential for the owners of upstream patents to block development of cumulative systems technologies (cf. Merges and Nelson, 1994). Johns Hopkins University's Curt Civin discovered an antibody (My-10) that selectively binds to an antigen, CD34, found on stem cells but not on more differentiated cells. In 1990, Hopkins was awarded a patent that claims all antibodies that recognize CD34. Baxter obtained an exclusive license. The chief rival was CellPro, a company founded in 1989 based on two key technologies: one a method for using selectively binding antibodies to enrich bone marrow stem cells or deplete tumor cells, and the other an unpatented antibody, 12-8, that also binds to CD34, although in a different class of antibodies from Civin's My-10 and recognizing a different epitope (binding site) on CD34. CellPro combined these two discoveries with other innovations and know-how to produce a cell separator instrument for use in cancer therapies, particularly bone marrow transplants.

---

[30]Promega and Roche have been in a long-running dispute over the right to distribute *Taq* for research uses, with Promega attacking the validity of Roche's patents. In December 1999, the patent on *Taq* was ruled unenforceable (for inequitable conduct). Roche appealed the ruling to the Court of Appeals for the Federal Circuit (CAFC). The rest of the case, including Promega's request for damages and the validity of the PCR patents, will be set aside until after the CAFC rules on the inequitable conduct case. Roche has also faced unfavorable rulings in Europe and Australia.

[31]The following account draws primarily from Bar-Shalom and Cook-Deegan (2002).

Baxter offered CellPro a nonexclusive license for $750,000 plus a 16 percent royalty.[32] CellPro felt this was uneconomic and, armed with a letter from outside counsel saying that CellPro's technology did not infringe and that the patent was probably invalid, decided to move forward with development and to sue to invalidate the patent. Although the jury ruled in CellPro's favor, the *Markman* decision reopened the case and the judge ruled for Baxter, assessing treble damages totaling $7.6 million, as well as $8 million in legal fees.[33] The court also ordered license terms similar to (though somewhat higher than) existing licenses, a royalty of over $1,000 per machine. CellPro lost the appeal and went bankrupt. Baxter allowed sales of CellPro's machine until its own instrument (which Baxter was developing all through this) received FDA approval. In the end, the technology did not prove to be widely effective, and more successful rival technologies were developed by others.

From our perspective, the main lesson of the CellPro case is that, to the degree that upstream patents are broadly interpreted, IP holders can use this broad claim to prevent others from engaging in the subsequent development needed to bring the patented technology to market. This is troubling when the patent owner or exclusive licensee cannot effectively develop the technology in a timely fashion, which was the case with Baxter, which was at least 2 years behind CellPro in bringing a product to market.[34]

Another case was that of the Harvard OncoMouse, licensed by Harvard exclusively to DuPont. The OncoMouse contained a recombinant activated oncogene sequence that permitted it to be employed both as an important model system for studying cancer and permitting early-stage testing of potential anticancer drugs. After years of negotiations, NIH and DuPont finally signed a memo of understanding in January 2000 that, among other things, permitted relatively unencumbered distribution of the technology from one academic institution to another, although under specific conditions.[35] Although this agreement was the cause of relief on the part of academic researchers, DuPont has just recently begun asserting its patent against selected institutions (Neighbour, 2002). The diffi-

---

[32]Baxter also licensed the patents to two other firms, for $750,000 and 8 percent royalties. These figures illustrate the cost for the license terms for a component of a therapeutic system, with the top of this scale probably at the high end, because CellPro was reluctant to take the terms and Nexell could not profitably produce a product under its terms.

[33]This provides an example of the scale for legal fees involved in such a case (see below).

[34]Bar-Shalom and Cook-Deegan (2002) also suggest that royalty stacking may have made the technology economically unfeasible. Hopkins had licensed to B-D, which in turn licensed to Baxter, which in turn licensed to others, with each taking a share of the rents. However, in this case, the license stacking was all on the same technology being passed from hand to hand. Thus, this was not a tragedy of the anticommons, but one of a proliferation of middlemen.

[35]In 1998, NIH announced an agreement with DuPont covering cre-lox technology (see MOU at http://ott.od.nih.gov/textonly/cre-lox.htm). In January, 2000, NIH announced an agreement covering OncoMouse (see MOU at http://ott.od.nih.gov/textonly/oncomous.htm).

culty is that although the initial press release suggested that these nonpaying rights to use the OncoMouse covered nonprofit recipients of NIH funding, the actual agreement stated that DuPont would make available similar rights to nonprofit NIH grantees "under separate written agreements." Because most universities have not asked for those rights, they stand outside of the agreement, and DuPont has begun to approach some of them, claiming that they are infringing Harvard's patent rights and must take a license from DuPont. The difficulty is that these new license agreements, although also nonpaying in principle, go well beyond the earlier understanding and make a series of stringent demands. Under the proposed agreement, for example, universities cannot use the technology in industry-sponsored research without the sponsor taking a commercial license, *notwithstanding the content or intent of the sponsored research* (Neighbour, 2002). It is unclear at this point, however, what success DuPont will have or how NIH and other institutions will respond.

The most visible recent controversy over access to IP covering a foundational biomedical discovery is the case of embryonic stem cell technology.[36] In brief, Geron funded the research of a University of Wisconsin developmental biologist, James Thompson, who in 1998 first isolated human embryonic stem cells and was issued a very broad patent. The Wisconsin Alumni Research Foundation (WARF), a university affiliate, held the patent and granted Geron exclusive rights to develop the cells into six tissue types that might be used to treat disease as well as options to acquire the exclusive rights to others. Another beneficiary of Geron support, Johns Hopkins University, also provided Geron with exclusive licenses on stem cell technology. In August 2001, WARF sued Geron—who had been trying to expand its rights to include an additional 12 tissue types—to be able to offer licensing rights to Geron's competitors. In January 2002, a settlement was reached that narrowed Geron's exclusive commercial rights to the development of only three types of cells—neural, heart, and pancreatic, gave it only nonexclusive rights to develop treatments based on three other cell types—bone, blood, and cartilage, and removed its option to acquire exclusive rights over additional cell types. Geron and WARF also agreed to grant rights free of charge to academic and government scientists to use the stem cell patents for research but not for commercial purposes.[37] Companies wishing to use the stem cells for research purposes would, however, have to license the patents. Thus it would appear that although WARF would like to license the technology broadly,

---

[36]This paragraph is based largely on two articles in *The New York Times*: Aug. 14, 2001, p. C2 and Jan. 10, 2002, p. C11.

[37]The WiCell Research Institute, set up by WARF, also provides cell lines to NIH and university researchers for $5,000 [essentially "cost"], using a standard materials transfer agreement (MTA) that includes no restrictions on publications nor reach-through claims to inventions using the cells. However, commercialization of those inventions may still require negotiating rights to the potentially blocking patents owned by WARF and licensed to Geron.

Geron still retains control over key application areas of the technology and may well decide to pursue those applications itself. Indeed, David Greenwood, CFO and senior VP of Geron, noted that Geron did not have to allow others to develop products in the three areas where it retained exclusive rights. It is unclear, however, whether Geron's now limited control of IP rights block others' research on stem cell technology. According to one respondent, infringement of Geron's IP is commonplace.[38] Moreover, scientific advances in both adult stem cell technology and the use of unfertilized eggs to spawn stem cells may weaken the constraints imposed by Geron's IP by broadening the access to the commercial development of noninfringing stem cell technology.

We have considered the question of research access to a small number of important upstream discoveries. Evidence from AUTM also suggests that, at least for licensing relationships between universities and small firms, access to relatively upstream discoveries—that is, the kind of discoveries that tend to originate from university labs—is commonly restricted. Specifically, in 1999, 90 percent of licenses to start-ups were exclusive, whereas only 39 percent of licenses to large firms were exclusive (AUTM, 2000). Similarly, in their study of licensing practices for genetic inventions, Henry et al. (2002) report that 68 percent of licenses granted by university and public labs were exclusive, whereas only 27 percent of licenses granted by firms were exclusive. However, only a minority of university-based discoveries are patented to begin with. Henry et al. (2002) [consistent with Mowery et al.'s prior (2001) results] find that only about 15 percent of university-based genetic discoveries are patented, with the vast majority going into the public domain without IP protection.

Even where universities employ restrictive licensing terms, however, it is not clear that such a practice diminishes follow-on discovery, at least when applied to smaller firms. One manager of a university-based start-up suggests that exclusive licensing to smaller biotech firms may actually advance follow-on discovery:

> The traditional way universities did this [technology transfer] would be to go license a large company. Those kinds of agreement [include a]...minimal up front [fee] and small royalty, 1-2 percent. What the experience has been then is often the large company will work on it for a while but if it doesn't look very promising, or they run into problems, which invariably they do...since they haven't invested much in it, they don't have a whole lot of motivation to stick with it. So, most of these licensing agreements that universities have done ended

---

[38]In response to the question of whether the patents keep others out, a scientist for a stem cell company responded: "No. People are infringing all over the place. None of the stem cell companies have the financial wherewithal to do anything about it. The conventional wisdom is, all these cells seem similar; we will patent and fight later. The first to market will win and then we can fight later. It is not clear who really owns what." See below for a discussion of infringement of research tools patents in general.

up going nowhere. The idea the university had, and other universities are beginning to do this, is to create small companies like us where the small company has every motivation to develop it because it is the only intellectual property that they have. The university then has more control over the situation because they are an equity owner. Hopefully the small company can develop the molecule to the point where some real value would be achieved, ...where we get somebody interested, and that somebody will take it over and eventually market it.

### *Restrictions on the Use of Targets*

In our interviews, we heard widespread complaints from universities, biotechnology firms and pharmaceutical firms over patentholders' assertion of exclusivity over an important class of research tools, namely "targets," which refers to any cell receptor, enzyme, or other protein implicated in a disease, thus representing a promising locus for drug intervention. Our respondents repeatedly complained about a firm excluding all others from exploiting its target (in the anticipation of doing so itself) or, similarly, a firm or university licensing the target exclusively. About one-third of respondents (representing all three types of respondents) voiced concerns over patents on gene targets (for example, the COX-2 enzyme patent, the CCR5 HIV receptor patent, and the hepatitis C protease patent).

Before considering the degree to which the assertion and licensing of IP on targets may be restricting their use in downstream research, we should recall that, to the extent that patents on targets do confer effective exclusivity, even over the ability of other firms to conduct research on a particular disease, this is the purpose of a patent—to allow temporary exclusivity. Responding to complaints about restricted access to their patented targets, a respondent from a pharmaceuticals firm stated: "Your competitors find out that you've filed against anything they might do. They complain, 'How can we do research?' I respond, 'It was not my intent for you to do research.'" Others also defended their rights to exclude rivals from their patented targets. More importantly, this right to assert exclusivity may confer a benefit in the form of increasing the incentives to do the research to discover the target to begin with, as well as incentives for follow-on investment to exploit the target. A key question, then, is whether those incentives can be protected while allowing reasonably broad access.[39]

Patents on targets, if broad in scope and exploited on an exclusive basis, may preclude the benefits of different firms with distinctive capabilities and percep-

---

[39]A related and important set of questions is how much incentive do patents actually provide (i.e., how effective are patents) and is this incentive necessary to bring forth the innovation, given the alternative means of capturing the rents from the innovation and given public subsidies for inventive activity (see Scherer, 2002, for a review of these issues).

tions pursuing different approaches to the problem (cf. Nelson, 1961; Cohen and Klepper, 1992). For example, big pharmaceuticals firms have libraries of compounds that might affect the target. These libraries vary by firm and are either kept secret or patented. Thus, narrowing access to the target entails a social cost. The following quote from a large biotech firm summarizes the issue:

> The problem is, a target is just that, a target, say, a receptor on a cell. If we did an exclusive license, and we've had that opportunity, the only compounds tested would be those in the chemical library of our licensee. Generally there is no chemical relation among the compounds that [act on this target]. The drugs work by occupying the reception site, for example. They [the licensee] throw all the compounds in their library and they may or may not have one that works well. The libraries vary a lot across firms. A lot are patented. Also large pharma companies have huge collections of compounds they've synthesized over the last 100 years that have never seen the light of day. With an exclusive license, the odds of finding an active drug, let alone the best, are not good. Therefore, we [the target owner] want the target technology broadly available. Broad licensing only makes economic sense in our view.

The problem of "limited lines of attack" may be greater when exclusive access to a set of targets is held by a smaller firm with limited capabilities, and, as noted above, much of the university licensing of biomedical innovations to small firms is on an exclusive basis. Although perhaps biased, a scientist from a large pharmaceutical firm described the broader capabilities of the large pharmaceutical firm to develop the potential of a target: "[Once the target had been identified], then, the power of the pharmaceutical company comes into play. You put an army of 50 molecular biologists and one-third of the medicinal chemists at [the firm] on this single problem."

In addition to the constraints imposed by firms' particular capabilities on the approaches taken to exploiting targets, there are also differences in firm strategies or approaches to drug development.[40] The following quotation from a scientist at a small start-up highlights this problem:

> Part of the problem that comes in here is that many of these firms are very specialized and many times somebody holds patents but they don't do all the

---

[40]An executive from a large pharmaceutical firm stated: "We all have access to the same body of literature, same collection of issued patents. Same kinds of research people taught in some of the very same universities. But, one person may look at his understanding of a disease and say, 'I think it's that this antibody is the problem and if I block that antibody that's going to cure this disease or treat the disease.' Someone else may say, 'It's not the antibody that's the problem, it's the enzyme that synthesizes the antibody,' or something like that. And take a completely different approach. The chemistries may be similar, may be completely different. Or you may have a compound that has multiple effects and one person may say, 'I want to focus on this effect of this drug.' And they'll start with that as a prototype, but then they'll start to make to modifications to enhance that effect and minimize the other effects. Let's take aspirin, for example. Aspirin is a cyclooxygenase inhibitor, it's an anti-inflamma-

applications feasible. So, what happens is they don't think about doing something and many times the royalty is so high that other companies, small companies that come up with ideas, may not be able to come in and negotiate the license deal. So, it becomes, by default, what happens now. It's not that the patent holder says the idea is great but I'm not going to let anybody do it. But, it never occurs to them.

Although limiting access to targets may well limit their exploitation, the question is how often this occurs. We do not have systematic data on the frequency with which this occurs. From interviews and secondary sources, however, we heard of a number of prominent examples of firms' being accused of asserting exclusivity over (or allowing only limited access to) a target. One case that has garnered a lot of attention is Myriad and its patents on a breast cancer gene (BRCA1). Myriad has been accused of stifling research because it has been unwilling to broadly license diagnostic use of its patents (Blanton, 2002). Myriad counters that over a dozen institutions had been licensed to do tests and that "Myriad's position is to not require a research license for anybody," while reserving the right to decide whether particular uses are research or commercial (Bunk, 1999; Blanton, 2002). Myriad sent a letter threatening a lawsuit to the University of Pennsylvania to stop them from performing genetic tests, arguing that this was commercial infringement (see below). Chiron has also developed a reputation for aggressively enforcing its patents on research targets. Chiron has filed suits against four firms that were doing research on drugs that block the hepatitis C virus (HCV) protease (in addition to filing suits against three firms doing diagnostics), and some have claimed that these suits are deterring others from developing HCV drugs (Cohen, 1999). Chiron responded to this claim by pointing out that it had licensed its patent to five pharmaceutical companies for drug development work (as well as at least five firms for diagnostic testing) and that the firms being sued had refused a license on essentially the same terms, which included significant up-front payments as well as "reach-through" royalties on the drug.

Also mentioned by our respondents was the case of telomerase as a potential target for cancer drugs. One university scientist observed: "I've asked heads of discovery why they were not using telomerase as a target. The response was, 'intellectual property.'" A scientist from a biotech firm suggested that Geron, the key IP owner, had been stymied in pursuing this because of the complexity of the

---

tory, it's an anti-febrile, but it also keeps platelets from sticking. One person may engineer compounds, starting with aspirin as a base, to make better anticoagulants. Another may go off toward the anti-inflammatory side. You're all starting with a limited body of knowledge as to how you think the disease functions, and how you think that's going to affect the course of that disease. Opinions can differ, based on your background, based on what you know of the literature, and based on what your experience is in the lab."

biology[41] and had redirected their efforts toward stem cell research, which looked more promising. Upon investigation, we found that Geron had indeed established a substantial patent position, with 56 U.S. patents related to telomerase. However, we also found that there is a great deal of research being done on telomerase in universities (see Figure 1) and that at least three other firms (Amgen, Novartis, Boehringer Ingelheim) are reported to be pursuing telomerase as a target (Marx, 2002b). In addition, Geron presented the results of three separate studies on telomerase-based anticancer projects at the April 2002 meetings of the American Association for Cancer Research. Furthermore, Geron has formed a number of nonexclusive licensing agreements for the exploitation of telomerase, typically with small biotech firms possessing complementary technology. Thus, although

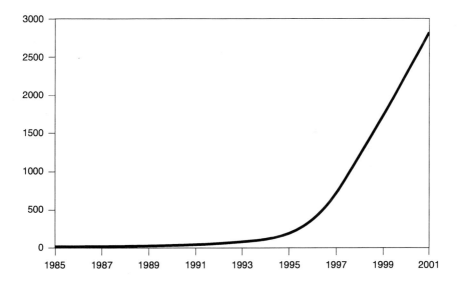

**FIGURE 1** Cumulative citations for "Telomerase" in MEDLINE.
Credit: C. Greider/Johns Hopkins University School of Medicine.
Source: Marx (2002a).

---

[41]Cal Harley of Geron predicted in 1994 that they would be in clinical trials within 4 years. They now hope to have a product in clinical trials in 2003, "if all goes well" (Marx, 2002b).

we again see some evidence of researchers being excluded, we do not find a failure to exploit the target.

Others have also complained about being blocked from working on targets because of restrictions imposed by patent holders. For example, *The New York Times* reported, "Peter Ringrose, chief scientific officer at Bristol-Myers, has said there are more than 50 proteins possibly involved in cancer that the company was not working on because the patent holders either would not allow it or were demanding unreasonable royalties" (*The New York Times*, Jan. 8, 2001, p. C2).

Ringrose's complaint was reported, however, in the context of the announcement of a licensing pact between Bristol-Myers and Athersys, which has a technique for producing proteins without isolating the corresponding gene, allowing use of the protein as a target without infringing patents on the gene, and therefore allowing circumvention of patents on genetic targets. This last point raises a question about the degree to which third-party patents on targets can actually impede firms' abilities to pursue R&D programs dedicated to particular therapies. We discuss this important issue below in the section on working solutions.

### *Costs and Delays*

In this section we consider the transactions costs associated with gaining access to one or multiple patents or responding to third-party assertions.[42] For instance, firms may avoid derailing in-house R&D projects but only by engaging in long and costly negotiations or litigation with IP holders. Firms may also invent around or conduct the R&D overseas, possibly at the cost of reducing R&D efficiency. Finally, IP holders may have to invest in monitoring the use of their IP, which, from a social welfare perspective, also constitutes a cost. Over a third of respondents (representing all three sectors) noted that dealing with research tool patents did cause delays and add to the cost of research.[43]

---

[42] Here we are only concerned with social costs, not the transfers of rents reflected in licensing fees.

[43] A respondent from a pharmaceutical firm expressed the firm's frustration:

> We do have frustration internally because we can't do what we consider basic research with a cloned gene, not selling the gene, just using it to make another discovery. To be cut off from that, it sits badly. Because, at the end of the day, you are cut off from tools, from making a breakthrough discovery. Because there is a patent on the human gene, you work with the guinea pig gene, but it is not the best approach. That's very frustrating. In a number of cases, we can't work with this protein or this gene and it slows things down. We are looking at ways to get around this. How to not infringe their IP. And, we are coming up with ways to do that, but it involves some labor and time.

Another biotech respondent stated:

> If there is a patent on manufacturing in different host cells, and certain others that don't have a royalty, then, on the last step, you don't make it in this cell, but over here. But that incurs some technical costs. It is a different system, you are not as familiar with the technology, but you go there because you don't want to pay the royalty.

University respondents referred to problems of negotiating MTAs (see below).

Litigation costs are likely to be a significant component of the social costs of the assertion and licensing of patents in biomedicine. Furthermore, biomedical patents are more likely to be litigated than are patents on other technologies (Lanjouw and Schankerman, 2001). Although estimates of litigation costs vary, estimates commonly ranged between $1 million-10 million for each side (see, for example, the discussion of the CellPro case above, where attorney fees were $8 million). One respondent from a biotech firm used the following comparison to put this into perspective:

> [XX lawsuit] cost $8 million per year and it was not done in a year. Think about $8 million, about what a biotech could do for $8 million. You could get a lot of science done. Depending on your burn rate, you could easily run a fully dedicated drug discovery program for $8 million. You could have afforded a reasonable number of people to work on that project for that year without any question whatsoever. It wouldn't surprise me that you could get in that position [because of a suit] that you would have to shut down a particular program.

In addition to these out-of-pocket expenses, we tried to estimate the opportunity cost of engaging in patent litigation. Out of the 16 industry respondents that addressed this issue, all but 1 suggested that litigation imposed a significant burden on the managers and scientists involved. In terms of actual work time, estimates were usually in terms of a few weeks over the course of a year for the individuals involved. Respondents also underscored the time spent worrying about the progress and outcome of the case. One respondent from a biotech described the process:

> Going to court is risky any time.... Patentability is complicated. You spend a lot of money educating the jury. You have to go searching through notebooks. If you do decide to sue, you have to be committed. The CEO, CFO are involved. You pull in business people to evaluate. Senior management and the particular inventors spend copious amounts of time on it; it is a huge distraction. They are in deposition, practicing for depositions, researching, responding to interrogatories, providing information. In a year, it costs a couple of man months. The CEO is in deposition for a week. My firm's experience has been that they want to ask everyone under the sun who was involved. I was tangentially involved and in deposition for a day and half. Duplicating all your files. Each page of your notebook, from 10 years ago, you have to find it, reproduce it. It is an enormous time sink and I think people underestimate it. Clients underestimate it. Even a winner may say, "If I knew then, I might say 'No'." Meantime, you are not doing science. Time not spent on new compounds, spent on what will they do in trial tomorrow.

About a third of our respondents addressed the question of negotiation delays or litigation, and nearly all of them felt that the process of sifting through a large number of potentially relevant patents and subsequent negotiations was very time consuming. One characterized the process as "complex, ongoing, and labor intensive," but a cost of doing business. Another stated: "All these patents makes research more expensive. It can slow it down, while you secure licenses." One biotechnology executive suggested that about a third of his firm's R&D projects suffer delays while licensing and related agreements are worked out. The above-cited quotation on reducing the stack of patents suggests that it commonly took about three or four weeks to sift through the patents potentially relevant to a project, often identifying somewhere between 5 and 20 that may be worth investigating intensively over the course of another three to six months. The respondent noted, however, that the research itself would typically be moving forward during this time. At this point, it would be typically determined that there are about three to six patents where agreements were required, and these negotiations could affect the progress and direction of the firm's R&D. The costs of negotiations as well as those for reviewing potentially relevant patents can be substantial in absolute terms. One attorney responsible for evaluating research tool IP from a large pharmaceuticals firm provided estimates for the time attorneys were occupied with evaluating the IP of third parties and the time associated with actual negotiations that implied a total of $2 million in annual expenses.

Another respondent from a large pharmaceuticals firm suggested that the transactions costs associated with biotech IP were especially high. He gave the following metric: Lawyers in the small molecule division (of this firm) are responsible for about eight projects each, whereas those in the biotech division can only handle about two projects each, because of the greater complexity of dealing with input technologies in biotech-based projects.

The question is, although perhaps high in absolute terms, do these transactions costs represent a significant expense? The answer depends on the firm. For large pharmaceutical firms, although the expense is by no means trivial, our respondents did not convey that it significantly affected their returns from drug development. For example, one executive responsible for biotech IP gave figures suggesting that the costs for evaluating and negotiating IP rights amounted to about one one-thousandth of the firm's total R&D budget. This same figure (in absolute terms), however, could represent a significant burden for a small firm, especially one with limited access to capital.

Although our respondents suggest that IP reviews and negotiations are costly and time consuming and that their complexity has increased, it is not clear whether these efforts have increased over the recent past. To address this question, we supplemented our interview data with data from the American Intellectual Property Law Association (AIPLA) and Biotechnology Industry Organization (BIO). The AIPLA's Report of Economic Survey (AIPLA, 1995; 1997; 2001) reports the number of responding attorneys working in the area of biotechnology and the median percentage of effort dedicated to biotechnology by each respondent. Assuming that the AIPLA's data are representative (which they may not be, with only a 18-20 percent response rate), they suggest slightly more than a 10 percent increase in the number of attorneys working on biotech between 1995 and 2001 and a 25 percent jump in the amount of time (at least per the median) that each attorney commonly dedicates to biotechnology. Therefore, there is roughly a 33 percent increase in resources devoted to what one might broadly construed as the "transaction costs" of filing, enforcing and contracting for patents. Ernst & Young LLP's Annual Biotechnology Industry Reports suggest, however, that in nominal terms, R&D expenditures by biotechnology firms have increased over 80 percent during the 1994-2000 period. If we use an annual R&D cost deflator of 5 percent, then real R&D has increased by about 40 percent. Therefore, attorney activity per R&D dollar is unlikely to have increased significantly in the recent past. Even allowing for some increase in attorneys' hourly fees,[44] these data suggest that the patenting of research tools has not itself dramatically increased demand for legal resources and, by extension, that the transaction costs have not increased disproportionately.

## Universities

There is particular concern among academic commentators about the effects of patenting on university research (Heller and Eisenberg, 1998; Eisenberg, 2001; Barton, 2000; Cook-Deegan and McCormack, 2001). We find only limited support for the idea that negotiations over rights stymie precommercial research conducted in universities. Industrial respondents all claim that university researchers, to the extent they are doing noncommercial work, are largely left alone. In fact, firms often welcome this research because it helps further develop knowledge of the patented technology. University researchers among our respondents confirm this claim. Also, many of the firms interviewed expressed the view that the negative publicity that an aggressive assertion of rights against a university would entail was not worth it. One university technology transfer officer reports that the university will indeed receive letters of notification of infringement. The respondent indicated that the typical response was effectively to ignore such letters and inform the IP holder that the university was engaged in research, did not intend to threaten the firm's commercial interests, and would not cease its research.[45] However, receiving such letters is not that common. For example, one respondent reported that in 15 years as a university administrator, overseeing 50 faculty members, he had never had a case of a professor coming to him with a notification letter.

There is a major exception to this norm of leaving university researchers alone, and that is the case of clinical research based on diagnostic tests using

---

[44]The American Intellectual Property Law Association (2001) survey, for example, suggests that the charges associated with filing a patent application increased in nominal terms about 25 percent between 1998 and 2000.

[45]Recently, however, this university did agree to engage in negotiations over the use of a research tool over which a firm had rights.

patented technologies.[46] Merz, Cho, and their colleagues have recently conducted several studies of the frequency with which clinical labs have been affected by patents on diagnostic tests. One study found that 25 percent of laboratory physicians reported abandoning a clinical test because of patents. They also reported royalty rates ranging from 9 percent for PCR to 75 percent for the human chorionic gonadotropin (hCG) patent. In a follow-up survey of 119 labs capable of performing hemochromatosis testing, they found that many had adopted the test immediately upon publication (Merz et al., 2002). When the patent issued a year later, it was licensed exclusively to SmithKline Beecham (SKB). Nearly all respondents in the Merz et al. study said they knew of the patent. About half had received letters from SKB. Twenty-six percent said they did not develop the genetic test for hemochromatosis, and another four percent said they abandoned the test, in part because of the patent. Much of the controversy around Myriad's use of its patent on BRCA1 revolves around this distinction between research and clinical practice. Myriad allows licensees to do tests provided that no fees are charged and the tests are not used for clinical purposes. Myriad also provides reduced fee diagnostics tests ($1,200 vs. $2,680) for NIH-funded projects (Blanton, 2002). However, according to Myriad's Gregory Critchfield, "If you give test results back to patients, it crosses over the line, and it's no longer a simple research test. [It] is really a very bright line" (Blanton, 2002). On the other hand, Merz argues that "There is no clear line to be drawn between clinical testing and research testing, because the state of the art of genetic tests is such that much more clinical study is necessary to validate and extend the early discovery of a disease gene. Thus, the restriction of physicians from performing clinical

---

[46]One controversial case was the diagnostic test for the Canavan disease gene mutation. Miami Children's Hospital held the patent and was charging a royalty of $12.50 per test, even though the doctor did the test himself. Washington University's Michael Watson was among those complaining that this royalty hurts research and patient care: "We would be happy to pay for some kind of test kit that is faster, better, cheaper. But they are trying to control manual testing, which is not appropriate" (Regalado, 2000, p. 55). This quotation reflects the opinion of many academics that they should not be forced to pay royalties for "do-it-yourself" technologies. One respondent from a clinical testing company said that the $12.50 royalty for the Canavan test was "substantial" [the cost of the test is reported to be $8 to $9 (Kotulak, 1999)]. Merz et al. (2002) also report royalties of $5 for Gaucher disease and $2 for a cystic fibrosis test, although only for labs doing more than 750 tests per year. Furthermore, they report that license stacking on a battery of tests for Ashkenazi Jewish patients can bring the royalty total to around $100, representing about 20 percent of the cost.

Another example is the case of the Bogart patent on the triple marker test for Down syndrome. Here, the patent owner demanded $5 per test, even though his patented test was only one of three that was needed to accurately determine the presence of Down syndrome. While the patent owner pointed out that labs routinely charge $75 for this test, critics noted that the direct costs for the test were also about $5, and that Medicaid reimbursement could be as low as under $10 in some states. This case was all the more controversial because the test was claimed to be widely known before it was patented.

testing will directly reduce the knowledge about these genes."[47] One of our respondents, a former medical school dean, echoed these remarks, saying that he had been shown letters of notification of infringement from medical school researchers and that programs had been stopped. He noted that the fact that the universities charge for these tests complicates the matter, but that clinical work is critical for the research process. Cho et al. (2003) report that about half of the diagnostic labs in their sample are also using the test results for clinical research. Cho also notes that sharing of test results within the clinical diagnostic communities is an important means of advancing clinical and scientific understanding of diseases.[48]

Thus, in some cases, firms are willing to assert their patents against universities that are doing diagnostic testing and charging a fee without licensing the patented tests, and at least some labs are stopping their testing as a result. However, the majority continue with the testing. So long as the university is not generating revenue based on the patented technology, universities appear to be largely left alone, although some firms will send letters.

## Materials Transfers

Eisenberg suggests that another significant cost of the patenting of research tools is that associated with the costs and delays in negotiating access to research materials, that is, those associated with materials transfer agreements (MTAs). Although our interviews did not focus on such transactions, to the extent we considered them the interviews also suggested that MTAs are a source of some concern and vexation. An academic researcher reinforced Eisenberg's findings when he said: "Things are becoming more bureaucratic. MTAs, they are crazy. Before, whenever someone wanted a plasmid from my lab, I would just send it. Now, the university says they own it and I have to go through the IP office. It goes back and forth between the two offices and it takes a long time. Before, we would just send it in the mail, and you would have it and could use it. Basic science is now becoming interested in 'value.' The university is particularly interested in value." Although the material being transferred may or may not be patented, the delays often involve negotiating an allocation of patent rights over the discoveries that may build on the material. Of those who addressed the issue of MTAs nearly all respondents from universities and from industry confirmed that when dealing with MTAs, especially those involving university technology transfer offices, delays could be substantial. For example, one industry respon-

---

[47]Testimony before House Judiciary Committee, 7/13/00, http://www.house.gove/judiciary/merz0713.htm.
[48]Personal communication, 1/24/2002.

dent suggested that about a third of the firm's projects involved some research agreement with a university and that such negotiations result in substantial delays (on the order of months). Some respondents noted that universities are learning and that the process is becoming smoother. NIH has developed a standard MTA for transfers between university or government labs. This MTA, which is endorsed by AUTM, is a single page and does not involve any reach-through claims.

Although materials transfers may indeed be problematic, the relevant question is the extent to which patenting per se has introduced impediments. Experimental biology has long been a competitive field in which scientists were somewhat reluctant to share materials with competing scientists (see Hagstrom, 1974; Sullivan, 1975; Campbell et al., 2002; but see McCain, 1991). Thus it is not clear whether the growing interest in "value" has changed the *willingness* of scientists to exchange information and materials. Campbell et al. (2002) find that about half of life science faculty report that their requests for materials or information have been denied at least once in the last three years, although 90 percent of requests were granted. About 12 percent of respondents say they have denied requests [Blumenthal et al. (1997) find a very similar result]. Furthermore, although 35 percent report that such withholding is increasing, 65 percent report that has stayed the same or decreased over the last 10 years. Also, although commercial value or industry sponsorship are important predictors of failing to share, the major reason given for not sharing was the effort required to actually produce the material or information, with concerns about scientific competition also being important, and with commercial concerns ranking at the bottom of the list. Blumenthal et al. (1997) have very similar findings, with protecting scientific lead and the expense or scarcity of the materials being the most important reasons for refusing to share materials or information and commercial concerns ranking at the bottom of the list. However, multivariate analyses by Campbell et al. (2003) and Blumenthal et al. (1997) do show that commercial concerns (such as patenting or industry sponsorship) are associated with refusing to share materials and results. These studies did not measure the impact of scientific competition on data sharing.

Walsh and Hong (2003) compare data from surveys of scientists in experimental biology, physics, and mathematics conducted in the 1960s (i.e., before the Bayh-Dole Act and the rise of patenting of academic science) with data from another survey done in 1998 using the same items. They find that secrecy (measured as the willingness to discuss one's current research with others) has indeed increased overall. Furthermore, secrecy has increased particularly in experimental biology, with only 14 percent of experimental biologists from the recent survey feeling safe to discuss their current work with all others (compared to 45 percent in the 1960s). Although it is difficult to eliminate the expectation of gains from some prospective, patent-based commercialization of some downstream discovery or drug as the cause of this increased secrecy, a multivariate analysis building on the models used by Campbell et al. (2002) and Blumenthal et al. (1997) shows that the primary predictor of secrecy is scientific competition. Industry-related activity has a mixed effect, with having applied for a patent showing no relation to secrecy, having industry funding related to greater secrecy, and having industry collaborators associated with less secrecy. Thus we have some evidence of an increasing reluctance to be open with scientific findings and materials. However, the results on whether this is due primarily to commercial concerns or to scientific competition are mixed.

What has clearly changed is that, when there is a willingness to share materials, processing that transfer has become more complicated, with the time increasing from days often to months (Eisenberg, 2001).[49] This negotiation can delay the research because the research on the material cannot go forward until the material is transferred. Three university scientists did tell us that they have had to abandon particular projects because of an inability to get materials, but in those cases, they moved on to other projects.[50] However, in each of these cases, the respondent suggested that this was the exceptional case and that for other projects getting research materials was routine. Campbell et al. (2002) find that 21 percent of respondents report abandoning a promising line of research and 24 percent report that their own publications were significantly delayed because of an inability to get access to others' materials or information. One solution reported by our respondents is that some firms and universities have a standard, take-it-or-leave-it agreement (cf. Eisenberg, 2001). This does reduce delays, but it probably also reduces the number of transactions. Also, several respondents (six from industry and three from universities) suggested that trust that comes from long-term relationships reduces the friction of these kinds of transactions, for example, the willingness to edit the standard agreement to take out offending clauses such as publication review clauses or reach-through rights (Bolton et al., 1994; Uzzi, 1996).[51] In fact, two of the scientists we interviewed said they routinely send materials without bothering with MTAs, although this would probably upset their technology transfer offices.

In addition, the availability of supply houses to provide licensed copies of patented research materials did facilitate access and distribution according to some

---

[49] Although Campbell et al. (2002) do not address MTAs in particular, it is possible that some of the "effort required to actually produce materials or information" (the major reason for not sharing) includes the bureaucratic procedures for dealing with MTAs.

[50] In one case, when he finally was able to get the reagent a year later, one respondent decided he did not have time to pursue that line and instead continued to pursue his current research, suggesting that the loss of that avenue of research was not critical.

[51] For an extensive discussion of the relation between norms and IP for biotech-related research see Rai (1999).

respondents. Our interviews also point to an intriguing possibility: This commercialization of research materials may actually increase access by creating market-based institutions for distributing them rather than relying on gift exchange among researchers. Several university scientists noted that the demand for important reagents can easily become overwhelming, and licensing these to a commercial firm was seen as a way of increasing, rather than limiting, access for the research community. They also noted that what is sometimes perceived by the requester as reduced access is often the result of a scarcity of materials or time (in addition to scientific competition) preventing compliance, rather than concerns over property rights (see also Campbell et al. 2002 and Blumenthal et al. 1997).[52]

Thus, to the degree that the patenting of biomedical discoveries may impose additional costs and delays in materials transfers, it is partly because Bayh-Dole and related acts have provided university administrations, and especially their technology transfer offices, a vested commercial interest in the disposition of intellectual property.

## WORKING SOLUTIONS: OVERCOMING THE ANTICOMMONS AND RESTRICTIONS ON ACCESS

Notwithstanding concerns about the proliferation of IP on research inputs and about the ability of rights holders to limit access to upstream discoveries and promising research targets, the problem was generally considered to be manageable. Firms reported a variety of private strategies and institutional responses that limited the adverse effects of the changing IP landscape. Although negotiations over IP and licensing fees surely affect access, and sometimes choice of projects, our conclusion is that patents on research tools do not yet pose the threat to research projects that they might given the number of patents and diversity of owners. In this section, we review the private strategies adopted by firms and universities and responses from government that allow research and commercialization to go forward despite the proliferation of biomedical intellectual property and claimants over the past decade or so.

One important reason why research tool patents tend not to interfere with research is that it is typically not that difficult to contract. As noted above, although the process of identifying the relevant patents is time consuming, the number of actual patents involved is often moderate, about a dozen or less. Licensing is routine in the drug industry. For example, from 1990 to 1997, there was an average of 379 licenses each year in the drug and chemicals industries (SIC28). For comparison, during this same period there were an average of only 276 li-

---

[52]For example, two of our university respondents (in one case asking for material and in the other case being asked) mentioned that the death of a cell line was the explanation for not complying.

censes per year in electronics (SIC36), where cross-licensing of patents is common (Arora et al., 2001).

Many of our responding firms suggested that if a research tool was critical, they would buy access to it. Several companies that had patents on targets noted that, in addition to trying to develop their own therapeutics, they include the liberal and broad licensing of those targets to others as part of their business model, reflecting a belief on the part of some holders of target patents that by giving several firms a nonexclusive license they increase the chances that one will discover a useful drug. We also observe that most of what might be called "general purpose" tools—tools that cut across numerous therapeutic and research applications that tend to be non-rival-in-use—tend to be licensed broadly. Thus many of the more fundamental (general purpose) research tools, such as genomics databases, DNA chips, recombinant DNA technology, PCR, etc., are made widely available through nonexclusive licenses. Incyte, for example, licensed its genomics database to over 20 pharmaceutical firms (who together account for about 75 percent of total private pharmaceutical R&D). They have also begun expanding their licensing program to include biotech firms and universities as well.[53] Similarly, *Taq* polymerase and thermal cyclers for PCR are available from a variety of authorized reagent and equipment vendors (Beck, 1998). Human Genome Sciences' semi-exclusive licensing of its databases to only about five firms reflects an exception to this pattern.

Liberal licensing practices are also encouraged to the extent that inventing around tool patents is feasible. Under such circumstances, patentholders are more willing to license on reasonable terms assuming the prospective user does not invent around to begin with. The ability to "invent around" puts an upper limit on the value of the rival's patent. Indeed, our respondents frequently noted their ability to invent around a patent as one component in their suite of solutions to blocking patents. Firms have also occasionally developed technologies that, it was claimed, made it possible to circumvent a number of the patents in the field.[54] Although some respondents argued that target patents were often unassailable, others claimed that for many important diseases (AIDS, many types of cancer, etc.) there are likely to be multiple approaches to the metabolic pathways. One

---

[53] Furthermore, Incyte's license requires users to "grant back" nonexclusive rights to use of genes discovered from its database, providing freedom to operate to firms in the network and creating what Incyte refers to as an "IP Trust."

[54] For example, Athersys, a Cleveland-based biotech firm, advertises its RAGE technology, which uses automated techniques to create protein expression libraries (i.e., activate and express every gene and therefore produce every protein) without using any knowledge about the location and structure of the corresponding gene. The company's website reports that some established pharmaceutical firms (Bristol-Myers Squibb and Pfizer) had licensed this technology (www.athersys.com).

university scientist considered the issue of whether a patent on a target protein could confer exclusive rights to working on a disease:

> I have never worked with a disease where one particular protein makes the only difference. A patent gets you exclusive rights to a class of drugs, but there may be other classes.... I could imagine a genetic disease where a single target was involved, but I don't think that the big medical problems fall into this case. Cancer, AIDS are my areas. AIDS is one hundredth as complex as cancer and even there a single protein is not the solution. Heart disease is as complicated as cancer. Cystic fibrosis is a good example of a single gene, single protein disease.

Thus the specifics of both the patent claims and the scientific understanding determine whether researchers are more or less able to invent around a given patent.

Aside from conventional methods for coming to terms, we find that firms have adopted a set of complementary strategies that create "working solutions" to address either a prospective anticommons (e.g., the need to license numerous tools) or a potentially blocking patent on one tool or discovery. These solutions include (in addition to licensing and inventing around) ignoring patents (sometimes invoking an informal research exemption), going offshore, creating public databases, and challenging patents in court. One pharmaceutical executive summarizes the range of strategies employed:

> If someone has a patent on genes, when the gene encodes a therapeutic product and they are ahead of us, we drop those projects. That is different than the case of a gene as a target for a small molecule screen. There we don't drop the project. If it is just an application, it is not until the patent issues that it is infringing. Lots of these patents are pretty thin. It is an issue whether it is valid. Third, you can do things offshore. Fourth, it may be available for license and fifth, they don't tend to enforce them.

These working solutions combine to create a free space in the patent landscape that allows research projects to proceed relatively unencumbered.

### Infringement and the "Research Exemption"

One solution to restrictive patents on upstream inventions is simply to ignore some or all of them. Several respondents noted that infringement of research tool patents is often hard to detect, facilitating such behavior. Thus, if research tool patents have created a minefield, they are mines with fairly insensitive triggers.

University researchers have a reputation for routinely ignoring IP rights in the course of their research (Seide and MacLeod, 1998). Respondents note that many research tools are "do-it-yourself" technologies and therefore they do not feel they should be required to pay royalties for the work. In fact, some strongly believe that these patented technologies were well-known in the scientific community and therefore the patents are not valid (see for example, Kornberg, 1995).

University researchers will often invoke a "research exemption," although the legal research exemption as construed by the Court of Appeals for the Federal Circuit has been quite narrow.[55] Some reagent suppliers facilitate this practice by supplying "unlicensed" (and less expensive) materials, also invoking the research exemption.[56] Promega, for example, sells *Taq* polymerase for about half of what many licensed vendors charge and asserts that many of its customers in university and government labs do not need a license under the experimental use exemption (Beck, 1998).

Many firms claim to be reluctant to enforce their patents against universities to the extent that the university is engaging in noncommercial research, because of the low damage awards and bad publicity that suing a university would entail. For example, William Haseltine of Human Genome Sciences said that they were ready to give academics access to data and reagents related to their patented CCR5 HIV receptor: "We would not block anyone in the academic world from using this for research purposes" (Marshall, 2000a). As one university technology transfer officer stated, "Asserting against a university doesn't make sense. First, there are no damages. You cannot get injunctive relief and/or damages. What have you gained? You've just made people mad. Also, these firms are consumers of technology as well. No one will talk to you if you sue. We all scratch each others' backs. You will become an instant pariah if you sue a university." Similarly, from the industry side, Leon Rosenberg of Bristol-Myers Squibb said, "Frankly, we all know it is not good form to sue researchers in academic institutions and stifle their progress" (NRC, 1997, Ch.6, p. 3). These quotations suggest that one limit on opportunism is being a member of a community with the members being able to sanction overly aggressive behavior (Rai, 1999). This vulnerability to such sanctions is based on the need to buy as well as sell technology, or, perhaps especially, to informally trade information. Indeed, there is a strong interest in developing trusting relationships with university researchers to encourage information sharing (for the general issue of trust and information sharing see Uzzi, 1996; for a discussion of the importance for industrial R&D of informal informa-

---

[55]Building on *Roche Products, Inc. v. Bolar Pharm. Co.* (Fed. Cir. 1984) and *Embrex v. Service Engineering Corp*, the current standard of the Court of Appeals for the Federal Circuit (CAFC) for what qualifies for a research exemption includes uses of patented inventions "for amusement, to satisfy idle curiosity, or for strictly philosophical inquiry." (*Embrex*) Although there is some question about whether the term "philosophical inquiry" may actually refer to scientific inquiry (cf. Wegner, 2002), the October 3, 2002 decision of the CAFC in *Madey v. Duke*, discussed below, corroborates that the CAFC has sustained a narrow interpretation of the research exemption.

[56]For example, one respondent from a university said that they buy limited quantities of a licensed peptide from a supply house, for $235 per milligram, which they use for benchmarking their experiments. They get the bulk of it, unlicensed, from another lab in the university that can make it for $55 per milligram.

tion sharing with universities see Cohen et al., 2002). A respondent at a biotech firm put it this way:

> We rely on lots of outside collaborations with academic labs. Our scientists want to feel on good terms with the academic community. If you start suing, it breaks down the good feeling. We give out our research tools for free, frequently. All we ask is, if you invent anything that is directly related to the tool, you allow us the freedom to practice.

We heard similar comments from those in universities, large pharmaceutical firms, and biotech firms. From the manager of a diagnostics firm, we also heard of a counter-example where a tool owner was not a member of the research community. He described the Canavan case in which the patent holder, Miami Children's Hospital, was charging $12 per test, which was considered high. When asked why Miami Children's was behaving differently, he responded:

> They are not a company or an organization whose purpose is to continue to do research. They have a product to license, but do not have to sustain relationships. If you license and screw someone, it will get back. This is one time. They hire an obnoxious lawyer who makes people sign nonmarket price deals. What do they have to lose? They are a hospital; patients come to them. They have different interests. They can get away with it. Michigan, or Hopkins, institutions that produce a lot of research, medical schools with a lot in the pipeline, they are looking for licensees, they to want to sustain a relationship. They are very reasonable. They know the business. Canavan is a one-off situation. They don't have a clue. They are not in the industry. Their lawyer, I will never see him again. Other institutions, we'll come across each other.

This quotation highlights the repeated game nature of many of these licensing negotiations, which tends to reduce opportunism by the players (as well as noting the problem with one-shot players).

A similar case is DuPont's recent aggressive assertion of its exclusively licensed OncoMouse patent against universities that did not follow the precise terms of a prior memo of understanding between DuPont and NIH. In commenting on this behavior, Neighbour (2002) mused about why DuPont would do such a thing now that they are out of the business of research in molecular biology. We would suggest that that may be the explanation. They have now ceased to be a part of that community and therefore have little to lose and revenue to gain when they sacrifice the goodwill of that community. Thus DuPont's behavior is consistent with the notion that a community of practice restrains the aggressive assertion of IP.

Several respondents noted that they actually welcomed universities using their patented technologies because if the university discovers a new use, the patentholder is best positioned to exploit the innovation. If the university becomes a competitor, however, firms feel they then have a right to assert their patents. As noted above, this is particularly evident in cases where university physicians use patent-protected discoveries as the basis for diagnostic tests (Merz et al., 2002; Blanton, 2002).

As a rule, universities do not assert their rights against one another.[57] The following quotation from a professor with over 25 patents summarizes the university researcher's perspective on whether he had any difficulties in gaining access to research tools from other academics:

> It is not a problem. I know this is a murky legal issue, and you should talk to patent lawyers, but in everyday practice, it is not murky. There is a concept of "academic use." If you have published it, I can use it for academic purposes. I've never heard of any case where someone was sued for using patented technologies. I don't know if it is solidly defensible in the law, but it is the practice. When I have a patented technology, academic colleagues would not even think of paying to use it.

Infringement of research tool patents by firms also appears to be pervasive. A third of the industrial respondents (and all nine university or government lab respondents) acknowledged occasionally using patented research tools without a license, and most respondents suggested that infringement by others is widespread.[58] The firms felt that much of their research would not yield commercially valuable discoveries, and thus they saw little need to spend money to secure the rights to use the input technology, particularly because it is very difficult to police such infractions. If the research looked promising, then they would get a license, if necessary.[59] Furthermore, at least a few industry respondents argued strongly that using a gene patent as a research tool did not infringe or that infringement was limited to that experiment per se and did not extend to the product discovered

---

[57]Some universities feel that they have the moral right to assert against another university, however, if it is commercializing their innovation. One academic stated: "Universities have a general agreement of sharing results freely. There is a research ethic. But if a university were making money off of our technology, then there would be trouble."

[58]One respondent stated, "Sometimes we take a license and sometimes we don't. I think there is a lot of infringement out there. The scientists are not telling their patent counsel." One respondent was explicit: "If you are confronted with a patent on a target you need, you have to decide what to do. You can infringe, and take the risk of getting sued. They would have to know your practices. If you keep it secret, then they may only find out when you release a product. Then they may know you used it. But the statute of limitations may have run out. Some research tool owners are very aggressive. If they get a hint you are using their tool, they sue. You take all this into account." Another stated, "I think all the firms in the industry take on some infringement risk, because the behavior in the industry is that you have to try a million things to find one that is promising. Once you identify the promising candidate, then you look into licensing the research tools or sequences you used."

[59]If this is true, it suggests that when firms do ask for a license, the patent owner ought to suspect that the tool has been useful in generating a valuable discovery. This knowledge may lead to asking for a high license fee. However, this urge is balanced by the recognition that this promising candidate still needs to get through risky clinical trials, and if the price is too high, the buyer may chose one of his other promising candidates.

(in part) by using the research tool, i.e., that the scope of research tool patent claims is quite limited. In addition, because many of these patents are of debatable validity, they felt that if a license were not available, they could challenge the patent in court. Finally, not only is use of a patented research tool hard to detect, but because of the long drug development process, the 6-year statute of limitations may expire before infringement is detected.

Consistent with this behavior, we also find that firms feel that it is not worth their while to assert their patents on all other firms that might be infringing. They may send a letter, offering terms, but will not aggressively pursue infringers on their marginal patents. Respondents point out that the cost of pursuing these cases greatly outweighs their value in most instances: "The average suit costs millions of dollars. The target is worth $100,000. Even with treble damages, it doesn't pay to sue." There is an additional cost, and that is the risk of the patent being invalidated by the court. These firms note, however, that they will aggressively defend a patent central to the firm's competitive performance. Barring that, with reference to research tool patents, there is a sense that the industry practices "rational forbearance" (NRC, 1997, Ch. 6).

Respondents also pointed out that patents are national but the research community is global. Thus another means of avoiding research tool patents is to use the patented technology offshore. Although similar to the solution of ignoring the patent, in that it involves using patented technologies without securing the rights, this case differs in that firms are not violating the legal rights of the patent owner, at least not until there a product developed and the firm tries to import the product. Furthermore, a district court decisions in 2001 (*Bayer AG v. Housey Pharmaceuticals*) suggests that even then the drug maker may not be liable for infringement (see, for example, Maebius and Wegner, 2001).[60]

In summary, by infringing (and informally invoking a research exemption), inventing around, going offshore, or invalidating patents in court, firms were able to greatly reduce the complexity of the patent landscape. These strategies, combined with licensing when necessary, provide working solutions to the potential problem that an increasingly complex patent landscape represents.

## Institutional Responses by Firms, NIH, USPTO, and Courts

In addition to these private responses to overcoming the barriers that patents might create, we have also observed firms (especially larger pharmaceutical

---

[60]Housey Pharmaceuticals had a patent on a method for screening potential drug candidates. Bayer allegedly used this method to discover a drug, but the alleged use was outside the United States. The district court ruled in favor of Bayer on the grounds that the Housey patent did not cover the drug or the method of making the drug, but only a method for finding substances worthy of further development, and that, therefore, Bayer did not infringe by selling the product in the United States. This ruling is an example of a court narrowing the reach of research tool patents (see below). However, it is still not clear how the CAFC will treat this issue (Maebius and Wegner, 2001).

firms), the courts, the NIH and the USPTO undertaking initiatives and policies that have had the effect (if not always the intent) of broadening and easing access to research tools. For example, with substantial public, private, and foundation support, public databases (e.g., GenBank or the Blueprint Worldwide Inc. venture to create a public "proteomics" database) and quasi-public databases (such as the Merck Gene Index and the SNPs Consortium) have been created, making genomic information widely available. Similarly, Merck has sponsored an $8 million program to create 150 patent-free transgenic mice to be made available to the research community at cost, without patent or use restrictions. According to our respondents, these efforts partly represent an attempt by large pharmaceutical firms to undercut the genomics firms' business model by putting genomic and other related information into publicly available databases and then competing on the exploitation of this shared information to develop drug candidates (cf. Marshall, 1999a; 2001).[61] These initiatives represent a partial return to the time before the genomics revolution, when publicly funded university researchers produced a body of publicly available knowledge that was then used by pharmaceutical firms to help guide their search for drug candidates.

NIH has also taken the lead in pressing for greater access to research tools. For example, since 1997, NIH has negotiated with DuPont to provide more favorable terms for transgenic mice for NIH and NIH-sponsored researchers (Marshall, 2000b). NIH has also begun a "mouse initiative" to sequence the mouse genome and create transgenic mice. One of the conditions of funding is that grantees forgo patenting on this research. NIH also pushed for broader access to stem cells, as well as for a simplified, one-page MTA without reach-through claims or publication restrictions. Scientific journals have also pushed for access to research materials. For example, biology journals have long made it a condition of publication that authors deposit sequences in public databases such as GenBank or Protein Data Bank (Walsh and Bayma, 1996). Similarly, when Celera published its human genome map findings, *Science*'s editors were able to gain for academics largely unrestricted access to Celera's proprietary database.[62] Thus large institutional actors have been able to act as advocates for university researchers to increase their access to necessary research tools.

---

[61] For example, the firms in the SNPs Consortium include Bayer, Bristol-Meyers Squibb, Glaxo Wellcome, Hoechst Marion Roussel, Monsanto, Novartis, Pfizer, Roche, SmithKline Beecham, and Zeneca. Each firm contributed $3 million, and Wellcome Trust added another $14 million to the effort. Also, financed by IBM, Canada's MDS, Inc. and the Canadian government, the Blueprint Worldwide database could pose a threat to the joint effort of Myriad Genetics, Hitachi, and Oracle to launch a $185 million effort to map protein interactions, along with for-profit efforts by universities to market protein databases (*Wall Street Journal*, 2001).

[62] Academics have the right to access the data at no charge, do searches and download segments up to 1 megabase, publish, and patent. They can download the whole database if the university signs an agreement not to redistribute the data. There are no reach-through provisions or restrictions on publication. *Science* also kept a copy of the database in escrow to ensure compliance.

Responding partly to concerns expressed by NIH, universities, and large pharmaceutical firms, the USPTO has also adopted new policies that diminish the prospect of an anticommons. Specifically, in January 2001, the USPTO adopted new utility guidelines that have effectively raised the bar on the patentability of tools, particularly ESTs. These guidelines are designed to reduce the number of "invalid" patents (cf. Barton, 2000).

Some of our respondents have suggested that recent court decisions have also mitigated potential problems due to research tool patents by limiting the scope of tool patents or, in some cases, invalidating them. Thus, although patentholders have the right to sue for infringement, the perception is that they are increasingly likely to lose such a suit. Cockburn et al. (2003) find that the CAFC went from upholding the plaintiff in about 60 percent of the cases to finding for the plaintiff in only 40 percent of the cases in recent years. One case that comes up frequently among our respondents is *University of California v. Eli Lilly and Co.* As noted above, the University of California tried to argue that its patent on insulin, based on work on rats, covered Lilly's human-based bioengineered insulin production process. The CAFC ruled that California did not in fact possess this claimed invention at the time of filing; therefore, the claim was not valid, and Lilly was not infringing. Another controversial case was over a transgenic mouse used to study Alzheimer disease. Mayo had been widely distributing the mice at nominal cost to academic researchers.[63] In 1999 Elan Pharmaceuticals sued for infringement and sent subpoenas to individual researchers across the country, demanding their lab notes. In 2000, a District Court judge dismissed the patent infringement suit by Elan Pharmaceuticals against Mayo Foundation, invalidating the patents on the grounds that their claims were covered by an earlier patent.[64] The case of Roche versus Promega over *Taq* (see above) is another example of the courts ruling against a research tool patent holder. As one respondent put it: "These are good times for a patent infringer and not great times for a patent holder." This seeming change in the court's attitude may represent a shift toward more freedom to conduct research without undue concern over research tool patents.

There remains, however, a great deal of uncertainty over how the courts will rule on the validity of research tool patents generally. One case, discussed above, that has been closely watched is the *Rochester v. Searle* case over COX-2 inhibitors. The critical issue in the case is whether knowledge of a drug target allows one to claim ownership over specific classes of drugs (i.e., how broadly do initial discovery claims extend over future developments building on those discover-

---

[63]However, they charged some pharmaceutical companies up to $850,000 for a breeding group (Dalton, 2000).

[64]On August 30, 2002, the CAFC reversed the summary judgment of invalidity based on anticipation and remanded the case back to the District Court for further proceedings. The District Court and the CAFC have not yet ruled on the breadth of the Elan patent claims.

ies). Although the district court recently dismissed Rochester's complaint of infringement (see footnote 16), the case will apparently be appealed, and the ultimate outcome may have important implications for how patents on key upstream discoveries affect subsequent drug development and commercialization.

## DISCUSSION AND CONCLUSION

In this chapter, we have considered two possible impacts of the patenting of research tools on biomedical research. First, we considered whether the existence of multiple research tool patents associated with a new product or process poses particular challenges for either research on or commercialization of biomedical innovations. Second, we examined whether restricted access to some upstream discovery—perhaps protected by only one patent—has significantly impeded subsequent innovation in the field. In brief, we find that the former issue—the "anticommons"—has not been especially problematic. The latter issue of access, at least to foundational upstream discoveries, has not yet impeded biomedical innovation significantly, but our interviews and prior cases suggest that the prospect exists and ongoing scrutiny is warranted.

The patenting of research tools has made the patent landscape more complex. As suggested by Heller and Eisenberg (1998), our interviews confirm that there are on average more patents and more patentholders than before involved in a given commercializable innovation in biomedicine, and many of these patents are on research tools. Despite this increased complexity, almost none of our respondents reported commercially or scientifically promising projects being stopped because of issues of access to IP rights to research tools. Moreover, although we do not have comparably systematic evidence on projects never undertaken, our interviews suggest that IP on research tools, although sometimes impeding marginal projects, rarely precludes the pursuit of more promising projects. Why? Industrial and university researchers have been able to develop "working solutions" that allow their research to proceed. These working solutions combine taking licenses (i.e., successful contracting), inventing around patents, going offshore, the development and use of public databases and research tools, court challenges and using the technology without a license (i.e., infringement), sometimes under an informal and typically self-proclaimed research exemption. In addition, the members of a research community (which includes both academic and commercial researchers) are somewhat reluctant to assert their IP against one another if that means they will sacrifice the goodwill and information sharing that comes with membership in the community. Changes in the institutional environment, particularly new USPTO guidelines and some shift in the courts' views toward research tool patents, as well as pressure from powerful actors such as NIH (stimulated perhaps by the early concerns articulating the anticommons problem) also appear to have further reduced the threat of breakdown. Finally, the very high technological opportunity in this industry means that firms can shift their research to areas less encumbered by intellectual property claims, and, therefore, the walling off of particular areas of research may not, under some circumstances, exact a high toll on social welfare.

Although stopped and stillborn projects are not especially evident, many of the working solutions to the IP complexity can impose social costs. Firms' circumvention of patents, the use of substitute research tools, inventing around or going offshore—although all privately rational strategies—constitute a social waste. Court challenges and even the contract negotiations themselves can also impose significant social costs. Litigation can be expensive and non-out-of-pocket costs, represented by the efforts devoted to the matter by researchers and management, can be substantial. Even when there is no court challenge, the negotiations can be long and complex and may impose costly delays. Disagreements can and have led to litigation, which is especially costly for small firms and universities. It is difficult to know, however, how much contracting costs in biomedicine reflect an enduring feature of IP in biomedicine and how much is transitional, arising from the uncertainty associated with the newness of the technology and uncertainties about the scope and validity of patent claims. Moreover, as new institutions (i.e., universities) and firms become owners of intellectual property, there is a costly period of adjustment as these new actors learn how to manage their IP effectively. The development of standard contracts and templates may be helpful in diminishing these adjustment costs, and funding agencies such as NIH can play an important role in developing and encouraging the use of such standards.

The second issue that we examined is the impact on biomedical innovation of restricted access to research tools. In thinking about the issue of access, it is helpful to distinguish research tools along two dimensions. First, it is obviously of interest how essential or "foundational" a research tool is for subsequent innovation, both in the sense of whether the tool is key to subsequent research and in the sense of the breadth of innovation that might depend upon its use. Is the research tool a key building block for follow-on research on a specific approach to a specific disease, is the tool key to advance in a broad therapeutic area, or might its application even cut across a range of therapeutic and diagnostic domains?

A second dimension of interest is the degree to which a research tool is rival-in-use. By "rival-in-use" we mean research tools that are primarily used to develop innovations that will compete with one another in the marketplace. For instance, in the case of a receptor that is specific to a particular therapeutic approach to a disease, if one firm finds a compound that blocks the receptor, it undermines the ability of another to profit from its compound that blocks the same receptor. The defining feature of research tools that are not rival-in-use is that the use of the research tool by one firm will not typically reduce others' profits from using it. Such tools include PCR, microarrays, cre-lox, and combinatorial libraries. From a social welfare perspective, a research tool that is not rival-in-use is like a public good in that it has a high fixed cost of development and zero or very low marginal cost in serving an additional user. Thus maximization of social welfare requires that the tool be made available to as large a set of users as possible.

We have observed that holders of IP on nonrival research tools often charge prices that permit broad access, at least among firms. In some of these cases, the IP holders have also charged higher prices to commercial clients and lower prices to university and other researchers who intended to use the tool largely for noncommercial purposes. From a social welfare perspective, such price discrimination expands the use of the tool and is welfare enhancing. There are, however, cases in which the IP holder cannot or does not develop a pricing strategy that allows low-value and academic projects access to the tool, as for instance in the case of DuPont's initial terms for the cre-lox technology or Affymetrix's initial terms for GeneChips. However, DuPont eventually bowed to pressure from NIH (although, as noted above, the issue is not entirely settled) and Affymetrix developed a university pricing system that greatly increased access (while others developed do-it-yourself microarrays).[65]

The concern with regard to IP access tends to be the greatest when a research tool is rival-in-use and is potentially key to progress in one or more broad therapeutic areas. When a foundational research tool is rival-in-use, the IP holders often either attempt to develop the technology themselves or grant exclusive licenses. As suggested above, exclusive exploitation of a foundational discovery is unlikely to realize the full potential for building on that discovery because no one firm can even conceive of all the different ways that the discovery might be exploited, let alone actually do so. Geron's exclusive license for human embryonic stem cell technology shows how restrictions on access to an important, broadly useful rival-use technology can potentially retard its development.[66] A more prosaic example is the pricing of licenses for diagnostic tests. Myriad's (and others') licensing practices show that, to the degree that a high price on a diagnostic test puts it out of the reach of clinics and hospitals involved in research that requires the test results, clinical research may be impeded, yielding long-term social costs.[67] The social welfare analysis of this situation is, however, not straightfor-

---

[65]We conjecture that it is exactly these non-rival-in-use technologies with many low-value uses that are likely to benefit from NIH intervention, if necessary, because there will be a large constituency of users who want access (including many researchers at NIH itself), most of the research community uses will be low value, and the cost to the patent owner of allowing these nonrival uses is low, because the high-value uses are not necessarily affected.

[66]Of course, President Bush's decision to deny federal funding to human embryonic cells lines created after August 9, 2001 limited the ability of researchers to invent around Geron's patents (Kotulak and Gorner, 2001).

[67]Here the difficulty is associated with the fact that the same activity that is rival-in-use (providing commercial diagnostic services) is also the (possibly non-rival-in-use) research use. The difficulty of separating these two activities in the American system of funding clinical research contributes to the problems associated with patents on diagnostic uses of genes.

ward. Even though knowledge, once developed, can be shared at little additional cost and may be best exploited through broad access, it does not follow that social welfare is maximized by mandating low-cost access if such access dampens the incentive to develop the research tool to begin with.

Many of the same kinds of "working solutions" that mitigate the prospect of an anticommons also apply to the issue of access for research. Our interviews suggest that a key "working solution," however, is likely infringement under the guise of a "research exemption." Firms and universities frequently ignore existing research tool patents, invoking a "research exemption" that is broader than the existing legal exemption and that is supported by norms of trust and exchange in the research community. As discussed above, such instances of possible infringement, especially on the part of universities, are tolerated by IP-holding firms, both for normative reasons and because of the high cost of enforcing rights through litigation, relative to the low payoff for stopping a low-value infringement. One can rationalize the failure of the IP holder to aggressively monitor infringement as a form of price discrimination, and, as suggested above, economic theory suggests that such price discrimination can improve social welfare.[68]

There are two central questions to ask when considering the effects of a given research tool patent on the progress of biomedical research. The first has to do with the specifics of the biology in question: Does current scientific knowledge provide us with many or few opportunities for modifying the biological system in question? As science progresses, we are likely to see an oscillation, with new discoveries opening promising but narrow shortcuts and further exploration of those discoveries uncovering a variety of lines of attack on the problem. Where there are many opportunities, the likelihood of a research tool patent impeding research is smaller. Here again, the Geron case provides an illustration, with the recent development of alternatives to the use of embryonic cells for exploiting the promise of stem cells mitigating the restrictive impact of Geron's control over embryonic stem cell technology.

The second question has to do with specifics of the legal rights in question, and was highlighted by Merges and Nelson (1990) and Scotchmer (1991): Does the scope of claims in this patent cover few or many of the research activities using this technology? As the USPTO and the courts become more familiar with a technology, uncertainty over the scope of patent claims should diminish. The eventual outcome of the *Rochester v. Searle/Pharmacia* COX-2 case, for example, is likely to have significance beyond the parties' considerable financial stakes. If the district court decision against Rochester is upheld, we are likely to see research proceed with reduced concern over upstream research tool patents, although one

---

[68] As long as the infringing uses do not reduce the value of the tool to the users with a high willingness to pay, such price discrimination is likely to be privately profitable as well.

should then consider the impact of that decision on the incentives for developing that class of upstream discovery.

Through a combination of luck and appropriate institutional response, we appear to have avoided situations where a single firm or organization using its patents has blocked research in one or more broad therapeutic areas. However, the danger remains that progress in a broad research area could be significantly impeded by a patentholder trying to reserve the area exclusively for itself. The question is whether something systematic needs to be done. One possibility that has been considered is a revision of the law providing for research exemptions to better reflect the current norms and practices of the biomedical research community (cf. Rai, 1999; Ducor, 1997, 1999). It is not easy to discern when research is commercial or noncommercial notwithstanding what kind of institution is doing the research (cf. OECD, 2002). Thus it is not apparent that society would benefit from a policy response as opposed to continued reliance on current ad hoc practices of *de facto* infringement under the informal rubric of the "research exemption." The viability of this latter approach may, however, be undermined by the recent October, 2002 CAFC decision in *Madey v. Duke* which effectively narrows the research exemption to exclude, in essence, any use of IP in the course of university research. The effect of this decision is not to make the unauthorized use of others' IP in academic biomedical research illegal; such uses, as suggested above, were already likely illegal in light of recent, pre-Madey interpretations of the research exemption. Rather, this decision will focus attention on such practices, sensitizing both faculty and university administrations to the possible illegality of—and liability for—such uses of IP. This could well chill some of the "offending" biomedical research that is conducted in university settings. Given the importance of this informal exemption for allowing open science to proceed relatively unencumbered, this outcome would be unfortunate. Thus, policymakers should ensure an appropriate exemption for research intended for the public domain.

We cannot, therefore, rule out future problems resulting from patents currently under review, court decisions, new shifts in technology, or even assertions of patents on foundational discoveries. Therefore, we anticipate a continuing need for the active defense of open science. Yet the social system we observe has appeared to develop a robust combination of working solutions for dealing with these problems. Recent history suggests that these solutions can take time and expense to work out, and the results may not be optimal from either a private or social welfare perspective, but research generally moves forward. It should also be recalled that patents benefit biomedical innovation broadly by providing incentives that have called forth enormous investment in R&D (cf. Arora et al., 2003), and that the research tools developed have increased the productivity of biomedical research (e.g., Henderson et al., 1999).

Thus, our conclusion is that the biomedical enterprise seems to be succeeding, albeit with some difficulties, in developing an accommodation that incorporates both the need to provide strong incentives to conduct research and development and the need to maintain free space for discovery. As technologies change and as court decisions such as *Madey v. Duke* emerge, these issues may need to be periodically revisited.

## REFERENCES

American Intellectual Property Law Association (AIPLA). (1995, 1997, 2001). *Report of the Economic Survey*. Washington, D.C.: AIPLA.

Arora, A. A., A. Fosfuri, and A. Gambardella. (2001). *Markets for Technology*. Cambridge: MIT Press.

Arora, A. A., M. Ceccagnoli, and W. M. Cohen. (2003). "R&D and the Patent Premium." NBER Working Paper 9431.

Association of University Technology Managers (AUTM). (2000). *AUTM Licensing Survey FY1999 Survey Summary*. Northbrook, IL: Association of University Technology Managers.

Bar-Shalom, A., and R. Cook-Deegan. (2002). "Patents and Innovation in Cancer Therapeutics: Lessons from CellPro." *Millbank Quarterly*, 80(4): 637-676.

Barton, J. (2000). "Intellectual Property Rights. Reforming the Patent System." *Science* 287: 1933-1934.

Beck, S. (1998). "Do You Have a License?: Products Licensed for PCR in Research Applications." *The Scientist* (June 8) 12(12): 21.

Bolton, M. K., R. Malmrose, and W. G. Ouchi. (1994). "The Organization of Innovation in the United States and Japan." *Journal of Management Studies* 31: 653-679.

Blanton, K. (2002). "Corporate Takeover." *Boston Globe* (24 February 2002): Magazine.

Blumenthal, D, E. G. Campbell, M. S. Anderson, N. Causino, and K. S. Louis. (1997). "Withholding Research Results in Academic Life Science. Evidence from a National Survey of Faculty." *JAMA* (16 Apr) 277(15): 1224-1228.

Bunk, S. (1999). "Researchers Feel Threatened by Disease Gene Patents." *The Scientist* (October 11)13(20): 7.

Campbell, E. G., B. R. Clarridge, M. Gokhale, L. Birenbaum, S. Hilgartner, N. A. Holtzman, and D. Blumenthal. (2002). "Data Withholding in Academic Genetics." *JAMA* (23/30 Jan) 287: 473-480.

Cho, M., S. Illangasekare, M. A. Weaver, D. G. B. Leonard, and J. F. Merz. (2003). "Effects of Patents and Licenses on the Provision of Clinical Genetic Testing Services." *Journal of Molecular Diagnostics* 5(1): 3-8.

Cockburn, I., R. Henderson, L. Orsenigo, and G. P. Pisano. (2000). "Pharmaceuticals and Biotechnology." In D.C. Mowery, ed., *U.S. Industry in 2000: Studies in Competitive Performance*. Washington, D.C.: National Academy Press.

Cockburn, I., and R. Henderson. (2000). "Publicly Funded Science and the Productivity of the Pharmaceutical Industry." In A. Jaffe, J. Lerner and S. Stern, eds., *Innovation Policy and the Economy*. Cambridge: MIT Press.

Cockburn, I., S. Kortum, and S. Stern. (2003)."Are All Patent Examiners Equal? Examiners, Patent Characteristics, and Litigation Outcomes." In W. Cohen and S. Merrill, eds., *Patents in the Knowledge-Based Economy*. Washington, D.C.: The National Academies Press.

Cohen, J. (1997). "The Genomics Gamble." *Science* 275 [7 Feb]: 767-772.

Cohen, J. (1999). "Chiron Stakes Out Its Territory." *Science* 285 [2 Jul]: 28.

Cohen, W. M., and S. Klepper. (1992). "The Tradeoff Between Firm Size and Diversity for Technological Progress." *Journal of Small Business Economics*, December.

Cohen, W. M., R. R. Nelson, and J. P. Walsh. (2000). "Protecting Their Intellectual Assets: Appropriability Conditions and Why U.S. Manufacturing Firms Patent (or Not)." NBER Working Paper 7552.

Cohen, W. C., A. Goto, A. Nagata, R. R. Nelson, and J. P. Walsh. (2002). "R&D Spillovers, Patents and the Incentives to Innovate in Japan and the United States." *Research Policy*, 30(9): 1349-1367.

Cook-Deegan, R. M., and S. J. McCormack. (2001). "Patents, Secrecy, and DNA." *Science* 293 (13 July): 217.

Dalton, R. (2000). "Patent Suit in Alzheimer's Rejected." *Nature* 405 [29 June]: 989.

Dasgupta, P., and E. Maskin. (1987). "The Simple Economics of Research Portfolios." *Economic Journal* 97(387): 581-595.

Doll, J. J. (1998). "The Patenting of DNA." *Science* 280 (1 May): 689-690.

Drews, J. (2000). "Drug Discovery: A Historical Perspective." *Science* 287 (17 March): 1960-1962.

Ducor, P. (1997). "Are Patents and Research Compatible?" *Nature* 387 (1 May): 13-14.

Ducor, P. (1999). "Research Tool Patents and the Experimental Use Exemption." *Nature Biotechnology* (17 October)17: 1027-1028.

Eisenberg, R. S. (2001). "Bargaining over the Transfer of Proprietary Research Tools: Is This Market Failing or Emerging?" In R. C. Dreyfuss, D. L. Zimmerman, and H. First, eds., *Expanding the Boundaries of Intellectual Property: Innovation Policy for the Knowledge Society*. Oxford: Oxford University Press.

Evenson, R., and Y. Kislev. (1973). "Research and Productivity in Wheat and Maize." *Journal of Political Economy* 81: 1309-1329.

Freire, M. (2002). "The Impact of Patenting and Licensing Practices on Research." Paper presented at OECD Workshop on Genetic Inventions, Intellectual Property Rights and Licensing Practices, January 24, 2002, Berlin.

Gambardella, A. (1995). *Science and Innovation: The US Pharmaceutical Industry in the 1980s*. Cambridge MA.: Cambridge University Press.

Hackett, L. B., and J. T. Totten. (1995). *Report on the United States Vaccine Industry*, commissioned by the Department of Health and Human Services and prepared by Mercer Management Consulting, June 14, 1995.

Hagstrom, W.O. (1974). "Competition in Science." *American Sociological Review* 39: 1-18.

Hall, B. H., and R. H. Ziedonis. (2001). "The Patent Paradox Revisited: An Empirical Study of Patenting in the U.S. Semiconductor Industry, 1979-1995." *RAND Journal of Economics* 32(1): 101-128.

Hamilton, J. O. (1997). "Stanford's DNA Patent 'Enforcer' Grolle Closes the $200M Book on Cohen-Boyer." *Signals Magazine*, November 25, 1997.

Heller, M. A. (1998). "The Tragedy of the Anticommons: Property in the Transition from Marx to Markets." *Harvard Law Review* 111(3): 621-688.

Heller, M. A., and R. S. Eisenberg. (1998). "Can Patents Deter Innovation? The Anticommons in Biomedical Research." *Science* 280 (1 May): 698-701.

Henderson, R., L. Orsenigo, and G. P. Pisano. (1999). "The Pharmaceutical Industry and the Revolution in Molecular Biology: Interactions among Scientific, Institutional and Organizational Change." In D. C. Mowery and R. R. Nelson, *Sources of Industrial Leadership: Studies of Seven Industries*. New York: Cambridge University Press.

Henry, M. R., M. K. Cho, M. A. Weaver, and J. F. Merz. (2002). "DNA Patenting and Licensing." *Science* 297 (23 August): 1279.

Hicks, D., T. Breitzman, D. Olivastro and K. Hamilton. (2001) "The Changing Composition of Innovative Activity in the US—A Portrait Based on Patent Analysis," *Research Policy* 30(4): 681-704.

Jewkes, J., D. Sawers, and R. Stillerman. (1958). *The Sources of Invention*. London: Macmillan.

Kash, D. E. (2000). "Patents in a World of Complex Technologies." Paper presented to NAS STEP Board Conference on Intellectual Property Rights: How Far Should They Be Extended?, Washington, D.C., February 2-3.

Kornberg, A. (1995). *The Golden Helix.* Sausalito, CA: University Science Books.

Kotulak, R. (1999). "Taking License with Your Genes: Biotech Firms Say They Need Protection." *Chicago Tribune* (12 September): 1.

Kotulak, R., and P. Gorner. (2001). "Stem Cell Limits Bring New Fears." *Chicago Tribune* (9 September).

Kryder, R. D., S. P. Kowalski, and A. F. Krattiger. (2000). "The Intellectual and Technical Property Components of pro-Vitamin A Rice (GoldenRice™): A Preliminary Freedom-To-Operate Review." ISAAA Brief No. 20-2000.

Kunin, S. G. (2000). "Patentability Issues in Biotechnology." Presentation to the International Property Rights Symposium, Tokyo.

Lanjouw, J. O., and M. Schankerman. (2001). "Enforcing Intellectual Property Rights." NBER Working Paper 8656.

Lerner, J. (1995). "Patenting in the Shadow of Competitors." *Journal of Law and Economics* 38(2): 463-495.

Levin, R. (1982). "The Semiconductor Industry," In R.R. Nelson, ed., *Government and Technical Progress: A Cross-Industry Analysis.* New York: Pergamon Press.

Levin, R., A. Klevorick, R. R. Nelson, and S. G. Winter. (1987). "Appropriating the Returns from Industrial R&D." *Brookings Papers on Economic Activity* 3:783-820.

Maebius, S., and H. Wegner. (2001). "Research Methods Patents: A Territoriality Loophole." *National Law Journal* (24 December): C3-C4

Malakoff, D., and R. F. Service. (2001). "Genomania Meets the Bottom Line." *Science* 291 (16 Feb): 1193-1203.

Mansfield, E. (1986). "Patents and Innovation: An Empirical Study." *Management Science* 32: 173-181.

Marshall, E. (1997). "Ethics in Science: Is Data-Hoarding Slowing the Assault on Pathogens?" *Science* 275 (7 Feb): 777-780.

Marshall, E. (1999a). "Drug Firms to Create Public Database of Genetic Mutations." *Science* 284 (16 April): 406-407.

Marshall, E. (1999b). "Do-It-Yourself Gene Watching." *Science* 286 (15 October): 444-447.

Marshall, E. (2000a). "Patent on HIV Receptor Provokes an Outcry." *Science* 287 (25 February): 1375-1377.

Marshall, E. (2000b). "Property Claims: A Deluge of Patents Creates Legal Hassles for Research." *Science* 288 (14 April): 255-257.

Marshall, E. (2001). "Bermuda Rules: Community Spirit, with Teeth." *Science* 291 (16 Feb): 1192.

Marx, J. (2002a). "Chromosome End Game Draws a Crowd." *Science* 295 (29 March): 2348-2351.

Marx, J. (2002b). "Tackling Cancer at the Telomeres." *Science* 295 (29 March): 2350.

McCain, K. W. (1991). "Communication, Competition, and Secrecy: The Production and Dissemination of Research-Related Information in Genetics." *Science, Technology, and Human Values* 16: 491-516.

Merges, R. (1994). "Intellectual Property Rights and Bargaining Breakdown: The Case of Blocking Patents." *Tennessee Law Review* 62(1): 74-106.

Merges, R. (1999). Remarks to the Board on Science, Technology, and Economic Policy, National Research Council, Workshop on Intellectual Property Rights Research. Berkeley, CA, October 19, 1999.

Merges, R. P., and R. R. Nelson. (1990). "On the Complex Economics of Patent Scope." *Columbia Law Review* 90(4): 839-916.

Merges, R. P., and R. R. Nelson. (1994). "On Limiting or Encouraging Rivalry in Technical Progress: The Effect of Patent Scope Decisions." *Journal of Economic Behavior and Organization* 25: 1-24.

Merz, J. F, D. G. Kriss, D. D. G. Leonard, and M. K. Cho. (2002). "Diagnostic Testing Fails the Test." *Nature* 415 (7 February): 577-579.
Mowery, D., R. Nelson, B. Sampat, and A. Ziedonis. (2001). "The Growth of Patents and Licensing by U.S. Universities." *Research Policy* 30: 99-119.
National Research Council. (1997). *Intellectual Property Rights and Research Tools in Molecular Biology.* Washington, D.C.: National Academy Press.
National Science Foundation. (1998). *Science and Engineering Indicators.* Washington: U.S. GPO.
Neighbour, A. (2002). "Presentation to the National Cancer Policy Board," Institute of Medicine, April 23.
Nelson, R. R. (1961). "Uncertainty, Learning, and the Economics of Parallel Research and Development Efforts." *Review of Economics and Statistics* 43: 351-364.
Nelson, R. R. (1982). "The Role of Knowledge in R&D efficiency." *Quarterly Journal of Economics* 97: 453-470.
OECD. (2002). "Genetic Inventions, IPRS and Licensing Practices." Working Party on Biotechnology, Directorate for Science, Technology and Industry.
Pollack, A. (2000). "Battling Searle, University Gets Broad Patent for New Painkiller." *The New York Times*, April 12, C1-2.
Rai, A. K. (1999). "Regulating Scientific Research; Intellectual Property Rights and the Norms of Science." *Northwestern University Law Review* 94(1): 77-152.
Regalado, A. (2000). "The Great Gene Grab." *Technology Review* September/October: 49-55.
Scherer, F.M., S. Herzstein, Jr., A. Dreyfoos, W. Whitney, O. Bachmann, C. Pesek, C. Scott, T. Kelly and J. Galvin. (1959). *Patents and the Corporation: A Report on Industrial Technology Under Changing Public Policy.* 2nd Edn. Boston, MA: Harvard University, Graduate School of Business Administration.
Scherer, F. M. (2002). "The Economics of Human Gene Patents." *Academic Medicine* 77: 1348-1366.
*Science.* (1997). "Genomics' Wheelers and Dealers." *Science* 275 (7 Feb): 774-775.
*Science.* (1998). "Biochip-Makers do Battle in Court." *Science* 279 (16 January): 311b.
*Science.* (2000). "A Cheaper Way to Buy Genomic Data." *Science* 288 (14 April 2000): 223.
Scotchmer, S. (1991). "Standing on the Shoulders of Giants: Cumulative Research and the Patent Law" *Journal of Economic Perspectives* 5(1): 29-41.
Scott, R. (2000). Testimony Before House Judiciary Committee, 7/13/00. http://www.house.gov/judiciary/scot0713.htm
Seide, R. K., and J. M. MacLeod. (1998). "Comment on Heller and Eisenberg." *ScienceOnline* http://www.sciencemag.org/feature/data/980465/seide.shl
Service, R. F. (1998). "DNA Analysis: Will Patent Fights Hold DNA Chips Hostage?" *Science* 282(October 16): 397.
Service, R. F. (2001). "Can Data Banks Tally Profits?" *Science* 291(16 Feb): 1203.
Shapiro, C. (2000). "Navigating the Patent Thicket: Cross Licenses, Patent Pools, and Standard-Setting." In A. Jaffe, J. Lerner and S. Stern, eds., *Innovation Policy and the Economy.* Cambridge: MIT Press.
Sullivan, D. (1975). "Competition in Bio-Medical Science: Extent, Structure, and Consequences." *Sociology of Education* 48: 223-241.
Thursby, J. G., and M. C. Thursby. (1999). "Purdue Licensing Survey: A Summary of Results." Unpublished manuscript, Krannert Graduate School of Management, Purdue University.
University of Rochester. (2003). "University to Appeal Ruling in Patent Case Against Pfizer and Pharmacia." Press release 3/5/2003.
USPTO. (2001). http://www.uspto.gov/web/offices/com/speeches/01-06.htm
USPTO. (n.d.). "Synopsis of application of written description guidelines." http://www.uspto.gov/web/menu/written.pdf
Uzzi, B. (1996). "The Sources and Consequences of Embeddedness for the Economic Performance of Organizations: The Network Effect." *American Sociological Review* 61(4): 674-698.

Wall Street Journal. (2001). "IBM and Others Are Financing A Public Database on Proteins." (30 May 2001).
Walsh, J. P., and T. Bayma. (1996). "Computer Networks and Scientific Work." *Social Studies of Science* 26: 661-703.
Walsh, J. P., and W. Hong. (2003). "Secrecy Is Increasing in Step with Competition," *Nature* (24 April) 422: 801-802.
Warcoin, J. (2002). "Intellectual Property and Development of Products in Biotechnology." Paper presented at OECD Workshop on Genetic Inventions, Intellectual Property Rights and Licensing Practices, January 24, 2002, Berlin.
Wegner, H. C. (2002). "The Right to Experiment with a Patented Invention." Paper presented to a meeting of the Bar Association of the District of Columbia PTC Section, Dec. 10, 2002, Washington, D.C.
Werth, B. (1994). *Billion-Dollar Molecule*. New York: Touchstone.
Whyte, W. F. (1984). *Learning from the Field*. Beverly Hills, CA: Sage.
Williamson, O. E. (1979). "Transaction Cost Economics: The Governance of Contractual Relations." *Journal of Law and Economics* 22: 233-261.